Communications
in Computer and Information Science          52

Ana Fred   Joaquim Filipe   Hugo Gamboa (Eds.)

# Biomedical Engineering Systems and Technologies

International Joint Conference, BIOSTEC 2009
Porto, Portugal, January 14-17, 2009
Revised Selected Papers

 Springer

Volume Editors

Ana Fred
Institute of Telecommunications
Technical University of Lisbon
Lisbon, Portugal
E-mail: afred@lx.it.pt

Joaquim Filipe
Departament of Systems and Informatics
Polytechnic Institute of Setúbal
Setúbal, Portugal
E-mail: j.filipe@est.ips.pt

Hugo Gamboa
Institute of Telecommunications
Technical University of Lisbon
Lisbon, Portugal
E-mail: hugo.gamboa@lx.it.pt

Library of Congress Control Number: 2010921115

CR Subject Classification (1998): J.3, K.4, I.2.9, J.2, J.7, C.3

| | |
|---|---|
| ISSN | 1865-0929 |
| ISBN-10 | 3-642-11720-1 Springer Berlin Heidelberg New York |
| ISBN-13 | 978-3-642-11720-6 Springer Berlin Heidelberg New York |

springer.com

© Springer-Verlag Berlin Heidelberg 2010
Printed in Germany

Typesetting: Camera-ready by author, data conversion by Scientific Publishing Services, Chennai, India
Printed on acid-free paper      SPIN: 12834969      06/3180      5 4 3 2 1 0

# Preface

This book contains the best papers of the Second International Joint Conference on Biomedical Engineering Systems and Technologies (BIOSTEC 2009), organized by the Institute for Systems and Technologies of Information Control and Communication (INSTICC), technically co-sponsored by the IEEE Engineering in Medicine and Biology Society (EMB), IEEE Circuits and Systems Society (CAS) and the Workflow Management Coalition (WfMC), in cooperation with AAAI and ACM SIGART.

The purpose of the International Joint Conference on Biomedical Engineering Systems and Technologies is to bring together researchers and practitioners, including engineers, biologists, health professionals and informatics/computer scientists, interested in both theoretical advances and applications of information systems, artificial intelligence, signal processing, electronics and other engineering tools in knowledge areas related to biology and medicine.

BIOSTEC is composed of three co-located conferences; each specializes in one of the aforementioned main knowledge areas, namely:

- BIODEVICES (International Conference on Biomedical Electronics and Devices) focuses on aspects related to electronics and mechanical engineering, especially equipment and materials inspired from biological systems and/or addressing biological requirements. Monitoring devices, instrumentation sensors and systems, biorobotics, micro-nanotechnologies and biomaterials are some of the technologies addressed at this conference.

- BIOSIGNALS (International Conference on Bio-inspired Systems and Signal Processing) is a forum for those studying and using models and techniques inspired from or applied to biological systems. A diversity of signal types can be found in this area, including image, audio and other biological sources of information. The analysis and use of these signals is a multidisciplinary area including signal processing, pattern recognition and computational intelligence techniques, amongst others.

- HEALTHINF (International Conference on Health Informatics) promotes research and development in the application of information and communication technologies (ICT) to healthcare and medicine in general and to specialized support for persons with special needs in particular. Databases, networking, graphical interfaces, intelligent decision support systems and specialized programming languages are just a few of the technologies currently used in medical informatics. Mobility and ubiquity in healthcare systems, standardization of technologies and procedures, certification and privacy are some of the issues that medical informatics professionals and the ICT industry in general need to address in order to further promote ICT in healthcare.

The joint conference, BIOSTEC, received 380 paper submissions from more than 55 countries in all continents. In all, 57 papers were published and presented as full papers, i.e., completed work (8 pages/30-minute oral presentation), 126 papers reflecting work-in-progress or position papers were accepted for short presentation, and another 63 contributions were accepted for poster presentation. These numbers, leading to a "full-paper" acceptance ratio below 15% and a total oral paper presentations acceptance ratio below 49%, show the intention of preserving a high-quality forum for the next editions of this conference. This book includes revised and extended versions of a strict selection of the best papers presented at the conference.

The conference included a panel and six invited talks delivered by internationally distinguished speakers, namely: Egon L. van den Broek, Pier Luigi Emiliani, Maciej Ogorzalek, Vimla L. Patel and Edward H. Shortliffe.

We must thank the authors, whose research and development efforts are recorded here. We also thank the keynote speakers for their invaluable contribution and for taking the time to synthesize and prepare their talks. The contribution of all Program Chairs of the three component conferences was essential to the success of BIOSTEC 2009. Finally, special thanks to all the members of the INSTICC team, whose collaboration was fundamental for the success of this conference.

November 2009
Ana Fred<br>Joaquim Filipe<br>Hugo Gamboa

# Organization

## Conference Co-chairs

| | |
|---|---|
| Ana Fred | IST - Technical University of Lisbon, Portugal |
| Joaquim Filipe | Polytechnic Institute of Setúbal / INSTICC |
| Hugo Gamboa | Institute of Telecommunications, Lisbon, Portugal |

## Program Co-chairs

| | |
|---|---|
| Luis Azevedo | Instituto Superior Técnico, Portugal (HEALTHINF) |
| Pedro Encarnação | Catholic Portuguese University, Portugal (BIOSIGNALS) |
| Teodiano Freire Bastos Filho | Federal University of Espírito Santo, Brazil (BIODEVICES) |
| Hugo Gamboa | Telecommunications Institute, Portugal (BIODEVICES) |
| Ana Rita Londral | ANDITEC, Portugal (HEALTHINF) |
| António Veloso | FMH, Universidade Técnica de Lisboa, Portugal (BIOSIGNALS) |

## Organizing Committee

| | |
|---|---|
| Sérgio Brissos | INSTICC, Portugal |
| Marina Carvalho | INSTICC, Portugal |
| Helder Coelhas | INSTICC, Portugal |
| Vera Coelho | INSTICC, Portugal |
| Andreia Costa | INSTICC, Portugal |
| Bruno Encarnação | INSTICC, Portugal |
| Bárbara Lima | INSTICC, Portugal |
| Raquel Martins | INSTICC, Portugal |
| Elton Mendes | INSTICC, Portugal |
| Carla Mota | INSTICC, Portugal |
| Vitor Pedrosa | INSTICC, Portugal |
| Vera Rosário | INSTICC, Portugal |
| José Varela | INSTICC, Portugal |

## BIODEVICES Program Committee

| | |
|---|---|
| Oliver Amft, Switzerland | Susana Borromeo, Spain |
| Rodrigo Varejão Andreão, Brazil | Enrique A. Vargas Cabral, Paraguay |
| Luciano Boquete, Spain | Ramón Ceres, Spain |

Fernando Cruz, Portugal
Pedro Pablo Escobar, Argentina
Marcos Formica, Argentina
Juan Carlos Garcia Garcia, Spain
Gerd Hirzinger, Germany
Jongin Hong,UK
Giacomo Indiveri, Switzerland
Bozena Kaminska, Canada
Rui Lima, Portugal
Ratko Magjarevic, Croatia
Dan Mandru, Romania
Manuel Mazo, Spain
Paulo Mendes, Portugal
Joseph Mizrahi, Israel
Raimes Moraes, Brazil
Pedro Noritomi, Brazil
Kazuhiro Oiwa, Japan

Evangelos Papadopoulos, Greece
Laura Papaleo, Italy
José Luis Martínez Pérez, Spain
Jose Luis Pons, Spain
Alejandro Ramirez-Serrano, Canada
Adriana María Rios Rincón, Colombia
Joaquin Roca-Dorda, Spain
Mario Sarcinelli-Filho, Brazil
Mohamad Sawan, Canada
Fernando di Sciascio, Argentina
Wouter Serdijn, The Netherlands
Jorge Vicente Lopes da Silva, Brazil
Amir M. Sodagar, USA
Ioan G. Tarnovan, Romania
Alexandre Terrier, Switzerland
Mário Vaz, Portugal
Chua-Chin Wang, Taiwan

## BIODEVICES Auxiliary Reviewers

Getúlio Igrejas, Portugal
Susana Palma, Portugal

Hugo Silva, Portugal

## BIOSIGNALS Program Committee

Andrew Adamatzky,UK
Oliver Amft, Switzerland
Peter Bentley,UK
C.D. Bertram, Australia
Paolo Bonato, USA
Tolga Can, Turkey
Rodrigo Capobiaco Guido, Brazil
Mujdat Cetin, Turkey
Marleen de Bruijne, Denmark
Dick de Ridder, The Netherlands
Suash Deb, India
Wael El-Deredy, UK
Eugene Fink, USA
Sebastià Galmés, Spain
Aaron Golden, Ireland
Cigdem Gunduz-Demir, Turkey
Bin He, USA
Huosheng Hu, UK
Helmut Hutten, Austria
Christopher James, UK
Yasemin Kahya, Turkey

Borys Kierdaszuk, Poland
Jonghwa Kim, Germany
Gunnar W. Klau, Germany
T. Laszlo Koczy, Hungary
Georgios Kontaxakis, Spain
Igor Kotenko, Russian Federation
Narayanan Krishnamurthi, USA
Arjan Kuijper, Germany
Vinod Kumar, India
Jason JS Lee, Taiwan
Kenji Leibnitz, Japan
Marco Loog, The Netherlands
Elena Marchiori, Netherlands
Martin Middendorf, Germany
Alexandru Morega, Romania
Mihaela Morega, Romania
Kayvan Najarian, USA
Tadashi Nakano, USA
Asoke Nandi,UK
Hasan Ogul, Turkey
Kazuhiro Oiwa, Japan

Oleg Okun, Sweden
Ernesto Pereda, Spain
Leif Peterson, USA
Vitor Pires, Portugal
Marcos Rodrigues, UK
Virginie Ruiz, UK
Heather Ruskin, Ireland
Maria Samsonova, Russian Federation
Carlo Sansone, Italy
Gerald Schaefer, UK
Dragutin Sevic, Serbia

Iryna Skrypnyk, Finland
Alan A. Stocker, USA
Junichi Suzuki, USA
Asser Tantawi, USA
Gianluca Tempesti, UK
Hua-Nong Ting, Malaysia
Anna Tonazzini, Italy
Duygu Tosun, USA
Bart Vanrumste, Belgium
Yuanyuan Wang, China
Didier Wolf, France

## BIOSIGNALS Auxiliary Reviewer

Francesco Gargiulo, Italy

## HEALTHINF Program Committee

Abdullah N. Arslan, USA
Osman Abul, Turkey
Abdullah N. Arslan, USA
Arnold Baca, Austria
Anna Divoli, USA
Adrie Dumay, The Netherlands
Mourad Elloumi, Tunisia
Alexandru Floares, Romania
Jose Fonseca, Portugal
David Greenhalgh, UK
Tiago Guerreiro, Portugal
Cigdem Gunduz-Demir, Turkey
Jin-Kao Hao, France
Chun-Hsi Huang, USA
Stavros Karkanis, Greece
Andreas Kerren, Sweden
Ina Koch, Germany
Georgios Kontaxakis, Spain
Athina Lazakidou, Greece
Feng-Huei Lin, Taiwan
Dan Mandru, Romania

Alice Maynard, UK
Boleslaw Mikolajczak, USA
Ahmed Morsy, Egypt
Chris Nugent, USA
Oleg Okun, Sweden
Chaoyi Pang, Australia
Leif Peterson, USA
Göran Petersson, Sweden
Axel Rasche, Germany
Marcos Rodrigues, UK
George Sakellaropoulos, Greece
Nickolas Sapidis, Greece
Boris Shishkov, Bulgaria
Iryna Skrypnyk, Finland
John Stankovic, USA
Adrian Tkacz, Poland
Ioannis Tsamardinos, Greece
Hassan Ugail, UK
Aristides Vagelatos, Greece
Athanasios Vasilakos, Greece
Jana Zvarova, Czech Republic

## Invited Speakers

Edward H. Shortliffe        Arizona State University, USA
Vimla L. Patel             Arizona State University, USA
Pier Luigi Emiliani        Institute of Applied Physics "Nello Carrara" (IFAC) of
                           the Italian National Research Council (CNR), Italy
Maciej Ogorzalek           Jagiellonian University, Poland
Egon L. Van Den Broek      University of Twente, The Netherlands

# Table of Contents

## Part II: BIOSIGNALS

## Part III: HEALTHINF

# Invited Papers

# Computational Intelligence and Image Processing Methods for Applications in Skin Cancer Diagnosis

Maciej Ogorzałek, Grzegorz Surówka, Leszek Nowak, and Christian Merkwirth

Department of Information Technologies, Faculty of Physics
Astronomy and Applied Computer Science, Jagiellonian University
ul. Reymonta 4, 30-059 Kraków, Poland
{Maciej.Ogorzalek,Grzegorz.Surowka}@uj.edu.pl
http://www.zti.uj.edu.pl

**Abstract.** Digital photography provides new powerful diagnostic tools in dermatology. Dermoscopy is a special photography technique which enables taking photos of skin lesions in chosen lighting conditions. Digital photography allows for seeing details of the skin changes under various enlargements and coloring. Computer-assisted techniques and image processing methods can be further used for image enhancement and analysis and for feature extraction and pattern recognition in the selected images. Special techniques used in skin-image processing are discussed in detail. Feature extraction methods and automated classification techniques based on statistical learning and model ensembling provide very powerful tools which can assist the doctors in taking decisions. Performance of classifiers will be discussed in specific case of melanoma cancer diagnosis. The techniques have been tested on a large data set of images.

**Keywords:** Medical diagnosis, Dermoscopy, Melanoma, Image analysis, Medical image processing, Feature extraction, Statistical learning, Classification, Ensembling.

## 1 Introduction

In the last decade there has been a significant rise in the number of cases of skin cancer noticeable world-wide. Early diagnosis of cancerous changes on the skin is of paramount importance for the possible therapy and prognosis of life for the patient. With the development of specialized equipment - the dermoscope or epiluminescence microscope the diagnosis process has been greatly facilitated and in training of dermatologists image atlas of reference pictures are widely available [7], [9], [31]. Inspite of all these developments there exists still a broad margin for mis-interpretations of skin lesion images [5]. The diagnosis is especially difficult in the early stages of cancer when the lesions are small. In the year 2000 the Second Consensus Meeting on Dermoscopy has been held and its main conclusions were that four algorithms, namely pattern analysis, ABCD rule [4], Menzies scoring method and 7-point check list are good ways of evaluation of skin lesions using dermoscopy. All four methods share some common concepts and allow for selection of specific features possible to be done with the aid of computer. Let us briefly review the underlying concepts.

A. Fred, J. Filipe, and H. Gamboa (Eds.): BIOSTEC 2009, CCIS 52, pp. 3–20, 2010.

The so-called ABCD rule introduced in [25], [30] allows for computation of so-called TDS (Total Dermoscopy Score) factor allowing quantitative characterization of the lesion. Four characteristic features of the image are taken into account:

1. **A**symmetry - The dermoscopic image is divided by two perpendicular axes positioned to produce the lowest possible asymmetry score. If the image shows asymmetric properties with respect to both axes with regard to colors and differential structures, the asymmetry score is 2. If there is asymmetry on one axis the score is 1. If asymmetry is absent with regard to both axes the score is 0. Most melanomas have an asymmetry score of 2 compared to about only 25% of benign melanocytic nevi. Because of its high (1.3) weighting, the assessment of asymmetry is crucial for the final TDS value.
2. **B**order The images of the lesions are divided into eighths and a sharp, abrupt cut-off of pigment pattern at the periphery within one eighth has a score 1. In contrast, a gradual, indistinct cut-off within one eighth has a score of 0. So, the maximum border score is 8, and the minimum score is 0. As a rule the border score in nevi is very low and in melanomas is predominantly between 3 and 8. Because of its low weight (0.1) the border score is not very relevant.
3. **C**olor - Six different colors: white, red, light-brown, dark-brown, blue-gray, and black, are counted for determining the color score. White is only counted if the area is lighter than the adjacent skin. When all six colors are present the maximum color score is 6; the minimum score is 1. Cancerous skin changes are usually characterized by three or more colors and in about 40% of melanomas even five or six colors are present.
4. **D**ifferential structure - Stolz proposed five features for evaluation of so-called differential structures: pigment network, structure-less or homogeneous areas, streaks, dots, and globules. Structure-less or homogenous areas must be larger than 10% of the lesion. Streaks and dots are counted only when more than two are clearly visible. For counting a globule only the presence of one single globule is necessary. The more structures are present in the picture, the higher the probability of the lesion being a cancer.

Once the ABCD values are evaluated for the image under consideration one can compute the $TDS$ factor:

$$TDS = A * 1.3 + B * 0.1 + C * 0.5 + D * 0.5 \tag{1}$$

$TDS < 4.75$ gives indication of a benign lesion, $4.75 < TDS < 5.45$ is nonconclusive while $TDS > 5.45$ gives strong indication that the lesion is cancerous.

Menzies [22], [23] proposed a scoring method based on inspection of the lesion image. Asymmetry of pattern, more then one color and presence of one to nine positive specific features are simply adding one point to the score. The positive features are blue-and-white vail,multiple brown dots, pseudopods, radial streaming, scar-like depigmentation, peripheral black dots/globules, multiple colors (five or six), multiple blue/gray dots, broad pigment network. One can easily notice that most of the features proposed by Menzies are present also in the ABCD rule and serve for TDS calculation.

The third method used by dermatologists is so-called seven point checklist. The criteria are divided into two categories: major criteria (assigned two points each) and minor criteria (assigned a single point each).Among the major criteria are

1. Atypical pigment network: Black, brown, or gray network with irregular meshes and thick lines
2. Blue-whitish veil: Confluent, gray-blue to whitish-blue diffuse pigmentation associated with pigment network alterations, dots/globules and/or streaks
3. Atypical vascular pattern: Linear-irregular or dotted vessels not clearly combined with regression structures and associated with pigment network alterations, dots/globules and/or streaks

Among the minor criteria one can list presence of:

1. Irregular Streaks: Irregular, more or less confluent, linear structures not clearly combined with pigment network lines
2. Irregular pigmentation: Black, brown, and/or gray pigmented areas with irregular shape and/or distribution
3. Irregular dots/globules: Black, brown, and/or gray round to oval, variously sized structures irregularly distributed within the lesion
4. Regression Structures: White areas (white scarlike areas) and blue areas (gray-blue areas, peppering, multiple blue-gray dots) may be associated, thus featuring so-called blue-whitish areas virtually indistinguishable from blue-whitish veil

Various properties of the images mentioned in the three scoring methods are depicted in Fig.1-3.

There has been a long discussion among medical specialists which of the three methods gives best results. Argenziano [1] and Johr [19] made a comparisons of different methods finding that the seven point checklist had sensitivity in the range of 95% and specificity of 75% while for pattern analysis these number were 91% and 90%

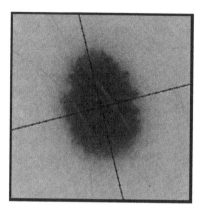 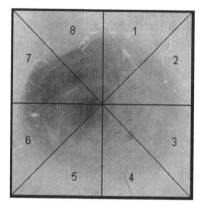

**Fig. 1.** Symmetry and border classification following the dermoscopic criteria. Symmetry against two main axes is considered plus border type between the lesion and normal skin in each of the eight sections.

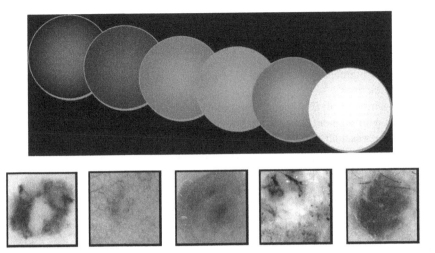

**Fig. 2.** Dermoscopic description considers presence of six main colorings in the lesion images as shown in the upper picture - ranging from black to white. In the bottom row typical images are shown in which each of the colorings could be identified.

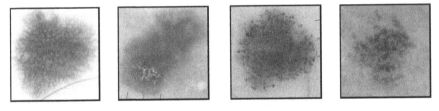

**Fig. 3.** Four types of patterns or structural elements visible in the dermoscopic images: colored net, smeared areas, condensed areas and dots

respectively going down to 85% and 66% respectively for the ABCD rule. The conclusion of the study was also that the seven-point rule is simplest to learn and even less-experienced doctors could obtain high diagnosis accuracy.

Collaboration of dermatologists, computer scientists and image processing specialists has led to significant automation of analysis of dermoscopic images improvement in their classification which serve as a diagnostic aid for doctors [12], [14], [15], [16] [29]. Several approaches to tele-diagnostic systems can be also mentioned when images to be evaluated and classified are sent via internet [18] or telephone [28]. There are several areas in dermoscopic image analysis where specific approaches, algorithms and methods from the image processing and computational intelligence toolkit could be used [3], [4], [6], [10].

## 2   Basic Image Processing

Digital photographs taken by dermoscopic cameras have typical properties and also problems known for any type of digital images. To be able to compare the features of various photographs taken possibly using different types of equipment with differently adjusted parameters, different sensor sensitivities and resolutions, under different lighting conditions one has to make their properties uniform. Often apart from the lesion under study the photograph contains also unwanted artefacts such as eg. hair or vigneting of the lens. Before any other analysis could be performed the acquired images have to be pre-processed. Pre-processing will include image normalization (size and color or gray shades), filtering and artefact removal.

All the medical diagnostic approaches stress the importance of the shape of the lesion and its geometric features such as borders. The shape of the lesion can be determined automatically by executing properly selected binarization of the image under consideration. It appears however that the oucome of the binarization process is very strongly dependent on the choice of the threshold. Figures 4 and 5 show typical problems associated with the choice of the threshold level. Misinterpretations are possible especially when the image exhibits multiple colorings or multiple gray shades. In such cases one can obtain very different shapes, borders and areas of the lesion depending on the selected threshold value.

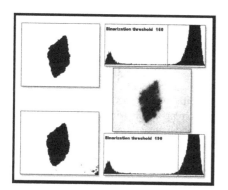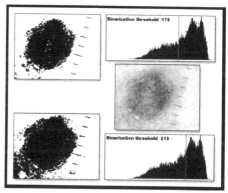

**Fig. 4.** Binarization of the image can lead to various interpretations based on the particular threshold selection. In the figure shown in the left the obtained result does not depend much on the threshold. For the image shown on the right there is a very strong dependence of the result on the selected binarization level.

A lot of information important for medical diagnosis is carried in the coloring of the image. Decomposition of color images into color components is one of the best known image analysis areas. Commonly two kinds of color decompositions are widely used namely RGB (additive) and CMY (subtractive) (or CMYK). Both are widely used in computer graphics and digital printing. Comparing the analysis done for obtaining the melanoma tests one can see that in the ABCD-scale existence of more then three out of six colorings gives a strong indication of cancerous lesion. As the scales proposed in medical diagnosis have to be simple enough not to cause confusion when used by any dermatologist the number of colors used is limited to six. Still different person will have a different perception of specific tints and colorings. Computerized methods provide a way to find the colorings which is user-independent. As the RGB decompositions are

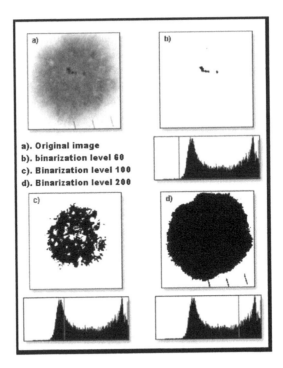

**Fig. 5.** Varying the threshold value one can single out from an image different structural components. For the binarization threshold equal 60 one can see the black spots present in the image. Choosing the binarization level 100 the fine pattern is becoming visible while for threshold level equal 200 the whole area of the lesion and its outer shape and boundaries can be determined.

readily available in any image processing package (eg. Matlab toolbox) one can provide thresholds for RGB levels to find the cancer specific colorings. Figures 6-8 show three examples of how the RGB thresholding could automatically extract white, black and grey-bluish area in the image.

In all the experiments the success depends highly on the proper selection of levels and thresholds. These in the experiments have been established in a heuristic way based

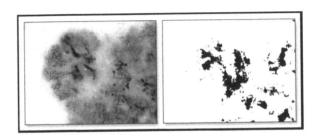

**Fig. 6.** Dangerous color components: white areas can be found by RGB thresholding setting $R > 110$, $G > R - 26$ and $B > R - 20$

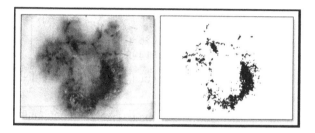

**Fig. 7.** Dangerous color components: black areas can be found by RGB thresholding setting $R < 90$, $G - 10 < B < G + 25B$ or $R/5 - 4 < G < R/2 + 12$, $B > R/4 - 3$

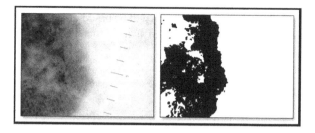

**Fig. 8.** Dangerous color components: grey-bluish areas can be found by RGB thresholding setting $R > 60$, $R - 46 < G < R + 15$ or $G > B - 30$, $R - 30 < B < R + 45$

on past experience and very close joint work with clinical dermatologists. The levels as shown in the experiments were adjusted to give best diagnostic results.

## 3 Engineering Decompositions of Images

Apart from common color decompositions used in art and computer graphics there exist also a number of different decompositions of the image signals used in TV/video applications. These specially created signal components carry important information and could possibly be used as additional features bringing more information about the lesion images. These components up to our knowledge have so far never been used in analysis of dermoscopic images. For the purpose of this study we used the HSV (Hue, Saturation and Value) decomposition. The V-value is directly the intensity of a particular pixel, S - determines the saturation $S = \frac{Max - Min}{Max}$, and Hue can be directly related to the RGB components as: $H = (\frac{G-B}{Max-Min})/6$ if $R = Max$, $H = (2 + \frac{B-R}{Max-Min})/6$ if $G = Max$ and $H = (4 + \frac{R-G}{Max-Min})/6$ if $B = Max$.

We also exploited another kind of decomposition widely used in Video/TV eg. in the NTSC standard – it is the YIQ or YUV space. These components are related to the RGB via a matrix transformation:

$$\begin{bmatrix} Y \\ I \\ Q \end{bmatrix} = \begin{bmatrix} 0.299 & 0.587 & 0.114 \\ 0.586 & -0.274 & 0.322 \\ 0.211 & -0.523 & 0.312 \end{bmatrix} \begin{bmatrix} R \\ G \\ B \end{bmatrix} \tag{2}$$

or

$$\begin{bmatrix} Y' \\ U \\ V \end{bmatrix} = \begin{bmatrix} 0.299 & 0.587 & 0.114 \\ -0.14713 & -0.28886 & 0.436 \\ 0.615 & -0.51499 & -0.10001 \end{bmatrix} \begin{bmatrix} R \\ G \\ B \end{bmatrix} \qquad (3)$$

Another very useful signal decomposition which we tested is the decomposition into luminance-chrominance components $YC_bC_r$:

$$\begin{bmatrix} Y \\ C_b \\ C_r \end{bmatrix} = \begin{bmatrix} 16 \\ 128 \\ 128 \end{bmatrix} + \begin{bmatrix} 0.25678824 & 0.50412941 & 0.09790588 \\ -0.1482229 & -0.29099279 & 0.43921569 \\ 0.43931569 & -0.36778831 & 0.07142737 \end{bmatrix} \begin{bmatrix} R \\ G \\ B \end{bmatrix} \qquad (4)$$

All the above-mentioned characterizations have been thoroughly tested on skin lesion images with the goal of finding those features which carry most information useful for diagnostic purposes. The process of diagnosis will be based on construction of best possible classifier which will be able to distinguish cancerous lesions taking into account the selected features of images.

## 4   How to Build a Good Classifier?

The approach we propose for building classifiers is to use statistical learning techniques for data-based model building. All classifiers we use are model-based. To construct an extremely efficient classifier we build ensembles of well trained but diverse models.

### 4.1   Model Types Used in Statistical Learning

There exist a vast variety of available models described in the literature which can be grouped into some general classes

- Global Models
  - Linear Models
  - Polynomial Models
  - Neural Networks (MLP)
  - Support Vector Machines
- Semi–global Models
  - Radial Basis Functions
  - Multivariate Adaptive Regression Splines (MARS)
  - Decision Trees (C4.5, CART)

- Local Models
  - k–Nearest–Neighbors
- Hybrid Models
  - Projection Based Radial Basis Functions Network (PRBFN)

Implementation of any of such modeling methods leads usually to solution of an optimization problem and further to operations such as matrix inversion in case of linear regression or minimization of a *loss function* on the training data or quadratic programming problem (eg. for SVMs).

## 4.2   Validation and Model Selection

The key for model selection is the Generalization error – how does the model perform on unseen data (samples) ? Exact generalization error is not accessible since we have only limited number of observations ! Training on small data set results in overfitting, causing generalization error to be significantly higher than training error. This is a consequence of mismatch between the capacity of the hypothesis space $\mathcal{H}$ (VC-Dimension) and the number of training observations. Any type of model constructed has to pass the validation stage – estimation of the generalization error using just the given data set. In a logical way we select the model with lowest (estimated) generalization error. To improve the generalization error typical remedies can be:

- Manipulating training algorithm (e.g. early stopping)
- Regularization by adding a penalty to the loss function
- Using algorithms with built-in capacity control (e.g. SVM)
- Relying on criteria like  BIC,  AIC,  GCV or  Cross Validation to select optimal model complexity
- Reformulating the loss function, e.q. by using an $\epsilon$-insensitive loss

## 5   Ensemble Methods

Building an **Ensemble** consists of averaging the outputs of several separately trained models

- Simple average $\bar{f}(\boldsymbol{x}) = \frac{1}{K}\sum_{k=1}^{K} f_k(\boldsymbol{x})$
- Weighted average $\bar{f}(\boldsymbol{x}) = \sum_k w_k f_k(\boldsymbol{x})$ with $\sum_k w_k = 1$

The ensemble generalization error is always smaller than the expected error of the individual models. An ensemble should consist of well trained but diverse models.

### 5.1   The Bias/Variance Decomposition for Ensembles

Our approach is based on the observation that the generalization error of an ensemble model can be improved if the predictors on which the averaging is done disagree and if their residuals are uncorrelated [21]. We consider the case where we have a given data set $D = \{(\mathbf{x}_1, y_1), \ldots, (\mathbf{x}_N, y_N)\}$ and we want to find a function $f(\mathbf{x})$ that approximates $y$ also for unseen observations of $\mathbf{x}$. These unseen observations are assumed to stem from the same but not explicitly known probability distribution $P(\mathbf{x})$. The expected generalization error $Err(\mathbf{x})$ given a particular $\mathbf{x}$ and a training set $D$ is

$$Err(\mathbf{x}) = E[(y - f(\mathbf{x}))^2 | \mathbf{x}, D] \tag{5}$$

where the expectation $E[\cdot]$ is taken with respect to the probability distribution $P$. The Bias/Variance Decomposition of $Err(\mathbf{x})$ is

$$Err(\mathbf{x}) = \sigma^2 + (E_D[f(\mathbf{x})] - E[y|\mathbf{x}])^2$$
$$+ E_D[(f(\mathbf{x}) - E_D[f(\mathbf{x})])^2] \tag{6}$$
$$= \sigma^2 + (\text{Bias}(f(x)))^2 + \text{Var}(f(x)) \tag{7}$$

where the expectation $E_D[\cdot]$ is taken with respect to all possible realizations of training sets $D$ with fixed sample size $N$ and $E[y|\mathbf{x}]$ is the deterministic part of the data and $\sigma^2$ is the variance of $y$ given $\mathbf{x}$. Balancing between the bias and the variance term is a crucial problem in model building. If we try to decrease the bias term on a specific training set, we usually increase the bias term and vice versa. We now consider the case of an ensemble average $\hat{f}(\mathbf{x})$ consisting of $K$ individual models

$$\hat{f}(\mathbf{x}) = \sum_{i=1}^{K} \omega_i f_i(\mathbf{x}) \qquad \omega_i \geq 0, \tag{8}$$

where the weights may sum to one $\sum_{i=1}^{K} \omega_i = 1$. If we put this into eqn. (6) we get

$$Err(\mathbf{x}) = \sigma^2 + \text{Bias}(\hat{f}(x))^2 + \text{Var}(\hat{f}(x)), \tag{9}$$

The bias term in eqn. (9) is the average of the biases of the individual models. The variance term of the ensemble could be decomposed in the following way:

$$\begin{aligned}
Var(\hat{f}) &= E\left[(\hat{f} - E[\hat{f}])^2\right] \\
&= E[(\sum_{i=1}^{K} \omega_i f_i)^2] - (E[\sum_{i=1}^{K} \omega_i f_i])^2 \\
&= \sum_{i=1}^{K} \omega_i^2 \left(E\left[f_i^2\right] - E^2\left[f_i\right]\right) \\
&\quad + 2\sum_{i<j} \omega_i \omega_j \left(E\left[f_i f_j\right] - E\left[f_i\right] E\left[f_j\right]\right),
\end{aligned} \tag{10}$$

where the expectation is taken with respect to $D$ and $\mathbf{x}$ is dropped for simplicity. The first sum in eqn. 10 gives the lower bound of the ensemble variance and contains the variances of the ensemble members. The second sum contains the cross terms of the ensemble members and disappears if the models are completely uncorrelated. The reduction of the variance of the ensemble is related to the degree of independence of the single models. This is a key feature of the ensemble approach.

There are several ways to increase model decorrelation. In the case of neural network ensembles, the networks can have different topology, different training algorithms or different training subsets [27,20]. For the case of fixed topology, it is sufficient to use different initial conditions for the network training [26]. Another way of variance reduction is Bagging, where an ensemble of predictors is trained on several bootstrap replicates of the training set [2]. When constructing k-Nearest-Neighbor models, the number of neighbors and the metric coefficients could be used to generate diversity.

Krogh et al. derive the equation $E = \bar{E} - \bar{A}$ which relates the ensemble generalization error $E$ with the average generalization error $\bar{E}$ of the individual models and the variance $\bar{A}$ of the model outputs with respect to the average output. When keeping the average generalization error $\bar{E}$ of the individual models constant, the ensemble generalization error $E$ should decrease with increasing diversity of the models $\bar{A}$. Hence we try to increase $A$ by using two strategies:

1. **Resampling:** We train each model on a randomly drawn subset of 80% of all training samples. The number of models trained for one ensemble is chosen so that usually all samples of the training set are covered at least once by the different subsets.
2. **Variation of Model Type:** We employ two different model types, which are linear models trained by ridge regression and k-nearest-neighbor (k-NN) models with adaptive metric.

# 6   Model Training and Cross Validation

In order to select models for the final ensemble we use a cross validation scheme for model training. As the models are initialized with different parameters (number of hidden units, number of nearest neighbor, initial weights, etc.), cross validation helps us to find a proper value for these model parameters.

The cross validation is done in several training rounds on different subsets of the entire training data. In every training round the data is divided in a training set and a test set. The trained models are compared by evaluating their prediction errors on the unseen data of the test set. The model with the smallest test error is taken out and becomes a member of the ensemble. This is repeated several times and the final ensemble is a simple average over its members. For example a $K$-fold cross validation training leads to an ensemble with $K$ members, where the weights in equ.(8) turn to $\omega_i = \frac{1}{K}$.

# 7   The ENTOOL Toolbox for Statistical Learning

The ENTOOL toolbox for statistical learning is designed to make state-of-the-art machine learning algorithms available under a common interface. It allows construction of single models or ensembles of (heterogenous) models. ENTOOL is Matlab-based with parts written in C++ and runs under Windows and Linux.

## 7.1   ENTOOL Software Architecture

Each model type is implemented as separate class in our simulator, all model classes share common interface. Exchange model types by exchanging constructor call. The system allows for automatic generation of ensembles of models. Models are divided into two brands:

1. Primary models like linear models, neural networks, SVMs etc.
2. Secondary models that rely on primary models to calculate output. All ensemble models are secondary models.

Each selected model goes through three phases: Construction, Training and Evaluation. In the construction phase topology of the model is specified. The model can't be used yet – it has now to be trained on some training data set $(x_i, y_i)$. After training, the model can be evaluated on new/unseen inputs $(x_n)$.

## 7.2  Primary Models Types

The user has a choice of various primary model types:

**ares.** Adaption of Friedman's MARS algorithm
**ridge.** Linear model trained by ridge regression
**perceptron.** Multilayer perceptron with iRPROP+ training
**prbfn.** Shimon Cohen's projection based radial basis function network
**rbf.** Mark Orr's radial basis function code
**vicinal.** k-nearest-neighbor regression with adaptive metric
**mpmr.** Thomas Strohmann's Mimimax Probability Machine Regression
**lssvm.** Johan Suykens' least-square SVM toolbox
**tree.** Adaption of Matlab's build-in regression/classification trees
**osusvm.** SVM code based on Chih-Jen Lin's libSVM
**vicinalclass.** k-nearest-neighbor classification

## 7.3  Secondary Models Types

The user has a choice of various secondary model types which can be used for ensembling or feature selection:

**ensemble.** Virtual parent class for all ensemble classes
**crosstrainensemble.** Ensemble class that trains models according to crosstraining scheme. Creates ensembles of decorrelated models.
**cvensemble.** Ensemble class that trains models according to crossvalidation/out-of-training scheme. Can be used to access OOT error.
**extendingsetensemble.** Boosting variant for regression.
**subspaceensemble.** Creates an ensemble of models where each single model is trained on a random subspace of the input data set.
**optimalsvm.** Wrapper that trains RBF osusvm/lssvm with optimal parameter settings ($C$ and $\gamma$)
**featureselector.** Does feature selection and trains model on selected subset

## 7.4  Experience in Ensembling

All the described methods have been implemented and are available for download from our web-site [11] which contains also manual and installation guide. The toolkit is under continuous development and new features and algorithms are being added. Also in the nearest future we will make available for users an on-line statistical learning service. The toolkit has been thoroughly tested on a variety of problems from ECG modeling, CNN training, financial time series, El Niño real data, physical measurements and many others [24].

## 7.5  Feature Selection

For the purpose of this study we did not adopt a simple strategy of verification of the ABCD rule using computerized methods which has been studied to some extent in [14] and has been the besis of some computer-assisted diagnostic systems [15], [12].

Among the color and signal decompositions several simple descriptors and also combined descriptors have been defined. Among these are lesion geometry descriptors which rely on basic geometric operations on images:

1. Estimated size (px)
2. symmetry (%)
3. Area of the lesion (%)
4. Area of the lesion (px)
5. Area of background (px)
6. Height (px)
7. Width (px)
8. Estimated borderline
9. Borders

Color components are also being used as in any of the medical approaches:

1. Average red
2. Average green
3. Average blue
4. White color (px)
5. Black color (px)
6. Gray-blue (px)
7. Grey-blue
8. Sum of color components

Various binary compositions of of colors and colors of the background can also be considered as characteristic features:

1. Binary sum of GBR
2. Binary RGB composition
3. Binary GRB composition
4. Binary RBG composition
5. Binary BGR composition
6. Binary BRG composition
7. Average red in background
8. Average green in background
9. Average blue in background
10. Sum of background color components

Apart from those commonly used characterizations we use also as features statistical properties of signal components:

1. Average luminance
2. Average $C_r$ component
3. Average comp. $C_b$
4. Average H
5. Average S
6. Average V
7. Average Y
8. Average I
9. Average Q
10. Average Y of background

11. Average I of background
12. Average Q of background
13. Average H of background
14. Average S of background
15. Average V of background
16. Average luminance of background
17. Average $C_r$ component of background
18. Average $C_b$ component of background

## 7.6  Results

The task of the dermatologist is to separate dangerous (cancerous) cases from normal ones. In many instances there is significant overlap in terms of the appearance of the image namely some patients developing cancer will have normal-looking dermoscopic lesion images while some patients in good health will have abnormal-looking diagnostic pictures. The population can be divided into several groups: those diagnosed as abnormal (in this group we have true positives (TP) and false positives (FP)) and those diagnosed as normal among whom are false negatives (FN) and true negatives (TN). For clinical tests often the notions of sensitivity and specificity of the tests are used. Sensitivity is the fraction of abnormal cases that are diagnosed as abnormal $Sen = \frac{TP}{TP+FN}$ while specificity is the fraction of normal cases that are diagnosed as normal $Spec = \frac{TN}{TN+FP}$. For displaying the results of classification so-called ROC curves are used (the name ROC stands for Receiver Operating Characteristics and was introduced by communication engineers for evaluation of performance of digital receivers receiving information over a noisy channel). A ROC curve is simply a plot of the true-positive fraction versus the false-positive fraction. A single threshold value chosen for the classifier will produce a single point on the ROC curve. Below we present two results showing the classifier performance when all the features were used for classification (Fig.9) and when the number of features has been reduced to 15 most significant ones ie. those that, which follows from many tests done, apparently carry most diagnostic information (Fig.10).

It is interesting to analyze the list of most significant features which give best classification results. On this list of 15 features we find:

1. Sum of color comp of the background
2. Average red
3. Average green
4. symmetry (%)
5. Average blue
6. Average Y
7. Average S of background
8. Average Y of background
9. Grey-blue,black and white
10. Average Q
11. Average green of background
12. Average luminance

13. Average V of background
14. Average Q of background
15. Average V

On this list we find standard features used by the dermatologists such as symmetry, grey-bluish, black and white components, coloring (RGB components) - which confirms that these features carry very important diagnostic information. It is interesting to

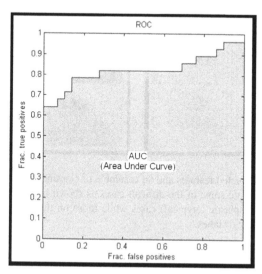

**Fig. 9.** ROC curve showing the quality of the classification when using all described features - AUC = 0.8246

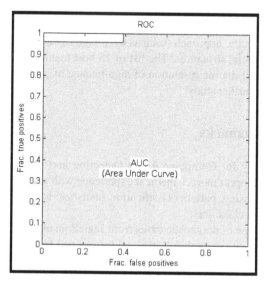

**Fig. 10.** ROC curve showing the quality of classification when using 15 best features as suggested by the sensitivity analysis - AUC = 0.9851

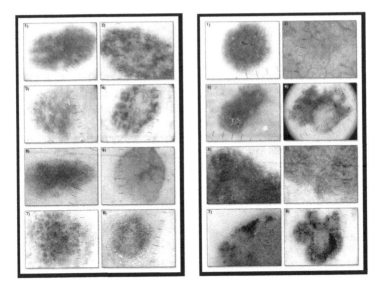

**Fig. 11.** Using 15 best selected features and an ensemble of classifiers using the approach it has been possible to diffrentiate some of the difficult cases as shown in the figure - the images on the left correspond to displastic (atypical) cases while those on the right are melanoma cases confirmed by histopatology studies.

notice that some of the video signal decompositions apparently are very important - average Y, Q and V and average luminance. Most striking conclusion however is that the color and signal components of the background play apparently very significant role. This could be understood that these features bring major contributions for the definition of borderline and also to some of the differential structures. It should be stressed that we did not use any specific approach (such as eg. image segmentation or texture analysis) to find the differential structures. The list of 15 best features has been found using sensitivity analysis and the interpretation of significance of individual features is still missing and requires further study.

## 8    Concluding Remarks

Proposed methodology for Computer Aided Detection and classification of skin lesions for diagnostic support merges medical experience with several cutting-edge technologies: image processing, pattern classification, statistical learning, ensembling techniques of model based classifiers.

For the large database of diagnostic cases from Jagiellonian University Dermatology Clinic which contains entries with full medical history and histopathology support the proposed approach proved to give excellent results giving correct classification of up to 98% of cases. It should be stressed that more thorough study is under way, and especially more melanoma cases have to be used in the training phase of the algorithms. Also as suggested in [32] and developed in [33] other methods such as wavelet based approaches could be combined with learning algorithms. It should be also noticed that the recognition of specific textures as listed in "D" of the ABCD scale has not been taken into account. We suspect that the signal decomposition approach adopted for the purpose of this study (and thus the signal components such as Y, Q, V) and use of averaged quantities contains this type of information.

**Acknowledgements.** The authors would like to thank dr. E. Żabińska and Ana Alekseenko from the Dermatology Clinic of the Jagiellonian University for enlightning discussion and help with preparation of the medical data. This work has been supported by research grant from the Jagiellonian University.

# References

1. Argenziano, G., Fabbrocini, G., Carli, P., DeGiorgi, V., Sammarco, P., Delfino, M.: Epiluminescence Microscopy for the Diagnosis of Doubtful Melanocytic Skin Lesions. Arch. Dermatl. 134, 1563–1570 (1998)
2. Breiman, L.: Bagging predictors. Machine Learning 24(2), 123–140 (1996), http://www.citeseer.ist.psu.edu/breiman96bagging.html
3. Burroni, M., Corona, R., Dell'Eva, G., Sera, F., Bono, R., Puddu, P., Perotti, R., Nobile, F., Andreassi, L., Rubegni, P.: Melanoma Computer-Aided Diagnosis: Reliability and Feasibility Study. Clin. Cancer Res. 10, 1881–1886 (2004)
4. Carli, P., DeGiorgi, V., Massi, D., Giannotti, B.: The ROle of Pattern Analysis and the ABCD rule of dermoscopy in the detection of histological atypia in Melanocytic Naevi. British J. Dematol. 143, 290–297 (2000)
5. Cassileth, B.R., Clark, W.H., Lusk, E.J., et al.: How well do physicians recognize melanoma and other problem lesions? J. Amer. Acad. Dermatol. 4, 555–560 (1986)
6. Chan, H.P., Sahiner, B., Wagner, R.F., Petrick, N.: Classifier Design for Computer-Aided Diagnosis: Effects of Finite Sample Size and on Mean Performance of Classical and Neural Network Classifiers. Med. Phys. 26, 2654–2668 (1999)
7. http://www.dermoscopy.org/atlas/base.htm
8. Dial, W.F.: ABCD rule aids in preoperative diagnosis of malignant melanoma. Cosmetic Dermatol. 8(3), 32–34 (1995)
9. Diepgen, T.L., Eysenbach, G.: Digital Images in Dermatology and the Dermatology Online Atlas on the World Wide Web. J. Dermatol. 25(12), 782–787 (1998)
10. Dreiseitl, S., Ohno-Machado, L., Kittler, H., Vinterbo, S., Billhards, H., Binder, M.A.: Comparison of Machine Learning Methods for the Diagnosis of Pigmented Skin Lesions. J. Biomed. Inform. 34, 28–36 (2001)
11. http://zti.if.uj.edu.pl/~merkwirth/entool.htm
12. Grammatikopoulos, G., Hatzigaidas, A., Papastergiou, A., Lazaridis, P., Zaharis, Z., Kampitaki, D., Tryfon, G.: Automated Malignant Melanoma Detection Using MATLAB. In: Proc. 5th Int. Conf. Data Networks, Communications and Computers, Bucharest, Romania, pp. 91–94 (2006)
13. Grin, C., Kopf, A., Welkovich, B., Bart, R., Levenstein, M.: Accuracy in the clinical diagnosis of melanoma. Arch. Dermatol. 126, 763–766 (1990)
14. Grzymala-Busse, P., Grzymala-Busse, J.W., Hippe, Z.S.: Melanoma prediction using data mining system LERS. In: Computer Software and Applications Conference, COMPSAC 2001, pp. 615–620 (2001)

15. Grzymala-Busse, J.W., Hippe, Z.S.: Postprocessing of rule sets induced from a melanoma data set. In: Proc. Computer Software and Applications Conference, COMPSAC 2002, pp. 1146–1151 (2002)
16. Hall, P.N., Claridge, E., Smith, J.D.: Computer Screening for Early Detection of Melanoma: Is there a Future? British J. Dermatol. 132, 325–328 (1995)
17. Hastie, T., Tibshirani, R., Friedman, J.: The Elements of Statistical Learning. Springer Series in Statistics. Springer, Heidelberg (2001)
18. Iyatomi, H., Oka, H., Hasimoto, M., Tanaka, M., Ogawa, K.: An Internet-based Melanoma Diagnostic System - Toward the Practical Application
19. Johr, R.H.: Dermoscopy: Alternative Melanocytic Algorithms - The ABCD Rule of Dermatoscopy, Menzies Scoring Method, and 7-Point Checklist, Clinics in Dermatology (Elsevier), vol. 20, pp. 240–247 (2002)
20. Krogh, A., Vedelsby, J.: Neural network ensembles, cross validation, and active learning. In: Tesauro, G., Touretzky, D., Leen, T. (eds.) Advances in Neural Information Processing Systems, vol. 7, pp. 231–238. MIT Press, Cambridge (1995), http://www.citeseer.ist.psu.edu/krogh95neural.html
21. Krogh, A., Sollich, P.: Statistical mechanics of ensemble learning. Physical Review E 55(1), 811–825 (1997)
22. Menzies, S.W.: Automated Epiluminescence Microscopy: Human vs Machine in the Diagnosis of Melanoma. Arch. Dermatol. 135, 1538–1540 (1999)
23. Menzies, S.W.: A method for the diagnosis of primary cutaneous melanoma using surface microscopy. Dermatol. Clin. 19, 299–305 (2001)
24. Merkwirth, C., Wichard, J., Ogorzałek, M.J.: Ensemble Modeling for Bio-medical Applications. In: Mitkowski, W., Kacprzyk, J. (eds.) Model. Dyn. in Processes & Sys. SCI, vol. 180, pp. 119–135. Springer, Heidelberg (2009)
25. Nachbar, F., Stolz, W., Merkle, T., Cognetta, A.B., Vogt, T., Landthaler, M., Bilek, P., Braun-Falco, O., Plewig, G.: The ABCD rule of dermatoscopy. High prospective value in the diagnosis of doubtful melanocytic skin lesions. Journal of the American Academy of Dermatology 30(4), 551–559 (1994)
26. Naftaly, U., Intrator, N., Horn, D.: Optimal ensemble averaging of neural networks. Network, Comp. Neural Sys. 8, 283–296 (1997)
27. Perrone, M.P., Cooper, L.N.: When Networks Disagree: Ensemble Methods for Hybrid Neural Networks. In: Mammone, R.J. (ed.) Neural Networks for Speech and Image Processing, pp. 126–142. Chapman and Hall, Boca Raton (1993)
28. Provost, N., Kopf, A.W., Rabinovitz, H.S., Stolz, W., De David, M., Wasti, Q., Bart, R.S.: Comparison of Conventional Photographs and Telephonically Transmitted Compressed Digitized Images of Melanomas and Dysplastic Nevi. Dermatology 196, 299–304 (1998)
29. Schmid-Saugeon, P., Guillod, J., Thiran, J.-P.: Towards a Computer-aided diagnosis System for Pigmented Skin Lesions. Computerized Medical Imaging and Graphics, 65–78 (2003)
30. Stolz, W., Riemann, A., Cognetta, A.B., Pillet, L., Abmayr, W., Hölzel, D., Bilek, P., Nachbar, F., Landthaler, M., Braun-Falco, O.: ABCD rule of dermatoscopy: a new practical method for early recognition of malignant melanoma. Eur. J. Dermatol. 7, 521–528 (1994)
31. Stolz, W., Braun-Falco, O., Bilek, P., Landthaler, M.: Farbatlas der Dermatoskopie. Blackwell Wiss.-Verl., Berlin (1993)
32. Surówka, G., Merkwirth, C., Żabińska-Płazak, Graca, A.: Wavelet based classification of skin lesion images. Bio-Algorithms and Med Systems 2(4), 43–50 (2006)
33. Surówka, G.: Supervised learning of melanotic skin lesion images. In: Human-Computer Systems Interaction. Backgrounds and Applications. Advances in Intelligent and Soft Computing. Springer, Heidelberg (2009) (in preparation)

# Affective Man-Machine Interface: Unveiling Human Emotions through Biosignals

Egon L. van den Broek[1], Viliam Lisý[2], Joris H. Janssen[3,4],
Joyce H.D.M. Westerink[3], Marleen H. Schut[5], and Kees Tuinenbreijer[5]

[1] Center for Telematics and Information Technology (CTIT), University of Twente
P.O. Box 217, 7500 AE Enschede, The Netherlands
vandenbroek@acm.org
[2] Agent Technology Center, Dept. of Cybernetics, FEE, Czech Technical University
Technická 2, 16627 Praha 6, Czech Republic
viliam.lisy@agents.felk.cvut.cz
[3] User Experience Group, Philips Research
High Tech Campus 34, 5656 AE Eindhoven, The Netherlands
{joris.h.janssen,joyce.westerink}@philips.com
[4] Dept. of Human Technology Interaction, Eindhoven University of Technology
P.O. Box 513, 5600 MB, Eindhoven, The Netherlands
j.h.janssen@tue.nl
[5] Philips Consumer Lifestyle Advanced Technology
High Tech Campus 37, 5656 AE Eindhoven, The Netherlands
{marleen.schut,kees.tuinenbreijer}@philips.com

**Abstract.** As is known for centuries, humans exhibit an electrical profile. This profile is altered through various psychological and physiological processes, which can be measured through biosignals; e.g., electromyography (EMG) and electrodermal activity (EDA). These biosignals can reveal our emotions and, as such, can serve as an advanced man-machine interface (MMI) for empathic consumer products. However, such a MMI requires the correct classification of biosignals to emotion classes. This chapter starts with an introduction on biosignals for emotion detection. Next, a state-of-the-art review is presented on automatic emotion classification. Moreover, guidelines are presented for affective MMI. Subsequently, a research is presented that explores the use of EDA and three facial EMG signals to determine neutral, positive, negative, and mixed emotions, using recordings of 21 people. A range of techniques is tested, which resulted in a generic framework for automated emotion classification with up to 61.31% correct classification of the four emotion classes, without the need of personal profiles. Among various other directives for future research, the results emphasize the need for parallel processing of multiple biosignals.

*That men are machines (whatever else they may be) has long been suspected; but not till our generation have men fairly felt in concrete just what wonderful psycho-neuro-physical mechanisms they are.*

William James (1893; 1842 – 1910)

A. Fred, J. Filipe, and H. Gamboa (Eds.): BIOSTEC 2009, CCIS 52, pp. 21–47, 2010.
© Springer-Verlag Berlin Heidelberg 2010

# 1  Introduction

Despite the early work of William James and others before him, it took more than a century before emotions became widely acknowledged and embraced by science and engineering. However, currently it is generally accepted that emotions cannot be ignored; they influence us, be it consciously or unconsciously, in a wide variety of ways [1]. Let us briefly denote four issues on how emotions influence our lives:

- long term physical well-being; e.g., Repetitive Strain Injury (RSI) [2], cardiovascular issues [3,4], and our immune system [5,6];
- physiological reactions / biosignals; e.g., crucial in communication [7,8,9,10];
- cognitive processes; e.g., perceiving, memory, reasoning [8,11]; and
- behavior; e.g., facial expressions [7,8,12].

As is illustrated by the three ways emotions influence us, we are (indeed) *psycho-neuro-physical mechanisms* [13,14], who both send and perceive biosignals that can be captured; e.g., electromyography (EMG), electrocardiography (ECG), and electrodermal activity (EDA). See Table 1 for an overview. These biosignals can reveal a plethora of characteristics of people; e.g., workload, attention, and emotions.

In this chapter, we will focus on biosignals that have shown to indicate people's emotional state. Biosignals form a promising alternative for emotion recognition compared to:

- facial expressions assessed through computer vision techniques [12,15,16]: recording and processing is notoriously problematic [16],
- movement analysis [15,17]: often simply not feasible in practice, and
- speech processing [12,18,19]: speech is often either absent or suffering from severe distortions in many real-world applications.

Moreover, biosignals have the advantage that they are free from social masking and have the potential of being measured by non-invasive sensors, making them suited for a wide range of applications [20,21]. Hence, such biosignals can act as a very useful interface between man and machine; e.g., computers or consumer products such as a mp3-player [22]. Such an advanced Man-Machine Interface (MMI) would provide machines with empathic abilities, capable of coping with the denoted issues.

In comparison to other indicators, biosignals have a number of methodological advantages as well. First of all, traditional emotion research uses interviews, questionnaires, and expert opinions. These, however, can only reveal subjective feelings, are very limited in explaining, and do not allow real time measurements: they can only be used before or after emotions are experienced [7,8,10]. Second, the recent progress in brain imaging techniques enables the inspection of brain activity while experiencing emotions; e.g., EEG and fMRI [11,28]. Although EEG techniques are slowly brought to practice; e.g., Brain Computer Interfacing (BCI) [29,30], these techniques are still very obtrusive. Hence, they are not usable in real world situations; e.g., for the integration in consumer products. As a way between these two research methods, psychophysiological (or bio)signals can be used [7,8,10,14]. These are not, or at least less, obtrusive, can be recorded and processed real time, are rich sources of information, and are relatively cheap to apply.

**Table 1.** An overview of common biosignals/ physiological signals and their features, as used for emotion analysis and classification

| physiology | features | unit | remark |
|---|---|---|---|
| cardiovascular activity [23] | heart rate (HR) | beats / min | |
| *through ECG or BVP* | SD IBIs | $s$ | heart rate variability (HRV) index |
| | RMSSD IBIs | $s$ | heart rate variability (HRV) index |
| | LF power (0.05Hz - 0.15Hz) | $ms^2$ | sympathetic activity |
| | HF power (0.15Hz - 0.40Hz) | $ms^2$ | parasympathetic activity |
| | VLF power ( $<$ 0.05Hz) | $ms^2$ | |
| | LF / HF | | |
| | pulse transit time (PTT) | $ms$ | |
| electrodermal activity (EDA) [24] | mean, SD SCL | $\mu S$ | tonic sympathetic activity |
| | number of SCRs | | rate phasic activity |
| | SCR amplitude | $\mu S$ | phasic activity |
| | SCR 1/2 recovery time | $s$ | |
| | SCR rise time | $s$ | |
| skin temperature [25] | mean, SD temp | $^\circ C(F)$ | |
| respiration [25] | respiration rate | | |
| | amplitude respirations | | |
| | respiratory sinus arrythmia | | |
| muscle activity | mean, SD corrugator supercilii | $\mu V$ | frowning |
| *through EMG* [26, 27] | mean, SD zygomaticus major | $\mu V$ | smiling |
| | mean, SD upper trapezius | $\mu V$ | |
| | mean, SD inter-blink interval | $ms$ | |

Legend: ECG: electrocardiogram; BVP: blood volume pulse; EMG: electromyogram; IBI: inter-beat interval; LF: low frequency; HF: high frequency; VLF: very low frequency; SCL: skin conductance level; SCR: skin conductance response; SD: standard deviation; RMSSD: root mean sum of square differences. See also Fig. 3 for plots of the three facial EMG signals and the EDA signal.

A number of prerequisites should be taken into account when using either traditional methods (e.g., questionnaires), brain imaging techniques, or biosignals to infer people's emotional state. In Van den Broek et al. (2009), these are denoted for affective signal processing (ASP); however, most of them also hold for brain imaging, BCI, and traditional methods. The prerequisites include:

1. the validity of the research employed,
2. triangulation; i.e., using multiple information sources (e.g., biosignals) and/or analysis techniques, and
3. inclusion and exploitation of signal processing knowledge ( e.g., determine the Nyquist frequencies of biosignals for emotion classification).

For a discussion on these topics, we refer to Van den Broek et al. (2009). Let us now assume that all prerequisites can be satisfied. Then, it is feasible to classify the biosignals in terms of emotions. In bringing biosignals-based emotion recognition to products, self-calibrating, and automatic classification is essential to make it useful for Artificial Intelligence (AI) [1,31], Ambient Intelligence (AmI) [20,32], MMI [7,33], and robotics [34,35].

In the pursuit toward empathic technology, we will describe our work on the automatic classification of biosignals. In the next section, we provide an overview of previous work. Section 3 provides an introduction to the classification techniques employed. Subsequently, in Sect. 4, we present the experiment in which we used four biosignals signals: three facial EMGs and EDA. After that, in Sect. 5, we will briefly introduce the preprocessing techniques employed. This is followed by Sect. 6 in which the classification results are presented. In Sect. 7, we reflect on our work and critically review it. Finally, in Sect. 8 we end with drawing the main conclusions.

## 2   Background

A broad range of biosignals are used in affective sciences; see Table 1. To enable processing of the signals, in most cases comprehensive sets of features have to be identified for each biosignal; see also Table 2. To extract these features, affective signals are processed in the time (e.g., statistical moments), frequency (e.g., Fourier), time-frequency (e.g., wavelets), or power domain (e.g., periodogram and autoregression) [36]. In Table 1, we provide a brief overview of the biosignals most often applied, including their best known features, with reference to their physiological source. In the next paragraph, we describe the signals and their psychological counterparts.

First, electrocardiogram (ECG; measured with electrodes on the chest) and blood volume pulse (BVP; measured with infra-red light around the finger or ear) can be used to derive heart beats. The main feature extracted from these heart beats is heart rate (HR; i.e., the number of beats per minute). HR is, however, not very useful in discriminating emotions as it is innervated by many different processes. Instead, the heart rate variability (HRV) provides better emotion information. HRV is more constant in situations where you are happy and relaxed, whereas it shows high variability in more stressful situations [20,55,56]. Second, respiration is often measured with a gauge band around the chest. Respiration rate and amplitude mediate the HRV and are, therefore,

**Table 2.** An overview of 20 studies on automatic classification of emotions, using biosignals / physiological signals

| information source | year | signals | parti-cipants | number of features | selection / reduction | classifiers | target | classification result |
|---|---|---|---|---|---|---|---|---|
| [37] Sinha & Parsons | 1996 | M | 27 | 18 | | LDA | 2 emotions | 86% |
| [9] Picard et al. | 2001 | C,E,R,M | 1 | 40 | SFS, Fisher | LDA | 8 emotions | 81% |
| [38] Scheirer et al. | 2002 | C,E | 24 | 5 | Viterbi | HMM | 2 frustrations | 64% |
| [39] Nasoz et al. | 2003 | C,E,S | 31 | 3 | | k-NN, LDA | 6 emotions | 69% |
| [40] Takahashi | 2003 | C,E,B | 12 | 18 | | SVM | 6 emotions | 42% |
| [41] Haag et al. | 2004 | C,E,S,M,R | 1 | 13 | | MLP | valence / arousal | 64 – 97% |
| [42] Kim et al. | 2004 | C,E,S | 175 | 10 | | SVM | 3 emotions | 78% |
| [43] Lisetti & Nasoz | 2004 | C,E,S | 29 | 12 | | k-NN, LDA, MLP | 6 emotions | 84% |
| [44] Wagner et al. | 2005 | C,E,R,M | 1 | 32 | SFS, Fisher | k-NN, LDA, MLP | 4 emotions | 92% |
| [45] Yoo et al. | 2005 | C,E | 6 | 5 | | MLP | 4 emotions | 80% |
| [46] Choi & Woo | 2005 | E | 1 | 3 | PCA | MLP | 4 emotions | 75% |
| [47] Healey & Picard | 2005 | C,E,R,M | 9 | 22 | Fisher | LDA | 3 stress levels | 97% |
| [34] Liu et al. | 2006 | C,E,M,S | 14 | 35 | | RT | 3 anxiety levels | 70% |
| [48] Rani et al. | 2006 | C,E,S,M,P | 15 | 46 | | k-NN, SVM, RT, BN | 3 emotions | 86% |
| [49] Zhai & Barreto | 2006 | C,E,S,P | 32 | 11 | | SVM | 2 stress levels | 90% |
| [50] Jones & Troen | 2007 | C,E,R | 13 | 11 | | ANN 5 valence levels | 5 arousal levels 26 / 57% | 31 / 62% |
| [51] Leon et al. | 2007 | C,E | 8 | 5 | DBI | AANN | 3 emotions | 71% |
| [52] Liu et al. | 2008 | C,E,S,M | 6 | 35 | | SVM | 3 affect states | 83% |
| [53] Katsis et al. | 2008 | C,E,M,R | 10 | 15 | | SVM, ANFIS | 4 affect states | 79% |
| [54] Yannakakis & Hallam | 2008 | C,E | 72 | 20 | ANOVA | SVM, MLP | 2 fun levels | 70% |
| [33] Kim & André | 2008 | C,E,M,R | 3 | 110 | SBS | LDA, DC | 4 emotions | 70 / 95% |

Signals: C: cardiovascular activity; E: electrodermal activity; R: respiration; M: electromyogram; S: skin temperature; P: pupil diameter. classifiers: MLP: MultiLayer Perceptron; HMM: Hidden Markov Model; RT: Regression Tree; BN: Bayesian Network; ANN: Artificial Neural Network; AANN: Auto-Associative Neural Network; SVM: Support Vector Machine; LDA: Linear Discriminant Analysis; k-NN: k-Nearest Neighbors; ANFIS: Adaptive Neuro-Fuzzy Inference System; DBI: Davies-Bouldin Index; PCA: Principal Component Analysis; SFS: Sequential Forward Selection; SBS: Sequential Backward Selection; DC: Dichotomous Classification.

often used in combination, which is called respiratory sinus arrhythmia (RSA) [25]. RSA is primarily responsive to relaxation and emotion regulation [57]. Third, electrodermal activity (EDA) measures the skin conductance of the hands or foots. This is primarily a response to increases in arousal. Beside the general skin conductance level (SCL), typical peaks in the signal, called skin conductance responses (SCRs), can be extracted. These responses are more event related and are valuable when looking at short timescales. Fourth, skin temperature, measured at the finger, is also responsive to increases in arousal, but does not have the typical response peaks as EDA has. Finally, electromyogram (EMG) measures muscle tension. In relation to emotions, this is most often applied in the face, where it can measure smiling and frowning [58,59].

When processing such biosignals some general issues have to be taken in consideration:

1. Biosignals are typically derived through non-invasive methods to determine changes in physiology [21] and, as such, are indirect measures. Hence, a delay between the actual physiological change and the recorded change in the biosignal has to be taken into account.
2. Physiological sensors are unreliable; e.g., they are sensitive to movement artifacts and to differences in bodily position.
3. Some sensors are obtrusive, preventing their integration in real world applications [20,22].
4. Biosignals are influenced by (the interaction among) a variety of factors [36,60]. Some of these sources are located internally (e.g., a thought) and some are among the broad range of possible external factors (e.g., a signal outside). This makes affective signals inherently noisy, which is most prominent in real world applications.
5. Physiological changes can evolve in a matter of milliseconds, seconds, minutes or even longer. Some changes hold for only a brief moment, while others can even be permanent. Although seldom reported, the expected time windows of change are of interest [20,22]. In particular since changes can add to each other, even when having a different origin.
6. Biosignals have large individual differences. On the one hand, this calls for methods and models tailored to the individual. It has been shown that personal approaches increase the performance of ASP [20,33,50]. On the other hand, generic features are of the utmost importance. Not in all situations, a system or product can be calibrated. Moreover, directing the quest too fast towards people's personal profiles could diminish the interest in generic features and, consequently, limit the progress in research towards them.

The features obtained from the biosignals (see Table 1) can be fed to pattern recognition methods (see Table 2); cf. [29]. These can be classified as: template matching, syntactic or structural matching, and statistical classification; e.g., artificial neural networks (ANN). The former two are not or seldom used in ASP, most ASP schemes use the latter.

Statistical pattern recognition distinguishes supervised and unsupervised (e.g., clustering) pattern recognition; i.e., respectively, with or without a set of (labeled) training data [61,62,63]. With unsupervised pattern recognition, the distance / similarity measure used and the algorithm applied to generate the clusters are key elements.

Supervised pattern recognition relies on learning from a set of examples (i.e., the training set). Statistical pattern recognition uses input features, a discriminant function (or network function for ANN) to recognize the features, and an error criterion in its classification process.

In the field of ASP, several studies have been conducted, using a broad range of signals, features, and classifiers; see Table 2 for an overview. Nonetheless, both the recognition performance and the number of emotions that the classifiers were able to discriminate are disappointing. Moreover, comparing the different studies is problematic because of:

1. The different settings the studies were applied in, ranging from controlled lab studies to real world testing;
2. The type of emotion triggers used;
3. The number of target states to be discriminated; and
4. The signals and features employed.

To conclude, there is a lack of general standards, low prediction accuracy, and inconsistent results. However, for affective MMI to come to fruition, it is eminent to deal with these issues. This illustrates the need for a well documented general framework. In this chapter, we set out to initiate its development, explore various possibilities, and apply it on a data set that will be introduced in the next section.

## 3  Techniques for Classification

In this section, we briefly introduce the techniques used in the research conducted, for those readers who are not familiar with them. Figure 1 presents the complete processing scheme of this research. The core of processing scheme consists of three phases, in which various techniques were applied.

First, analysis of variance (ANOVA) and principal component analysis (PCA) are introduced that enabled the selection of a subset of features for the classification of the emotions. Second, the classification was done using k-nearest neighbors (k-NN), support vector machines (SVM), and artificial neural networks (ANN), which will be briefly introduced later in this section. Third and last, the classifiers were evaluated using leave-one-out cross validation (LOOCV), which will be introduced at the end of this section.

### 3.1  Analysis of Variance (ANOVA)

Analysis of variance (ANOVA) is a statistical test to determine whether or not there is a significant difference between the means of several data sets. ANOVA examines the variance of data set means compared to within class variance of the data sets themselves. As such, ANOVA can be considered as an extension of the t-test, which can only be applied on one or two data sets. We will sketch the main idea here. For a more detailed explanation, we refer to Chapter 6 of [64].

ANOVA assumes that the data sets are independent and randomly chosen from a normal distribution. Moreover, it assumes that all data sets are equally distributed. These

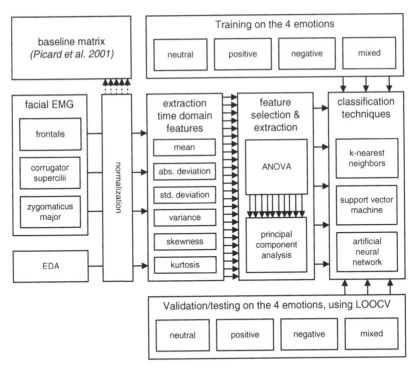

Legend: EMG: electromyography EDA: electrodermal activity; ANOVA of variance; LOOCV: leave-one-out cross validation.

**Fig. 1.** The complete processing scheme, as applied in the current research

assumptions usually hold with empirical data. Moreover, the test is fairly robust against limited violations.

Assume we have $D$ data sets. For each data set $d$, the sum $t_d$ and mean $\bar{s}_d$ of all samples are defined as:

$$t_d = \sum_{i=0}^{S-1} x_{id} \quad \text{and} \quad \bar{s}_d = \frac{t_d}{s_d}$$

where $x_{id}$ denotes one data sample and $s_d$ denotes the number of samples of data set $d$. Subsequently, the grand sum $T$ and the total number of data samples $S$ can be defined as:

$$T = \sum_{d=0}^{D-1} t_d \quad \text{and} \quad S = \sum_{d=0}^{D-1} s_d.$$

The total sum of squares $SS$ (i.e., the quadratic deviation from the mean) can be written as the sum of two independent components:

$$SS_H = \sum_{d=0}^{D-1} \frac{t_d^2}{s_d^2} - \frac{T^2}{S} \quad \text{and} \quad SS_E = \sum_{d=0}^{D-1} \sum_{i=0}^{S-1} x_{id}^2 - \sum_{d=0}^{D-1} \frac{t_d^2}{s_d^2},$$

where indices $H$ and $E$ denote hypothesis and error, as is tradition in social sciences. Together with $S$ and $D$, these components define the ANOVA statistic:

$$F(D - 1, S - D) = \frac{S - D}{D - 1} \cdot \frac{SS_H}{SS_E},$$

where $D - 1$ and $S - D$ can be considered as the degrees of freedom.

The hypothesis that all data sets were drawn from the same distribution is violated if

$$F_\alpha(D - 1, S - D) < F(D - 1, S - D),$$

where $F_\alpha$ denotes the ANOVA statistic that accompanies chance level $\alpha$, considered to be acceptable. Often $\alpha$ is chosen as either $0.05, 0.01$, or $0.001$. If $\alpha < 0.05$ the data sets are assumed to be different.

## 3.2 Principal Component Analysis (PCA)

Through principal component analysis (PCA), the dimensionality of a data set of interrelated variables can be reduced, preserving its variation as much as possible. The variables are transformed to a new set of uncorrelated but ordered variables: the principal components. The first principal component represents, as much as possible, the variance of the original variables. Each succeeding component represents the remaining variance, as much as possible. For a brief introduction on PCA, we refer to Chapter 12 of [64].

Suppose we have a set of data, each represented as a vector $x$, which consists of $n$ variables. Then, the principal components are defined as a linear combination $\alpha \cdot x$ of the variables of $x$ that preserves the maximum of the (remaining) variance, denoted as:

$$\alpha \cdot x = \sum_{i=0}^{n-1} \alpha_i x_i,$$

where $\alpha = (\alpha_0, \alpha_1, \ldots, \alpha_{n-1})^T$. The variance covered by each principal component $\alpha \cdot x$ is defined as:

$$var(\alpha \cdot x) = \alpha \cdot C\alpha,$$

where $C$ is the covariance matrix of $x$.

To find all principal components, we need to find the maximized $var(\alpha \cdot x)$ for them. Hereby, the constraint $\alpha \cdot \alpha = 1$ has to be taken into account. The standard approach to do so is the technique of Lagrange multipliers. We maximize

$$\alpha \cdot C\alpha - \lambda \left( \sum_{i=0}^{n-1} \alpha_i^2 - 1 \right) = \alpha \cdot C\alpha - \lambda(\alpha \cdot \alpha - 1),$$

where $\lambda$ is a Lagrange multiplier. Subsequently, we can derive that $\lambda$ is an eigenvalue of $C$ and $\alpha$ is its corresponding eigenvector.

Once obtained the vectors $\alpha$, a transformation can be made that maps all data $x$ to its principal components:

$$x \rightarrow (\alpha_0 \cdot x, \alpha_1 \cdot x, \ldots, \alpha_{n-1} \cdot x)$$

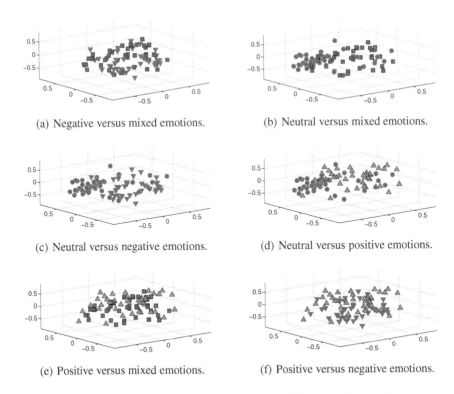

(a) Negative versus mixed emotions.          (b) Neutral versus mixed emotions.

(c) Neutral versus negative emotions.         (d) Neutral versus positive emotions.

(e) Positive versus mixed emotions.           (f) Positive versus negative emotions.

**Fig. 2.** Visualization of the first three principal components of all six possible combinations of two emotion classes. The emotion classes are plotted per two to facilitate the visual inspection.The plots illustrate how difficult it is to separate even two emotion classes, where separating four emotion classes is the aim. However, note that the emotion category neutral can be best separated from the other three categories: mixed, negative, and positive emotions, as is illustrated in b), c), and d).

Note that the principal components are sensitive to scaling. In order to tackle this problem, the components can be derived from the correlation matrix instead of the covariance matrix. This is equivalent to extracting the principal components in the described way after normalization of the original data set to unit variance.

PCA is also often applied for data inspection through visualization, where the principal components are chosen along the figure's axes. Figure 2 presents such a visualization: for each set of two emotion classes, of the total of four, a plot denoting the first three principal components is presented.

### 3.3   k-Nearest Neighbors (k-NN)

k-nearest neighbors (k-NN) is a very intuitive, simple, and often applied machine learning algorithm. It requires only a set of labeled examples (i.e., data vectors), which form the training set.

Now, let us assume that we have a training set $x^l$ and a set of class labels $C$. Then, each new vector $x_i$ from the data set is classified as follows:

1. Identify $k$ vectors from $x^l$ that are closest to vector $x_i$, according to a metric of choice; e.g., city block, Euclidean, or Mahalanobis distance.
2. Class $c_i$ that should be assigned to vector $x_i$ is determined by:

$$c_i = \operatorname*{argmax}_{c \in C} \sum_{i=0}^{k-1} w_i \gamma(c, c_i^l),$$

where $\gamma(.)$ denotes a boolean function that returns 1 when $c = c_i^l$ and 0 otherwise and

$$w_i = \begin{cases} 1 & \text{if } \delta(x_i, x_i^l) = 0; \\ \frac{1}{d(x_i, x_i^l)^2} & \text{if } \delta(x_i, x_i^l) \neq 0, \end{cases}$$

where $\delta(.)$ denotes the distance between vectors $x_i$ and $x_i^l$. Note that, if preferred, the factor weight can be simply eliminated by putting $w_i = 1$.
3. If there is a tie of two or more classes $c \in C$, vector $x_i$ is randomly assigned to one of these classes.

The algorithm presented applies to k-NN for weighted, discrete classifications, as will be applied in the current research. However, a simple adaptation can be made to the algorithm, which enables continuous classifications. For more information on these and other issues, we refer to the various freely available tutorials and introductions that have been written on k-NN.

### 3.4  Support Vector Machine (SVM)

Using a suitable kernel function, a support vector machine (SVM) ensures the division of a set of data into two classes, with respect to the shape of the classifier and misclassification of the training samples. The main idea of SVM can be best explained with the example of a binary linear classifier.

Let us define our data set as:

$$D = \{(x_i, c_i) | x_i \in \mathbb{R}^d, c_i \in \{-1, 1\}\} \text{ for } i = 0, 1, \ldots, N - 1,$$

where $x_i$ is a vector with dimensionality $d$ from the data set, which has size $N$. $c_i$ is the class to which $x_i$ belongs. To separate two classes, we need to formulate a separating hyperplane $w \cdot x = b$, where $w$ is a normal vector of length 1, $x$ is a feature vector, and $b$ is a constant.

In practice, it is often not possible to find such a linear classifier. In this case, the problem can be generalized. Then, we need to find $w$ and $b$ so that we can optimize

$$c_i(w \cdot x_i + b) \leq \xi_i,$$

where $\xi_i$ represents the deviation (or error) from the linearly separable case.

To determine an optimal plane, the sum of $\xi_i$ must be minimized. The minimization of this parameter can be solved by Lagrange multipliers $\alpha_i$. From the derivation of this method, it is possible to see that often most of the $\alpha_i$s are equal to $0$. The remaining relevant subset of the training data $x$ is denoted as the support vectors. Subsequently, the classification is performed as:

$$f(x) = \mathrm{sgn}\left( \sum_{i=0}^{S-1} c_i \alpha_i x \cdot x_i + b \right),$$

where $S$ denotes the number of support vectors.

For a non-linear classification problem, we can replace the dot product by a non-linear kernel function. This enables the interpretation of algorithms geometrically in feature spaces non-linearly related to the input space and combine statistics and geometry. A kernel can be viewed as a (non-linear) similarity measure and induce representations of the data in a linear space. Moreover, the kernel implicitly determines the function class, which is used for learning [63].

The SVM introduced here classified samples in two classes. In the case of multiple classes, two approaches are common: 1) for each class, a classifier can be build that separates that class from the other data and 2) for each pair of classes, classifiers can be build. With both cases, voting paradigms are used to assign the data samples $x_i$ to classes $c_i$. For more information on SVM, [62,63] can be consulted.

## 3.5  Artificial Neural Networks (ANN)

Artificial neural networks (ANN) are inspired by their biological counterparts. Often, ANN are claimed to have a similar behavior as biological neural networks. Although ANN share several features with biological neural networks (e.g., noise tolerance), this claim is hardly justified; e.g., a human brain consists of roughly $10^{11}$ brain cells, where an ANN consists of only a few dozens of units.

Nevertheless, ANN have proved their use for a range of pattern recognition and machine learning applications.

Moreover, ANN have a solid theoretical basis [61,62].

ANN consist of a layer of input units, one or more layers of hidden units, and a layer of output units. These units are connected with a weight $w_{ij}$, which determines the transfer of unit $u_i$ to unit $u_j$. The activation level of a unit $u_j$ is defined as:

$$a_j(t+1) = f(a_j(t), i_j(t)),$$

where $t$ denotes time, $f(.)$ is the activation function that determines the new activation based on the current state $a(t)$ and its effective input, defined as:

$$i_j(t) = \sum_{i=0}^{U_j-1} a_i(t) w_{ij}(t) + \tau_j(t),$$

where $\tau_j(t)$ is a certain bias or offset and $U_j$ denotes the number of units from which a unit $u_j$ can receive input. Note that at the input layer of a ANN, the input comes from the environment; then, $i$ is the environment instead of another unit.

On its own, each neuron of an ANN can only perform a simple task. In contrast, a network of units can approximate any function. Moreover, ANN cannot only process input, they can also learn from their input, either supervised or unsupervised. Although various learning rules have been introduced for ANN, most can be considered as being derived from Hebb's classic learning rule:

$$\Delta w_{ij} = \eta a_i a_j,$$

which defines the modification of the weight of connection $(u_i, u_j)$. $\eta$ is a positive constant. Its rationale is that $w_{ij}$ should be increased with the simultaneous activation of both units and the other way around.

Various ANN topologies have been introduced. The most important ones are recurrent and feed-forward networks, whose units respectively do and do not form a directed cycle through feedback connections. In the current research, a feed-forward network is applied: the classic multilayer perceptron (MLP), as is more often used for emotion recognition purposes; see also Table 2. It incorporated the often adopted sigmoid-shaped function applied to $f(.)$:

$$\frac{1}{1 + e^{-a_j}}$$

Throughout the 60 years of their existence, a broad plethora of ANN have been presented, varying on a range of aspects; e.g., their topology, learning rules, and the choice of either synchronous or asynchronously updating of its units. More information on ANN can be found in various introductions on ANN.

### 3.6  Leave-One-Out Cross Validation (LOOCV)

Assume we have a classifier that is trained, using a part of the available data set: the training data. The training process optimizes the parameters of a classifier to make it fit the training data. To validate the classifier's performance, an independent sample of the same data set has to be used [61,62].

Cross validation deviates from the general validation scheme since it enables the validation of a classifier without the need of an explicit validation set. As such, it optimizes the size of the data set that can be used as training data.

Various methods of cross validation have been introduced. In this section, we will introduce leave-one-out cross validation (LOOCV), a frequently used method to determine the performance of classifiers. LOOCV is typically useful and, consequently, used in the analysis of (very) small data sets. It has been shown that LOOCV provides an almost unbiased estimate of the true generalization ability of a classifier. As such, it provides a good model selection criterion.

Assume we have a classifier (e.g., k-NN, a SVM, or an ANN) of which we want to verify its performance on a particular data set. This data set contains (partly) data samples $x_i$ with known correct classifications $c_i^l$. Then, classifier's performance can be determined through LOOCV, as follows:

1. $\forall_i$ train a classifier $C_i$ with the complete data set $x$, except $x_i$.
2. $\forall_i$ classify data sample $x_i$ to a class $c_i$, using classifier $C_i$.

3. Compute the average error of the classifier through

$$\mathcal{E} = \frac{1}{D} \operatorname*{argmax}_{c \in C} \sum_{i=0}^{D-1} \gamma(c_i, c_i^l),$$

where $D$ denotes the number of data samples and $\gamma(.)$ denotes a boolean function, which returns 1 if $c_i = c_i^l$ and 0 otherwise. Note that $\frac{1}{D}$ can be omitted from the formula if no comparisons are made between data sets (with different sizes).

Instead of one data sample $x_i$, this validation scheme also allows a subset of the data to be put aside. Such a subset can, for example, consist of all data gathered of one person. This enables an accurate estimation of the classification error $\mathcal{E}$ on this unknown person.

The processing scheme as presented here can be adapted in various ways. For example, in addition to the boolean function $\gamma(.)$, a weight function could be used that expresses the resemblance between classes. Hence, not all misclassifications would be judged similarly.

All results reported in this chapter are determined through LOOCV, if not specified in another way. For more information on cross validation, LOOCV in particular, we refer to [62].

## 4   Recording Emotions

We conducted an experiment in which the subjects' emotions were elicited, using film fragments that are known to be powerful in eliciting emotions in laboratory settings; see also [58,59,65]. As biosignals, facial EMG and EDA were recorded. These are known to reflect emotions [66]; see also both Table 1 and Table 2. The research in which the data was gathered is already thoroughly documented in both [58] and [59]. Therefore, we will only provide a brief summary of it.

### 4.1   Participants

In the experiment, 24 subjects (20 females) participated (average age 43 years). Mainly females were solicited to participate since we expected a more and stronger facial emotion expression of females [67]. Consequently, a relative small number of males participated. The biosignal recordings of three subjects either failed or were distorted. Hence, the signals of 21 subjects remained for classification purposes.

### 4.2   Equipment and Materials

We selected 8 film fragments (120 sec. each) for their emotional content. For specifications of these film fragments, see [58,59]. The 8 film fragments were categorized as being neutral or triggering positive, negative, or mixed (i.e., simultaneous negative and positive; [68]) emotions; hence, 2 film fragments per emotion category. This categorization was founded on Russell's valence-arousal model, introduced in [69]. Note that the existence of mixed emotions, the way to determine them, and the method to analyze ratings of the possible mixed emotions is still a topic of debate; e.g., [20,22,68].

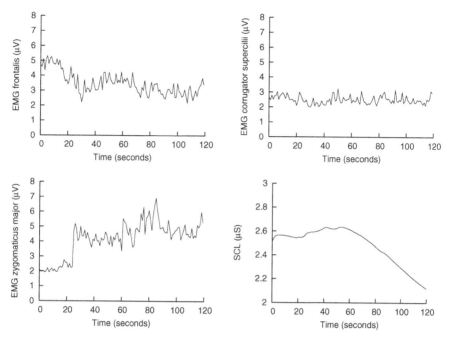

**Fig. 3.** Samples of the electromyography (EMG) in $\mu V$ of the frontalis, the corrugator supercilii, and the zygomaticus major as well as of the electrodermal activity (EDA) in $\mu V$, denoted by the skin conductance level (SCL). All these signals were recorded in parallel, with the same person.

A TMS International Porti5-16/ASD system was used for the biosignal recordings, which was connected to a PC with TMS Portilab software[1]. Three facial EMGs were recorded: the right corrugator supercilii, the left zygomaticus major, and the left frontalis muscle. The EMG signals were high-pass filtered at 20 Hz, rectified by taking the absolute difference of the two electrodes, and average filtered with a time constant of 0.2 sec. The EDA was recorded using two active skin conductivity electrodes and average filtering with a time constant of about 2 sec. See Fig. 3 for samples of the three EMG signals and the EDA signal.

### 4.3 Procedure

After the participant was seated, the electrodes were attached and the recording equipment was checked. The 8 film fragments were presented to the participant in pseudo-random order. A plain blue screen was shown between the fragments for 120 seconds. This assured that the biosignals returned to their baseline level, before the next film fragment was presented.

After the viewing session, the electrodes were removed. Next, the participants answered a few questions regarding the film fragments viewed. To jog their memory, representative print-outs of each fragment were provided.

---

[1] URL of TMS Portilab software: http://www.tmsi.com/

# 5    Preprocessing

The quest towards self-calibrating algorithms for consumer products and for AmI and AI purposes gave some constraints to processing the signals. For example, no advanced filters should be needed, the algorithms should be noise-resistant, and should (preferably) also be able to handle corrupt data. Therefore, we chose to refrain from advanced preprocessing schemes and applied basic preprocessing. Figure 1 presents the complete processing scheme as applied in the current research.

## 5.1    Normalization

Humans are known for their rich variety in all aspects, this is no different for their biosignals. In developing generic classifiers, this required the normalization of the signals. This was expected to boost its performance significantly [48].

For each person, for all his signals, and for all their features separately, the following normalization was applied:

$$x_n = \frac{x_i - \bar{x}}{\sigma},$$

where $x_n$ is the normalized value, $x_i$ the recorded value, $\bar{x}$ the global mean, and $\sigma$ the standard deviation.

Normalization of data (e.g., signals) has been broadly discussed. This has resulted in a variety of normalization functions; e.g., see [24,61,62].

## 5.2    Baseline Matrix

In their seminal article, Picard, Vyzas, and Healey (2001) introduced a baseline matrix for processing biosignals for emotion recognition. They suggested that this could tackle problems due to variation both within (e.g., inter day differences) and between participants. Regrettably, Picard et al. (2001) did not provide evidence for its working.

The baseline matrix requires biosignals recordings while people are in a neutral state. Regrettably, such recordings were not available. Alternatively, one of both available neutral film fragments was chosen [58,59].

In line with Picard et al. (2001), the input data was augmented with the baseline values of the same data set. A maximum performance improvement of 1.5% was achieved, using a k-NN classifier. Therefore, the baseline matrix was excluded in the final processing pipeline.

## 5.3    Feature Selection

To achieve good classification results with pattern recognition and machine learning, the set of input features is crucial. This is no different with classifying emotions [7,8,10]. As was denoted in Sect. 2, biosignals can be processed in the time, frequency, time-frequency, and power domain.

For EMG and EDA signals, the time domain is most often employed for feature extraction; see also Table 1. Consequently, we have chosen to explore a range of features from the time domain: mean, absolute deviation, standard deviation (SD), variance, skewness, and kurtosis. Among these are frequently used features (i.e., mean and SD)

**Table 3.** The best feature subsets from the time domain, for k-nearest neighbor (k-NN) classifier with Euclidean metric. They were determined by analysis of variance (ANOVA), using normalization per signal per participant. EDA denotes the electrodermal activity or skin conductance level.

| feature | EDA | facial electromyography (EMG) | | |
| --- | --- | --- | --- | --- |
| | | frontalis | corrugator supercilii | zygomaticus |
| mean | | | | o |
| absolute deviation | | | | o |
| standard deviation (SD) | | o | | o |
| variance | | o | | o |
| skewness | o | | o | o |
| kurtosis | | | o | |

**Table 4.** The recognition precision of the k-nearest neighbors (k-NN) classifier, with $k = 8$ and the Euclidean metric. The influence of three factors is shown: 1) normalization, 2) analysis of variance (ANOVA) feature selection (FS), and 3) Principal Component Analysis (PCA) transform.

| normalization | no *fs* | ANOVA *fs* (10 features) | ANOVA *fs* & PCA (5 components) |
| --- | --- | --- | --- |
| no | 45.54% | | |
| yes | 54.07% | 60.71% | 60.80% |

and rarely used, but promising, features (i.e., skewness and kurtosis) [58,59]; see also Table 3.

To define an optimal set of features, a criterion function should be defined. However, no such criterion function was available in our case. Thus, an exhaustive search in all possible subsets of input features (i.e., $2^{24}$) was required to guarantee an optimal set [70]. To limit this enormous search space, an ANOVA-based heuristic search was applied.

For both the normalizations, we performed feature selection based on ANOVAs. We selected the features with ANOVA $\alpha \leq 0.001$ (see also Sect. 3), as this led to the best precision. The features selected for each of the biosignals are presented in Table 3.

The last step of preprocessing was PCA; see also Sect. 3. The improvement of the PCA was small compared to feature selection solely. However, it was positive for normalization; see also Table 4. Figure 2 presents for each set of two emotion classes, of the total of four, a plot denoting the first three principal components. As such, the six resulting plots illustrate the complexity of separating the emotion classes.

## 6 Classification Results

This section reports the results of the three classification techniques applied: k-nearest neighbors (k-NN), support vector machines (SVM), and artificial neural networks

**Table 5.** Confusion matrix of the k-NN classifier of EDA and EMG signals for the best reported input preprocessing, with a cityblock metric and $k = 8$

| | | real | | | |
| --- | --- | --- | --- | --- | --- |
| | | neutral | positive | mixed | negative |
| classified | neutral | 71.43% | 19.05% | 9.52% | 14.29% |
| | positive | 9.52% | 57.14% | 9.52% | 21.43% |
| | mixed | 4.76% | 4.76% | 64.29% | 11.90% |
| | negative | 14.29% | 19.05% | 16.67% | 52.38% |

(ANN); see also Sect. 3. In all cases, the features extracted from the biosignals were used to classify participants' neutral, positive, negative, or mixed state of emotion; see also Fig. 2. For the complete processing scheme, we refer to Fig. 1.

### 6.1   k-Nearest Neighbors (k-NN)

For our experiments, we have used MATLAB[2] and a k-NN implementation, based on SOM Toolbox 2.0[3]. Besides the classification algorithm described in Sect. 3.3, we have used a modified version, more suitable for calculating the recognition rates. Its output was not the resulting class, but a probability of classification to each of the classes. This means that if there is a single winning class, the output is 100% for the winning class and 0% for all the other classes. If there is a tie of multiple classes, the output is divided among them and 0% is provided to the rest. All the recognition rates of the k-NN classifier reported in the current study were obtained by using this modified algorithm.

A correct metric is a crucial part of a k-NN classifier. A variety of metrics provided by the `pdist` function in MATLAB[2] was applied. Different feature subsets appeared to be optimal for different classes. Rani et al. (2006) denoted the same issue in their empirical review; cf. Table 3. The results of the best preprocessed input with respect to the four emotion classes (i.e., neutral, positive, negative, and mixed) is 61.31%, with a cityblock metric and $k = 8$; cf. Table 4.

Probability tables for the different classifications given a known emotion category are quite easy to obtain. They can be derived from confusion matrices of the classifiers by transforming the frequencies to probabilities. Table 5 presents the confusion matrix of the k-NN classifier used in this research, with a cityblock metric and $k = 8$.

### 6.2   Support Vector Machines (SVM)

We have used MATLAB[2] environment and a SVM and kernel methods toolbox[4], for experimenting with SVMs. We used input enhanced with the best preprocessing, described in the previous section. It was optimized for the k-NN classifier; however, we

---

[2] MATLAB online: http://www.mathworks.com/products/matlab/

[3] The MATLAB SOM Toolbox 2.0 is available through:
http://www.cis.hut.fi/projects/somtoolbox

[4] The SVM and kernel methods toolbox is available through:
http://asi.insa-rouen.fr/enseignants/ arakotom/toolbox/

expected it to be a good input also for more complex classifiers, including SVM. This assumption was supported by several tests with various normalizations. Hence, the signals were normalized per person, see also Sect. 5. After feature selection, the first 5 principal components from the PCA transformation were selected, see also Sect. 3.

The kernel function of SVM characterizes the shapes of possible subsets of inputs classified into one category [63]. Being SVM's similarity measure, the kernel function is the most important part of an SVM; see also Sect. 3. We applied both a polynomial kernel, with dimensionality $d$, defined as:

$$k_P(x_i, x^l) = (x_i \cdot x^l)^d$$

and a Gaussian (or radial basis function) kernel, defined as:

$$k_G(x_i, x^l) = \exp\left(-\frac{|x_i - x^l|^2}{2\sigma^2}\right),$$

where $x_i$ is a feature vector that has to classified and $x^l$ is a feature vector assigned to a class (i.e., the training sample) [63].

A Gaussian kernel ($\sigma = 0.7$) performed best with 60.71% correct classification. However, a polynomial kernel with $d = 1$ had a similar classification performance (58.93%). All the results were slightly worse than with the k-NN classifier.

### 6.3   Artificial Neural Networks (ANN)

We have used a multi-layer perceptron (MLP) trained by a back-propagation algorithm that was implemented in the neural network toolbox of MATLAB[2]; see also Sect. 3. It used gradient descent with moment and adaptive training parameter. We have tried to recognize only the inputs that performed best with the k-NN classifier.

In order to assess what topology of ANN was most suitable for the task, we conducted small experiments with both 1 and 2 hidden layers. In both cases, we did try 5 to 16 neurons within each hidden layer. All of the possible $12 + 12 \times 12$ topologies were trained, each with 150 cycles and tested using LOOCV.

The experiments using various network topologies supported the claim from [71] that bigger ANN do not always tend to over fit the data. The extra neurons were simply not used in the training process. Consequently, the bigger networks showed good generalization capabilities but did not outperform the smaller ones. A MLP with 1 hidden layer of 12 neurons showed to be the optimal topology.

An alternative method for stopping the adaptation of the ANN is using validation data. For this reason, the data set was split into 3 parts: 1 subject for testing, 3 subjects for validation, and 17 subjects for training. The testing subject was completely removed from the training process at the beginning. The network was trained using 17 randomly chosen training subjects. At the end of each training iteration, the network was tested on the 3 validation subjects.

This procedure led to a 56.19% correct classification of the four emotion classes.

## 6.4  Reflection on the Results

Throughout the last decade, various studies have been presented with similar aims. Some of these studies reported good results on the automatic classification of biosignals that should unveil people's emotions; see Table 2. For example, Picard et al. (2001) reports 81% correct classification on the emotions of one subject [9]. Haag et al. (2004) reports 64%–97% correct classification, using a band function with bandwidth 10% and 20%. This study was conducted on one subject. This study reports promising results but also lacks the necessary details needed for its replication [41]. More recently, Kim and André (2008) reported a recognition accuracy of 95% and 70% for subject-dependent and subject-independent classification. Their study included three subjects [33].

In comparison with [9,41,33], this research incorporated data of a large number of people (i.e., 21), with the aim to develop a generic processing framework. At first glance, with average recognition rates of 60.71% for SVM and 61.31% for k-NN and only 56.19% for ANN, its success is questionable. However, the classification rates differ among the four emotion categories, as is shown in Table 5, which presents the confusion matrix of the results of the k-NN classifier. Neutral emotional states are recognized best, with a classification rate of 71.43%. Negative emotional states are the most complex to distinguish from the other three emotion categories, as is marked by its 52.38% correct classification rate. The complexity of separating the four emotion classes from each other is illustrated in Fig. 2.

Taking in consideration the generic processing pipeline (see also Fig. 1) and the limitations of other comparable research (cf. Table 2), the results reported in this chapter should be judged as (at least) reasonably good. Moreover, a broad range of improvements are possible. One of them would be to question the need of identifying specific emotions, using biosignals for MMI. Hence, the use of alternative, rather rough categorizations, as used in the current research, should be further explored.

With pattern recognition and machine learning, preprocessing of the data is crucial. This phase could also be improved for the biosignals used in the current study. First of all, we think that the feature selection based on an ANOVA was not sufficient for more complex classifiers such as neural networks. The ANOVA tests gathered the centers of random distributions that would generate the data of different categories; hereby assuming that their variances were the same. However, a negative result of this test is not enough to decide that a feature did not contain any information. As an alternative for feature selection, the k-NN classifier could be extended by a metric that would weigh the features, instead of omitting the confusing or less informative features.

Taken it all together, the quest towards affective MMI continues. Although the results presented are good compared to related work, it is hard to estimate whether or not the classification performance is sufficient for embedding of affective MMI in real world applications. However, the future is promising with the rapidly increasing amount of resources allocated for affective MMI and the range of improvements that are possible. This assures that the performance on classification of emotions will achieve the necessary further improvements.

# 7   Discussion

This chapter has positioned *men as machines* in the sense that they are *psycho-neuro-physical mechanisms* [13]. It has to be said that this is a far from new position; it is already known for centuries, although it was rarely exploited in application oriented research. However, in the last decade interest has increased and subareas evolved that utilized this knowledge. This chapter concerns one of them: affective MMI; or as Picard (1997) coined it: affective computing.

A literature overview is provided of the work done so far, see also Table 1 and Table 2. In addition, some guidelines on affective MMI are provided; see Sects. 1 and 2. To enable the recognition of these emotions, they had to be classified. Therefore, a brief description was provided of the classification techniques used (Sect. 3). Next, a study is introduced in which three EMG signals and people's EDA were measured (see also Fig. 3), while being exposed to emotion inducing film fragments; see Sect. 4. See Fig. 1 for an overview of the processing scheme applied in the current research. Subsequently, preprocessing and the automatic classification of biosignals, using the four emotion categories, were presented in Sect. 5 and Sect. 6.

Also in this research, the differences among participants became apparent. They can be denoted on four levels; see also Sect. 1. People have different physiological reactions on the same emotions and that people experience different emotions with the same stimuli (e.g., music or films). Moreover, these four levels interact [7,8,14]. Although our aim was to develop a generic model, one could question whether or not this can be realized. Various attempts have been made to determine people's personal biosignals-profile; e.g., [9,14,33,48]. However, no generally accepted standard has been developed so far.

In pursuit to generic affective MMI processing schemes, the notion of time should be taken into consideration, as was already denoted in Sect. 2. This can help to distinguish between emotions, moods, and personality [20,72,73]:

1. Emotion: A short reaction (i.e., a matter of seconds) to the perception of a specific (external or internal) event, accompanied by mental, behavioral, and physiological changes [7,10].
2. Moods: A long lasting state, gradually changing, in terms of minutes, hours, or even longer. They are experienced without concurrent awareness of their origin and are not object related. Moods do not directly affect actions; however, they do influence our behavior indirectly [7,10,74].
3. Personality: People's distinctive traits and behavioral and emotional characteristics. For example, introvert and extrovert persons express their emotions in distinct ways. Additionally, also self-reports and physiological indicators / biosignals will be influenced by people's personality trait [19,75].

With respect to processing the biosignals, the current research could be extended by a more detailed exploration of the time windows; e.g., with a span of 10 seconds [7,8,10,22]. Then, data from different time frames can be combined and different, better suitable normalizations could be applied to create new features. For example, information concerning the behavior of the physiological signals could be more informative than only the integral features from a large time window. Studying short time frames could also provide a better understanding on the relation between emotions and their physiological correlates / biosignals, see also Table 1.

Other more practical considerations should also be noted. The advances made in wearable computing and sensors facilitates (affective) MMI; e.g., [21]. Last years, various prototypes have been developed, which enable the recording of physiological signals; e.g., [76]. This enables the recordings of various biosignals in parallel. In this way, an even higher probability of correct interpretation can be achieved [7,8,20].

Affective MMI can extent consumer products [22]. For example, a mp3-player could sense its listener's emotions and either provide suggestions for other music or automatically adapt its playing list to these emotions. In addition, various other applications have been proposed, mockups have been presented, and implementations have been made. Three examples of these are clothes with wearable computing, games that tweak its behavior and presentation depending on your emotions, and lighting that reacts on or adapts to your mood.

Affective signal processing (ASP) could possibly bring salvation to AI [1,20]. With understanding and sensing emotions, true AI is possibly (and finally) within reach. Current progress in biomedical and electrical engineering provide the means to conduct affective MMI in an unobtrusive manner and, consequently, gain knowledge about our natural behavior, a prerequisite for modeling it. As AI's natural successor, for AmI [20], even more than for AI, emotions play a crucial role in making it a success. Since AmI was coined by Emile Aarts [32], this has been widely acknowledged and repeatedly stressed; e.g., [20,32].

An extension of MMI is human-robot interaction. With robotics, embodiment is a key factor. Potentially, robots are able to enrich their communication substantially through showing some empathy from time to time. As with AI and AmI, this requires sensing and classification of emotions, as can be conveniently done through biosignals [34,35].

Of interest for affective MMI are also the developments in brain-computer interfacing (BCI) [29,30]. In time, affective BCI will possibly become within science's reach. Affective BCI, but also BCI in general, could advance AI, AmI, and human-robot interaction. Slowly this becomes acknowledged, as is illustrated by a workshop on affective BCI, as was held at the IEEE 2009 International Conference on Affective Computing and Intelligent Interaction[5]. With affective BCI, again both its scientific foundation and its applications will be of interest.

Without any doubt affective MMI has a broad range of applications and can help in making various areas more successful. Taking it all together, the results gathered in this research are promising. However, the correct classification rate is below that what is needed for reliable affective MMI in practice. Providing the range of factors that can be improved, one should expect that the performance can be boosted substantially. That this is not already achieved is not a good sign; perhaps, still some essential mistakes are made. One of the mistakes could be the computationally driven approach. A processing scheme that is founded on or at least inspired by knowledge from both biology, in particular physiology, and psychology could possibly be more fruitful . . .

[5] The IEEE 2009 International Conference on Affective Computing and Intelligent Interaction: http://www.acii2009.nl/

# 8   Conclusions

Affective MMI through biosignals is perhaps the ultimate blend of biomedical engineering, psychophysiology, and AI. However, in its pursuit, various other disciplines (e.g., electrical engineering and psychology) should not be disregarded. In parallel, affective MMI promotes the quest towards its scientific foundation and screams for its application [7,8,10]. As such, it is next generation science and engineering, which truly bridges the gap between man and machine.

As can be derived from this chapter, still various hurdles have to be taken in the development of a generic, self-calibrating, biosignal-driven classification framework for affective MMI. The research and the directives denoted here could help in taking some of these hurdles. When the remaining ones will also be taken; then, in time, the common denominators of people's biosignals can be determined and their relation with experienced emotions can be further specified. This would mark a new, biosignal-driven, era of advanced, affective MMI.

**Acknowledgements.** The authors thank Leon van den Broek (Radboud University Nijmegen, The Netherlands / University of Utrecht, The Netherlands), Frans van der Sluis (University of Twente, The Netherlands), and Marco Tiemann (Philips Research, The Netherlands) for their reviews of this book chapter. Furthermore, we thank the editors for inviting us to write a chapter for their book.

# References

1. Picard, R.W.: Affective Computing. MIT Press, Boston (1997)
2. van Tulder, M., Malmivaara, A., Koes, B.: Repetitive strain injury. The Lancet 369(9575), 1815–1822 (2007)
3. Schuler, J.L.H., O'Brien, W.H.: Cardiovascular recovery from stress and hypertension factors: A meta-analytic view. Psychophysiology 34(6), 649–659 (1997)
4. Frederickson, B.L., Manusco, R.A., Branigan, C., Tugade, M.M.: The undoing effect of positive emotions. Motivation and Emotion 24(4), 237–257 (2000)
5. Ader, R., Cohen, N., Felten, D.: Psychoneuroimmunology: Interactions between the nervous system and the immune system. The Lancet 345(8942), 99–103 (1995)
6. Solomon, G.F., Amkraut, A.A., Kasper, P.: Immunity, emotions, and stress with special reference to the mechanisms of stress effects on the immune system. Psychotherapy and Psychosomatics 23(1-6), 209–217 (1974)
7. Fairclough, S.H.: Fundamentals of physiological computing. Interacting with Computers 21(1-2), 133–145 (2009)
8. Mauss, I.B., Robinson, M.D.: Measures of emotion: A review. Cognition and Emotion 23(2), 209–237 (2009)
9. Picard, R.W., Vyzas, E., Healey, J.: Toward machine emotional intelligence: Analysis of affective physiological state. IEEE Transactions on Pattern Analysis and Machine Intelligence 23(10), 1175–1191 (2001)
10. van den Broek, E.L., Janssen, J.H., Westerink, J.H.D.M., Healey, J.A.: Prerequisits for Affective Signal Processing (ASP). In: Encarnação, P., Veloso, A. (eds.) Biosignals 2009: Proceedings of the International Conference on Bio-Inspired Systems and Signal Processing, Porto – Portugal, pp. 426–433 (2009)

11. Critchley, H.D., Elliott, R., Mathias, C.J., Dolan, R.J.: Neural activity relating to generation and representation of galvanic skin conductance responses: A functional magnetic resonance imaging study. The Journal of Neuroscience 20(8), 3033–3040 (2000)

12. Zeng, Z., Pantic, M., Roisman, G.I., Huang, T.S.: A survey of affect recognition methods: Audio, visual, and spontaneous expressions. IEEE Transactions on Pattern Analysis and Machine Intelligence 31(1), 39–58 (2009)

13. James, W.: Review: La pathologie des emotions by Ch. Féré. The Philosophical Review 2(3), 333–336 (1893)

14. Marwitz, M., Stemmler, G.: On the status of individual response specificity. Psychophysiology 35(1), 1–15 (1998)

15. Gunes, H., Piccardi, M.: Automatic temporal segment detection and affect recognition from face and body display. IEEE Transactions on Systems, Man, and Cybernetics – Part B: Cybernetics 39(1), 64–84 (2009)

16. Whitehill, J., Littlewort, G., Fasel, I., Bartlett, M., Movellan, J.: Towards practical smile detection. IEEE Transactions on Pattern Analysis and Machine Intelligence 31(11), 2106–2111 (2009)

17. Daly, A.: Movement analysis: Piecing together the puzzle. TDR – The Drama Review: A Journal of Performance Studies 32(4), 40–52 (1988)

18. Ververidis, D., Kotropoulos, C.: Emotional speech recognition: Resources, features, and methods. Speech Communication 48(9), 1162–1181 (2006)

19. Van den Broek, E.L.: Emotional Prosody Measurement (EPM): A voice-based evaluation method for psychological therapy effectiveness. Studies in Health Technology and Informatics (Medical and Care Compunetics) 103, 118–125 (2004)

20. van den Broek, E.L., Schut, M.H., Westerink, J.H.D.M., Tuinenbreijer, K.: Unobtrusive Sensing of Emotions (USE). Journal of Ambient Intelligence and Smart Environments 1(3), 287–299 (2009)

21. Gamboa, H., Silva, F., Silva, H., Falcão, R.: PLUX – Biosignals Acquisition and Processing (2010), http://www.plux.info (Last accessed January 30, 2010)

22. van den Broek, E.L., Westerink, J.H.D.M.: Considerations for emotion-aware consumer products. Applied Ergonomics 40(6), 1055–1064 (2009)

23. Berntson, G.G., Bigger, J.T., Eckberg, D.L., Grossman, P., Kaufmann, P.G., Malik, M., Nagaraja, H.N., Porges, S.W., Saul, J.P., Stone, P.H., van der Molen, M.W.: Heart rate variability: Origins, methods, and interpretive caveats. Psychophysiology 34(6), 623–648 (1997)

24. Boucsein, W.: Electrodermal activity. Plenum Press, New York (1992)

25. Grossman, P., Taylor, E.W.: Toward understanding respiratory sinus arrhythmia: Relations to cardiac vagal tone, evolution and biobehavioral functions. Biological Psychology 74(2), 263–285 (2007)

26. Fridlund, A.J., Cacioppo, J.T.: Guidelines for human electromyographic research. Psychophysiology 23(5), 567–589 (1986)

27. Reaz, M.B.I., Hussain, M.S., Mohd-Yasin, F.: Techniques of EMG signal analysis: detection, processing, classification and applications. Biological Procedures Online 8(1), 11–35 (2006)

28. Grandjean, D., Scherer, K.R.: Unpacking the cognitive architecture of emotion processes. Emotion 8(3), 341–351 (2008)

29. Lotte, F., Congedo, M., Lécuyer, A., Lamarche, F., Arnaldi, B.: A review of classification algorithms for EEG-based brain-computer interfaces. Journal of Neural Engineering 4(2), R1–R13 (2007)

30. Bimber, O.: Brain-Computer Interfaces. IEEE Computer 41(10) (2008); [special issue]

31. Minsky, M.: The Emotion Machine: Commonsense Thinking, Artificial Intelligence, and the Future of the Human Mind. Simon & Schuster, New York (2006)

32. Aarts, E.: Ambient intelligence: Vision of our future. IEEE Multimedia 11(1), 12–19 (2004)

33. Kim, J., André, E.: Emotion recognition based on physiological changes in music listening. IEEE Transactions on Pattern Analysis and Machine Intelligence 30(12), 2067–2083 (2008)

34. Liu, C., Rani, P., Sarkar, N.: Human-robot interaction using affective cues. In: Proceedings of the 15th IEEE International Symposium on Robot and Human Interactive Communication (RO-MAN 2006), Hatfield, UK, pp. 285–290. IEEE Computer Society, Los Alamitos (2006)

35. Rani, P., Sims, J., Brackin, R., Sarkar, N.: Online stress detection using psychophysiological signals for implicit human-robot cooperation. Robotica 20(6), 673–685 (2002)

36. Cacioppo, J.T., Tassinary, L.G., Berntson, G.: Handbook of Psychophysiology, 3rd edn. Cambridge University Press, New York (2007)

37. Sinha, R., Parsons, O.A.: Multivariate response patterning of fear. Cognition and Emotion 10(2), 173–198 (1996)

38. Scheirer, J., Fernandez, R., Klein, J., Picard, R.W.: Frustrating the user on purpose: A step toward building an affective computer. Interacting with Computers 14(2), 93–118 (2002)

39. Nasoz, F., Alvarez, K., Lisetti, C.L., Finkelstein, N.: Emotion recognition from physiological signals for presence technologies. International Journal of Cognition, Technology and Work 6(1), 4–14 (2003)

40. Takahashi, K.: Remarks on emotion recognition from bio-potential signals. In: Proceedings of the IEEE International Conference on Systems, Man and Cybernetics, Palmerston North, New Zealand, October 5-8, vol. 2, pp. 1655–1659 (2003)

41. Haag, A., Goronzy, S., Schaich, P., Williams, J.: Emotion recognition using bio-sensors: First steps towards an automatic system. In: André, E., Dybkjær, L., Minker, W., Heisterkamp, P. (eds.) ADS 2004. LNCS (LNAI), vol. 3068, pp. 36–48. Springer, Heidelberg (2004)

42. Kim, K.H., Bang, S.W., Kim, S.R.: Emotion recognition system using short-term monitoring of physiological signals. Medical & Biological Engineering & Computing 42(3), 419–427 (2004)

43. Lisetti, C.L., Nasoz, F.: Using noninvasive wearable computers to recognize human emotions from physiological signals. EURASIP Journal on Applied Signal Processing 2004(11), 1672–1687 (2004)

44. Wagner, J., Kim, J., André, E.: From physiological signals to emotions: Implementing and comparing selected methods for feature extraction and classification. In: Proceedings of the IEEE International Conference on Multimedia and Expo. (ICME), Amsterdam, The Netherlands, July 6-8, pp. 940–943 (2005)

45. Yoo, S.K., Lee, C.K., Park, J.Y., Kim, N.H., Lee, B.C., Jeong, K.S.: Neural network based emotion estimation using heart rate variability and skin resistance. In: Wang, L., Chen, K., S. Ong, Y. (eds.) ICNC 2005. LNCS, vol. 3610, pp. 818–824. Springer, Heidelberg (2005)

46. Choi, A., Woo, W.: Physiological sensing and feature extraction for emotion recognition by exploiting acupuncture spots. In: Tao, J., Tan, T., Picard, R.W. (eds.) ACII 2005. LNCS, vol. 3784, pp. 590–597. Springer, Heidelberg (2005)

47. Healey, J.A., Picard, R.W.: Detecting stress during real-world driving tasks using physiological sensors. IEEE Transactions on Intelligent Transportation Systems 6(2), 156–166 (2005)

48. Rani, P., Liu, C., Sarkar, N., Vanman, E.: An empirical study of machine learning techniques for affect recognition in human-robot interaction. Pattern Analysis & Applications 9(1), 58–69 (2006)

49. Zhai, J., Barreto, A.: Stress detection in computer users through noninvasive monitoring of physiological signals. Biomedical Science Instrumentation 42, 495–500 (2006)

50. Jones, C.M., Troen, T.: Biometric valence and arousal recognition. In: Thomas, B.H. (ed.) Proceedings of the Australasian Computer-Human Interaction Conference (OzCHI), Adelaide, Australia, pp. 191–194 (2007)

51. Leon, E., Clarke, G., Callaghan, V., Sepulveda, F.: A user-independent real-time emotion recognition system for software agents in domestic environments. Engineering Applications of Artificial Intelligence 20(3), 337–345 (2007)

52. Liu, C., Conn, K., Sarkar, N., Stone, W.: Physiology-based affect recognition for computer-assisted intervention of children with Autism Spectrum Disorder. International Journal of Human-Computer Studies 66(9), 662–677 (2008)

53. Katsis, C.D., Katertsidis, N., Ganiatsas, G., Fotiadis, D.I.: Toward emotion recognition in car-racing drivers: A biosignal processing approach. IEEE Transactions on Systems, Man, and Cybernetics–Part A: Systems and Humans 38(3), 502–512 (2008)

54. Yannakakis, G.N., Hallam, J.: Entertainment modeling through physiology in physical play. International Journal of Human-Computer Studies 66(10), 741–755 (2008)

55. Task Force: Heart rate variability: Standards of measurement, physiological interpretation, and clinical use. European Heart Journal 17(3), 354–381 (1996)

56. Ravenswaaij-Arts, C.M.A.V., Kollee, L.A.A., Hopman, J.C.W., Stoelinga, G.B.A., Geijn, H.P.: Heart rate variability. Annals of Internal Medicine 118(6), 436–447 (1993)

57. Butler, E.A., Wilhelm, F.H., Gross, J.J.: Respiratory sinus arrhythmia, emotion, and emotion regulation during social interaction. Psychophysiology 43(6), 612–622 (2006)

58. van den Broek, E.L., Schut, M.H., Westerink, J.H.D.M., van Herk, J., Tuinenbreijer, K.: Computing emotion awareness through facial electromyography. In: Huang, T.S., Sebe, N., Lew, M., Pavlović, V., Kölsch, M., Galata, A., Kisačanin, B. (eds.) ECCV 2006 Workshop on HCI. LNCS, vol. 3979, pp. 52–63. Springer, Heidelberg (2006)

59. Westerink, J.H.D.M., van den Broek, E.L., Schut, M.H., van Herk, J., Tuinenbreijer, K.: 14. In: Computing emotion awareness through galvanic skin response and facial electromyography. Philips Research Book Series, vol. 8, pp. 137–150. Springer, Dordrecht (2008)

60. Cacioppo, J., Tassinary, L.: Inferring psychological significance from physiological signals. American Psychologist 45(1), 16–28 (1990)

61. Mitchell, T.M.: Machine Learning. The McGraw-Hill Companies, Inc., Columbus (1997)

62. Bishop, C.M.: Pattern Recognition and Machine Learning. Information Science and Statistics. Springer Science+Business Media, LLC, New York (2006)

63. Schölkopf, B., Smola, A.J.: Learning with kernels: Support Vector Machines, Regularization, Optimization, and Beyond. In: Adaptive Computation and Machine Learning. The MIT Press, Cambridge (2002)

64. Rencher, A.C.: Methods of Multivariate Analysis, 2nd edn. Wiley Series in Probability and Statistics. John Wiley & Sons, Inc., New York (2002)

65. Rottenberg, J., Ray, R.R., Gross, J.J.: 1. In: Emotion elicitation using films, pp. 9–28. Oxford University Press, New York (2007)

66. Kreibig, S.D., Wilhelm, F.H., Roth, W.T., Gross, J.J.: Cardiovascular, electrodermal, and respiratory response patterns to fear- and sadness-inducing films. Psychophysiology 44(5), 787–806 (2007)

67. Kring, A.M., Gordon, A.H.: Sex differences in emotion: Expression, experience, and physiology. Journal of Personality and Social Psychology 74(3), 686–703 (1998)

68. Carrera, P., Oceja, L.: Drawing mixed emotions: Sequential or simultaneous experiences? Cognition & Emotion 21(2), 422–441 (2007)

69. Russell, J.A.: A circumplex model of affect. Journal of Personality and Social Psychology 39(6), 1161–1178 (1980)

70. Cover, T.M., van Campenhout, J.M.: On the possible orderings in the measurement selection problem. IEEE Transactions on Systems, Man, and Cybernetics SMC-7(9), 657–661 (1977)

71. Lawrence, S., Giles, C.L., Tsoi, A.: What size neural network gives optimal generalization? Convergence properties of backpropagation. Technical Report UMIACS-TR-96-22 and CS-TR-3617 (April 1996)

72. Barrett, L.F.: Valence as a basic building block of emotional life. Journal of Research in Personality 40, 35–55 (2006)

73. Russel, J.A., Barrett, L.F.: Core affect, prototypical emotional episodes, and other things called emotion: Dissecting the elephant. Journal of Personality and Social Psychology 26(5), 805–819 (1999)
74. Gendolla, G.H.E.: On the impact of mood on behavior: An integrative theory and a review. Review of General Psychology 4(4), 378–408 (2000)
75. Cooper, C.L., Pervin, L.A.: Personality: Critical concepts in psychology, 1st edn. Critical concepts in psychology. Routledge, New York (1998)
76. Lukowicz, P.: Wearable computing and artificial intelligence for healthcare applications. Artificial Intelligence in Medicine 42(2), 95–98 (2008)

# Part I
# BIODEVICES

# On-Chip Biosensors Based on Microwave Detection for Cell Scale Investigations

Claire Dalmay[1], Arnaud Pothier[1], M. Cheray[2], Fabrice Lalloué[2],
Marie-Odile Jauberteau[2], and Pierre Blondy[1]

[1] XLIM – UMR 6172 Université de Limoges/CNRS, 123 avenue Albert Thomas
87060 Limoges CEDEX, France
`claire.dalmay@xlim.fr`
[2] Homéostasie cellulaire et Pathologies – EA 3842 Université de Limoges
2 rue du Dr Marcland, 87025 Limoges CEDEX, France

**Abstract.** This paper presents original label free bio sensors allowing the study of electrical properties of human cells and so potentially cell identification and discrimination. Proposed biosensors are based on planar devices operating at microwave frequencies and fabricated using a standard microelectronic process. Actually, their microscopic sensitive areas allow an improved detection at the cell scale which represents a significant progress in the study of many biological phenomenon. In this paper, biosensor detection capabilities are demonstrated on only few biological cells analysis up to one single cell interacting with the sensor. Fabricated micro-sensors can be used to determine cell intrinsic electrical impedance at microwave frequencies allowing a label free approach to accurately discriminate biological cells.

**Keywords:** Biosensor, electrical bio-impedance, microelectronics, RF planar devices.

## 1 Introduction

In recent years, biosensors known a great interest as there is an important need for efficient tools that can quickly and accurately analyse biological elements like bio molecules or cells. Current optical and chemical bio-detection techniques can effectively analyse biological systems but present some drawbacks; especially their requirement in using specific labels to enhance the signal generation. These labeled methods make the sample preparation more complex, expensive and time consuming. In addition, the sample can be largely chemically altered prior analysis. On the other hand, electronic detection techniques are very interesting as they allow the development of label free methods [1]. Thanks to microelectronic technologies, a significant improvement of electrical sensor detection performance can be expected since miniaturized biosensors become able to really work at the cell scale. Especially, the capability to reach the one single cell measurement should represent a major advancement in the understanding of many biological mechanisms [2].

A. Fred, J. Filipe, and H. Gamboa (Eds.): BIOSTEC 2009, CCIS 52, pp. 51–63, 2010.
© Springer-Verlag Berlin Heidelberg 2010

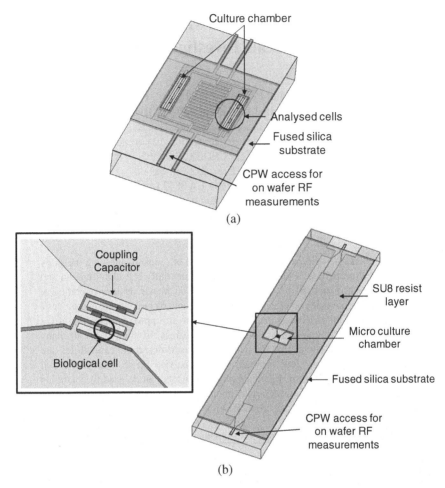

**Fig. 1.** Proposed RF biosensors concept based on stop band resonator (a) and two poles filter designs (b)

In this paper is detailed the development of label free biosensors allowing detection of a very small concentration of cells until a single one. The developed bio detection method is based on an electronic technique which allows evaluating intracellular medium impedance in the gigahertz frequency domain. At such high frequencies, the cell membrane is almost transparent making easier to probe its content. In addition, size of RF device working at these frequencies is generally small (typically few millimeters) and seems well adapted to interact with small size objects.

Actually, the cell electrical impedance, as said its permittivity and conductivity, are influenced by different parameters like the cell type, its morphology but also the cell physiological state. As example, tumoral cells are well known to present a larger conductivity and permittivity than normal cells [3-4]. Hence, individual cell electrical properties measurement represents a complementary tool able to allow efficient cell identification.

Two designs of biosensors are presented in this paper, both based on a coplanar microwave technology implementation as illustrated on figure 1. They have been especially designed to work with a tiny biological sample concentration insuring accurate analysis with a significant sensitivity.

The first section of the paper is dedicated to the biosensor RF design, then the detection capabilities of micro biosensors are demonstrated using a specific experimental protocol that we have especially developed for these measurements. To finish, experimental results performed on glial cells coming from human main nervous system are presented and discussed.

## 2  Biosensor RF Design: From Few Cells to Single Cell Detection

To develop biosensors allowing analysis until the single cell sample, the sensor sensitivity has to be especially enhanced. That is why a resonant RF structure has been preferred to wide band devices since they are intrinsically much more sensitive to small parameter changes [5-6] and so more appropriated for accurate characterization on few biological samples. On the other hand, the analysis spectrum is limited to a narrow frequency band around the sensor resonant frequency, requiring several resonant structures with different resonant frequencies if wider band investigations are needed.

In the present case, the sensor sensitivity will mainly rely on its capability to exhibit an obvious change in its RF response induced by the biological sample interaction with the sensor. Using a resonant structure, we expect to measure a significant sensor resonant frequency shift linked to the intrinsic electrical impedance of analyzed cells.

In the following sections two different biosensors concepts are introduced both based on resonant RF devices. We will see that following this approach accurate RF characterization of biological cell until the single one become then possible.

### 2.1  Compact RF Band Stop Resonator Biosensor

This micro-biosensor design uses a compact meandered inductor and an inter-digital capacitor association that behaves as a band stop resonator. Indeed, this device presents a resonant frequency at which almost no RF signal can be transmitted across the sensor since all the energy is reflected to its input. This kind of microwave resonator can be modeled as a simple RLC circuit and its resonance frequency $Fr$ is directly relied on its inductor L and capacitor C values as described on equation (1).

$$Fr = \frac{1}{2\pi\sqrt{LC}} \qquad (1)$$

The frequency of operating is then fixed by design functions of the meandered inductor and interdigital capacitor size and geometries.

This coplanar band stop resonator design is a good candidate for biosensing, since as illustrated on figure 2, the electromagnetic (EM) field distribution at the resonant frequency is strongly concentrated in the capacitive part of the device which consequently represents a highly sensitive interaction area. Indeed, a very small

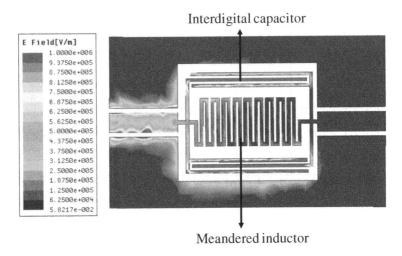

**Fig. 2.** RF electromagnetic field distribution at resonance frequency

concentration of biological media brought on this specific area will meaningfully disturb the EM field distribution and induce a significant sensor RF response shift (figure 3). Actually, the own electrical impedance of analyzed cells will locally increase the sensor capacitance and shift down the resonant frequency (2). The resulting frequency shift will be all the more significant if many cells are present on the sensor and if most of them are located in gaps between metallic lines where the electromagnetic field is stronger.

$$Fr'= \frac{1}{2\pi\sqrt{LC//C_{cell}}} < Fr \tag{2}$$

As shown on figure 1.a, some culture chambers allowing the cell growth on the sensor have been introduced to localize the analysis area to the most sensitive parts of the device. They have been especially patterned to make the frequency shift independent of cell location in these gaps. Consequently, considering one cell type, the measured frequency shift only relies on the intrinsic cell properties and on the number of cells which interact with the resonator.

As the sensor sensitivity is concerned, two main parameters have been especially optimized to enhance its capabilities to analyze a small number of cells. First, the interaction between the EM field and cells has been optimized considering gaps between metallic lines in the same order of magnitude with analyzed cell sizes. In the present sensor design, 10 μm gaps have be used. Secondly, the resonator intrinsic quality factor has been maximized. This resonator property is mainly limited by its resistive loss (i.e. the amount of RF signal that is dissipated because of the metallic line resistivity) that directly limits the magnitude of RF signal reflected at the resonator input. Consequently, the quality factor controls how the resonant frequency peak is narrow. The sensor frequency sensitivity is so highly dependent of this resonator property since a thin peak helps small frequency shift detections (figure 3).

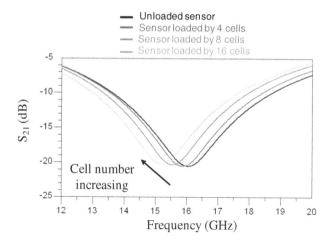

**Fig. 3.** Electromagnetic simulation of the cell number influence on the sensor RF response considering a cell permittivity of 42 and a cell conductivity of 0.1 S.m$^{-1}$. *The $S_{21}$ parameter represents the RF signal attenuation through the resonator.*

As shown on figure 3, full wave 3D simulations have been performed using Ansoft HFSS (electromagnetic simulator based on finite element method) to evaluate the cell EM influence on the band stop resonator sensor frequency response. In these simulations, cells are modeled using uniform dielectric elements with a specific permittivity and conductivity. One can see that the detection of a low number of cells (at least less than ten) seems possible with a good accuracy. Since the sensor sensitive area is mainly capacitive, it is also quite sensitive to small permittivity change making also suitable cell discrimination [7]. Nevertheless, these simulations have also highlighted that this band stop resonator topology still exhibits a too moderated unloaded quality factor (limited to 15-20) insufficient to reach the single cell detection. Indeed a tradeoff between a compact/high value meandered inductor and a low loss resonator performance has had to be found involving a narrow metallic lines design.

## 2.2 Ultra Sensitive Biosensor Based on a RF Filter Design Approach

To improve the detection capability and allow measurement up to the one single cell, another RF biosensor design approach has been followed considering a higher frequency selectivity device. This biosensor is then based on two resonant structures coupled together. In this case, two half wavelength microstrip resonators have been considered, able to achieve quality factor at least higher than 80. These resonators are coupled thanks to an interdigital capacitor as illustrated on the figure 1b. This capacitor allows the RF signal transmission from one resonator to the other one and in our case will be used as the sensor sensitive analysis area. This makes sense seeing the EM field distribution along the filter (figure 5) where most of electric field is concentrated in this capacitor. As done on the previous sensor design, to maximize the cell interaction with the device, the analysis area has been limited to the filter coupling capacitor with a polymer resist.

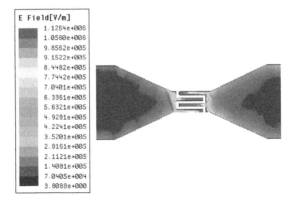

**Fig. 4.** RF electromagnetic field distribution in the coupling capacitor at the filter center frequency

Actually, the proposed filter design is a two pole bandpass filter in which the RF signal can be transmitted only on a restricted frequency band. But unlike conventional RF filter topologies; this device has been especially designed to present a strong ripple in its bandpass bandwidth. This approach allows differentiating clearly each resonant peak induced by the odd and even EM modes propagating along the filter. As shown on figure 5, the benefit of this approach is that only one of these modes is sensitive to the coupling capacitance change, the other one can be used as a reference.

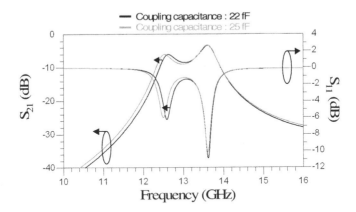

**Fig. 5.** Simulated response of the proposed RF biosensor. The frequency shift follows the arrows. *The $S_{21}$ parameter represents the RF signal attenuation through the filter whereas $S_{11}$ parameter represents the amount of signal reflected to the filter input.*

As a result, the bio detection principle consists in detecting inter-resonator coupling increase induces by the few biological cells attending on the interdigital capacitor. Since the global capacitance increase is very low (in the range of tens of femto Farads if only 5 cells are considered), this coupling capacitor design has been optimized setting its value as low as 20 fF to enhance the filter sensitivity. And finally, the biosensor center frequency can be set independently to the inter-resonator coupling, simply adjusting the two resonators length. In the presented case, the filter has been designed to operate at 13GHz.

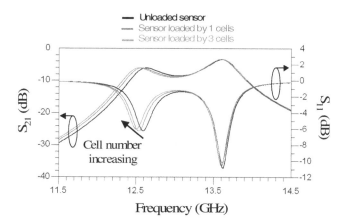

**Fig. 6.** Electromagnetic simulation of the cell number influence on the filter response considering a cell permittivity of 42 and a cell conductivity of 0.1 $S.m^{-1}$

As presented on figure 6, full wave 3D simulations performed using HFSS have permitted to validate the biosensor design. Following the same cell modeling, HFSS simulations show that one single cell detection is possible since the induced inter resonator's coupling increase generates a frequency response shift that seems to be detected.

## 2.3  Fabrication Process Details

Both RF biosensors presented on figure 7 are fabricated using standard microelectronic process based on classical UV photolithography techniques. Since for experimentation we perform an "on-chip" cell culture directly on the sensor surface, the fabrication process is mainly based on biocompatible materials for an optimal cell adhesion and growth between the sensor metallization gaps. Hence a fused silica substrate (glass substrate without impurities) has been preferred for its lower loss properties in the RF frequency domain and its transparency making easier the observation of cells throw it. Metal lines are made of gold for its superior conductivity properties. A SU8 polymer has been used to create a thick biocompatible layer to delimitate the micro-culture chambers. In addition, this layer favored the cell migration and adhesion on the fused silica substrate into the chambers.

The fabrication process is a two mask level and starts with the fused silica substrate cleaning using with a RCA bath ($H_2SO_4$, $H_2O_2$ and deionized water solution). A Chromium/Gold layer is then evaporated and thin gold lines are then patterned. These lines are next electroplating up to a thickness of 8 μm in order to maximize the surface interaction with cells. Then, micro-culture chambers are created on the sensor surface using a 20μm thick SU8 polymer. Sensors are finally diced and cleaned using autoclave sterilization techniques.

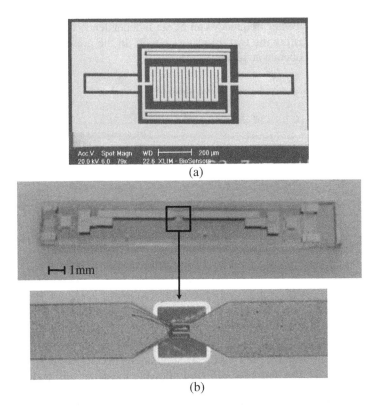

**Fig. 7.** Photograph of the fabricated biosensors stop band resonator (a) and band pass filter (b) design

## 3 Experimental Protocol for in Vitro RF Characterization

During characterizations, to ensure the integrity of cells a support biological media in which cells are protected is usually required. But most of support medium commonly used in biology are aqueous saline solutions which induce very strong losses for RF signal especially when they cover a large RF sensor surface. Actually in this configuration, the biological media alters dramatically the device RF performances inducing a strong degradation of the sensor quality factor and its detection capabilities.

For this study, we have developed a specific experimental protocol allowing a simple test procedure in which cells are grown directly on the sensor chip and measured in air without any biological media. This experimental protocol is illustrated on the figure 8 and has been used in the same conditions for cell growth on both resonator and filter sensor designs.

Initially, cells are grown at 37°C in a humidified 5% $CO_2$ - 95% air incubator in dedicated medium for cell culture (Minimum Essential Medium (MEM)). Once at confluence, cells are collected, put in another culture media solution and counted.

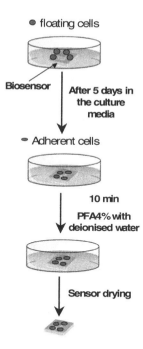

**Fig. 8.** Experimental protocol process

During the same time, sensors are deposited in 6-well plates and submerged in MEM solution. Then a defined cells concentration is seeded in each well plate. Few days are required to allow a sufficient cell adhesion on the sensor surface. When the number of adherent cells in the capacitor gaps is sufficient, sensors are put out the culture media and washed with Phosphate Buffered Saline (PBS). Then they are washed in deionized water to wipe out contamination from the culture media. Then, a paraformaldehyde 4 % bath (PFA) allows definitively fixing cells and so preventing from degradation during the measurement sequence. Since most of cell adhesion occurs preferentially on the silica substrate, most of fixed cells are located only between gold lines in the culture micro-chamber. Actually, the number of cells interacting with the sensor is controlled adjusting the time of culture.

After the PFA bath, cells are no longer alive, but they have kept their original form, their intracellular content and their electrical properties as in living conditions. Finally, sensors are washed again, left in deionized water and dried just before measurement.

Experimentations presented in this paper were led with a glioblastoma cell type named U87, some cancerous glial cell coming from human central nervous system. Actually, these cells constitute the most aggressive and frequent form of cerebral tumors (figure 9) and are the interest of intensive biological studies nowadays. They present good adherence properties and their size with typical diameter of 10-20 µm well matched our sensors metallic line gap spacings.

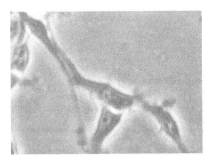

**Fig. 9.** Photograph of U87 cells grown on fused silica substrate

## 4  Results and Discussion

All characterizations are performed using classical microwave measurement techniques with on wafer probing in ambient air condition as it allows quick and successive measurements on different sensors.

The measurements are performed using a HP 8722ES calibrated microwave network analyser and consist in recording over frequency both reflected ($S_{11}$) and transmitted microwave signal attenuation ($S_{21}$) through biosensors.

First, each sensor is calibrated measuring their unloaded frequency response as a reference. The cell culture is then performed and once loaded with cells; sensors are dried and measured again following the same procedure. Due to own cell impedance effect on the sensor, a frequency shift can be detected. HFSS 3D EM simulations allow extracting the cells intrinsic electrical properties related to the observed frequency shift. Indeed, by fitting simulations data with measured one and modelling cells as homogenous, source-free and linear dielectric volumes, cell global permittivity can be extracted with a good accuracy on a narrow frequency band around the sensor resonance frequency (figures 10 &12). The cell conductivity can theoretically be also extracted, but the achieved accuracy for instance on this parameter still need to be improved.

### 4.1  Stop Band Resonator Sensor Experiments

Hence as shown on Figure 10, using a biosensor designed following the stop band resonators approach, a meaningful frequency response change has been observed while it was loaded with only 8 glial cells (figure 11). Indeed, this sensor which initially resonated at 16 GHz, shifted down to 15.63 GHz after the cell adhesion on the device.

As a result, we have extracted for U87 glial-cells an effective permittivity value of 42 ± 2 at 16 GHz and 20°C. As biological cell content is mainly dominated by water, the permittivity value is so highly dependent of the measurement frequency and the corresponding temperature. Nevertheless, these results are in good agreement with previous analysis that has been detailed in [7] and can also be compared to the effective permittivity of pure water which is close to 45 at 16 GHz and 20°C.

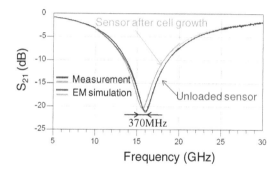

**Fig. 10.** Biosensor measured response before and after the U87 cell adhesion

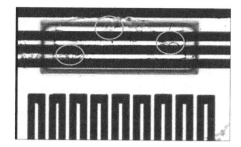

**Fig. 11.** Photograph of the sensor after the cell adhesion. U87 cells are located with circles.

With this RF biosensors design, measurements performed have also confirm that the effect of less than 4 to 6 cells on the biosensor cannot be observed clearly, limited by the frequency selectivity of the resonator. However, we have demonstrated the biosensor capability to be sensitive at least to less than ten cells. And using full wave simulations, we have succeeded to accurately model this effect, as illustrated on the figure 10, allowing us extracting individual cell electrical impedance.

## 4.2 Band Pass RF Filter Sensors Measured Capabilities

One single cell measurement has been successfully done using the RF filter biosensor design. As illustrated on Figure 12, with only one cell attending in the sensor capacitor gaps, a 13 MHz frequency shift has been observed.

Based on these preliminary results, at 20°C the extracted U87 global permittivity is estimate to $50 \pm 3$ at 12.5 GHz. As expected, this value is higher than the one obtained previously at 16GHz; mainly due to the frequency dependence of the water electrical impedance. Hence at 20°C, pure effective permittivity water is about 53 at 12.5 GHz and against 45 at 16 GHz [8]. Others experiments in which 3 cells were involved also confirm this trend and the frequency shift obtained are in agreement with the extracted permittivity value for this cell type.

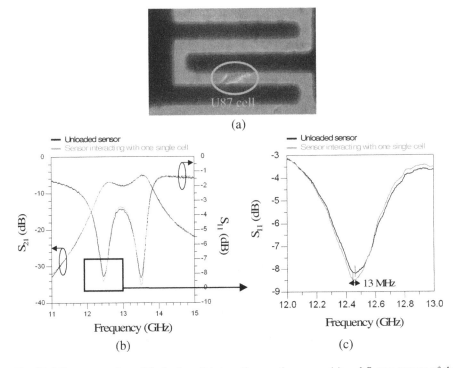

**Fig. 12.** Microscope view of 1 single cell interaction on the sensor (a) and S-parameters of the biosensor before and after the U87 cell growth (b & c)

## 5  Conclusions

An original label free bio-sensing approach for cellular analysis at radio frequencies has been introduced and validated. The detection feasibility has been proved on human cancerous cells coming from the central nervous system. We have seen that thanks to their sub millimetric size used sensors are able to work at the cell scale with a very limited number of biological cells until a single one. And for the first time RF characterisation on one single cell have been successfully achieved. This demonstrates the high sensitivity capabilities of the proposed RF micro-sensors concept and shows the promising potentiality of such biosensors to cell discrimination and analysis.

## References

1. Kim, Y.I., Park, T.S., Kang, J.H., Lee, M.C., Kim, J.T., Park, J.H., Baik, H.K.: Biosensors for label free detection based on RF and MEMS technology. Sensors and Actuators, 592–599 (January 2006)
2. Lu, X., Huang, W.H., Wang, Z.L., Cheng, J.K.: Recent developments in single-cell analysis. Analytica Chimica Acta 510, 127–138 (2004)

3. Blad, B., Baldetorp, B.: Impedance spectra of tumour tissue in comparison with normal tissue; a possible clinical application for electrical impedance tomography. Physiol.Meas. 17, A 105–A 115 (1996)
4. Vander Vorst, A., Rosen, A., Kotsuka, Y.: RF/ Microwave interaction with biological tissues. IEEE Press, Los Alamitos (2006)
5. Denef, N., Moreno-Hagelsieb, L., Laurent, G., Pampina, R., Foultier, B., Remacle, J., Flandre, D.: RF detection of DNA based on CMOS inductive and capacitive sensors. In: EUMW Conference Digest, September 2004, pp. 669–672 (2004)
6. Kim, Y.-I., et al.: Development of LC resonator for label-free biomolecule detection. Sens. Actuators A: Phys. (2007)
7. Dalmay, C., Pothier, A., Blondy, P., Lalloué, F., Jauberteau, M.-O.: Label free biosensors for human cell characterization using radio and microwave frequencies. In: IEEE MTT-S International Microwave Symposium Digest, IMS (2008)
8. Buchner, R., Barthel, J., Stauber, J.: The dielectric relaxation of water between 0°C and 35°C. Chemical Physics Letters 306, 57–63 (1998)

# Improvements of a Brain-Computer Interface Applied to a Robotic Wheelchair

André Ferreira[1], Teodiano Freire Bastos-Filho[1], Mário Sarcinelli-Filho[1],
José Luis Martín Sánchez[2], Juan Carlos García García[2], and Manuel Mazo Quintas[2]

[1] Department of Electrical Engineering, Federal University of Espirito Santo (UFES)
Av. Fernando Ferrari, 514 — 29075-910 Vitória-ES, Brazil
{andrefer,tfbastos,mario.sarcinelli}@ele.ufes.br
[2] Department of Electronics, University of Alcala (UAH)
Carretera Madrid-Barcelona, Km 33600 — 28871 Alcalá de Henares, Spain
{jlmartin,jcarlos,mazo}@depeca.uah.es

**Abstract.** Two distinct signal features suitable to be used as input to a Support-Vector Machine (SVM) classifier in an application involving hands motor imagery and the correspondent EEG signal are evaluated in this paper. Such features are the Power Spectral Density (PSD) components and the Adaptive Autoregressive (AAR) parameters. The best result (an accuracy of 97.1%) is obtained when using PSD components, while the AAR parameters generated an accuracy of 91.4%. The results also demonstrate that it is possible to use only two EEG channels (bipolar configuration around $C_3$ and $C_4$), discarding the bipolar configuration around $C_z$. The algorithms were tested with a proprietary EEG data set involving 4 individuals and with a data set provided by the University of Graz (Austria) as well. The resulting classification system is now being implemented in a Brain-Computer Interface (BCI) used to guide a robotic wheelchair.

**Keywords:** Adaptive Autoregressive Parameters, Power Spectral Density components, Support-Vector Machines, Brain-Computer Interfaces, Robotic Wheelchair.

## 1 Introduction

A Brain-Computer Interface (BCI) is a system that includes a way of acquiring the signals generated by the brain activity, a method/algorithm for decoding such signals and a subsystem that associates the decoded pattern to a behavior or action [1]. The BCI and its inherent challenges, involving areas such as signal processing, machine learning and neurosciences, have been the focus of several important research groups. The results of this new technology could be applied to improve the quality of life of many people affected by neuromotor disfunctions caused by diseases, like amyotrophic lateral sclerosis (ALS), or injuries, like spinal cord injury.

A basic structure of a BCI, according to the previous definition, is presented in Fig. 1. This paper is related to the phases of *feature extraction* and *feature translation* or *classification*, both indicated in the figure. The objective here is to evaluate Power Spectral Density (PSD) components and Adaptive Autoregressive parameters as inputs for a Support-Vector Machine (SVM) classifier. The SVM is supposed to be able to distinguish two mental tasks related to hands motor imagery, based on these two features

A. Fred, J. Filipe, and H. Gamboa (Eds.): BIOSTEC 2009, CCIS 52, pp. 64–73, 2010.

extracted from the EEG signal. Two data sets (a proprietary one acquired in the University of Alcala and one provided by the University of Graz) are used to evaluate the implemented algorithms. Configurations of three EEG channels (bipolar around $C_3$, $C_z$ and $C_4$) and two EEG channels (bipolar around $C_3$ and $C_4$) are tested.

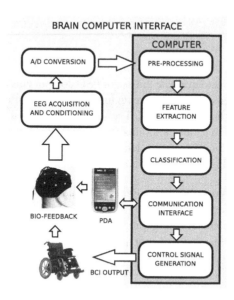

**Fig. 1.** Brain-Computer Interface available at UFES

The preliminaries and the results of such evaluation are hereinafter presented as follows: Section 2 contextualizes this work, introducing some previous works involving a robotic wheelchair commanded by a BCI; the methodology used to reach the objective is explained in Section 3, where the feature extraction and the classifier are described in details. The results obtained with the two data sets aforementioned and some comments are presented in Section 4, which is followed by Section 5, where the main conclusions of this work are highlighted.

## 2  Background

A robotic wheelchair commanded through a BCI is being developed at the Federal University of Espirito Santo, Brazil. The users of such BCI can select movements to be executed by the wheelchair from a set of options presented in the screen of a PDA connected to the BCI, as illustrated in Fig. 1.

A drawback of this approach is the need of eye-closing to generate the desired pattern, in this case an ERS [2]. An user who is not able to close the eyes for a while to select an option of movement, for example, will not get any profit using the current version of the BCI implemented in the wheelchair. In order to overcome such problem, other EEG information should be used.

In such a context, hands motor imagery is being tested here, in connection to a SVM-based classifier, to check the possibility of using this approach to implement a BCI to be used to command the robotic wheelchair aforementioned. The idea underlying this study is to use imaginary hand movements, instead of eye-closing, to generate recognizable EEG patterns.

## 3   Method

The focus of this paper is to evaluate the use of PSD components and AAR parameters, associated to EEG signals acquired in the region of the motor cortex of the human brain, as inputs of a classifier based on a SVM. The system is supposed to classify two different mental tasks related to hands motor imagery, aiming at allowing to implement a BCI to be used to command a robotic wheelchair [2]. In order to perform such evaluation, the following methodology was carried out:

1. evaluate two different approaches: PSD-SVM and AAR/RLS[1]-SVM, according to the sketch of Fig. 2;
2. evaluate different channel configurations: $[C_3\ C_z\ C_4]$ and $[C_3\ C_4]$[2];
3. PSD approach: evaluate for different time intervals (3-5s, 4-6s, 5-7s, 6-8s and 7-9s);
4. AAR/RLS approach: evaluate for different Classification Times (CT) [3]. The CTs used are 3s, 4s, 5s, 6s, 7s and 8s;
5. evaluate the algorithms using the proprietary UAH dataset and search for the best configuration (feature extractor and SVM classifier);
6. apply such configuration to the Graz dataset and evaluate the results.

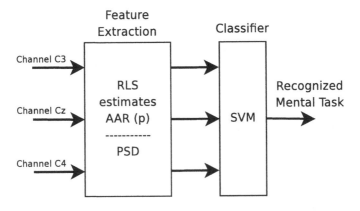

**Fig. 2.** A representation of the systems being evaluated

---

[1] Recursive Least Squares.
[2] Actually, the channels are bipolar, with electrodes placed around these positions, as shown in Fig. 3.

### 3.1   Graz Dataset

The Graz dataset was provided by the Department of Medical Informatics, University of Graz (Austria), during the BCI Competition 2003. It is named *Data set III* and is related to motor imagery. In this paper, 140 trials of this dataset, and the respective labels, were used, 70 related to left hand motor imagery and 70 related to right hand motor imagery. Each trial lasts 9 seconds, with a sampling rate of 128 Hz, resulting in 1152 samples/channel/trial. The data was obtained using a bipolar configuration around the positions $C_3$, $C_z$ and $C_4$, according to the 10-20 International System, as presented in Fig. 3. In the same figure (on the right side), an illustration of the protocol used during the experimental phase is presented. After the 2 initial seconds, a beep sounds and a cross is presented in the center of the screen, calling the subject's attention to the beginning of the experiment. One second later ($t = 3\ s$), an arrow pointing left or right is presented to the operator, suggesting which mental task should be accomplished, and lasts for 6 seconds (until $t = 9\ s$). The data was filtered, keeping the spectrum ranging from 0.5 Hz to 30 Hz, and visual feedback was used (more details can be found in [4]).

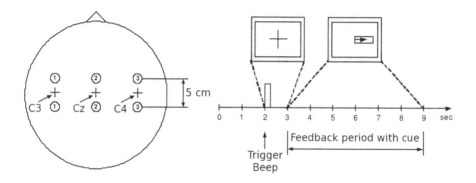

**Fig. 3.** Electrodes Placement and the experimental protocol associated to the Graz dataset

### 3.2   UAH Dataset

Experiments similar to those described in Section 3.1 were accomplished at the University of Alcala (UAH), Spain. The mental tasks are the same of the Graz dataset, also related to hands motor imagery. The dataset was recorded from 4 normal subjects in different sessions. Each session corresponds to 60 trials (half to each one of the two mental tasks considered) and each trial was 9 s long, resulting in 9 minutes/session. Three subjects participated in 3 sessions and one subject participates in 4 sessions, thus resulting in 780 trials. The bio-signal amplifier *g.BSamp* and the subsystem *g.16sys* compound the *g.tec* system used to record the EEG data, the software being implemented in Matlab. The data was also filtered to keep only the spectrum from 0.5 Hz and 30 Hz, but the volunteer had no visual feedback. Table 1 presents the experiments carried out in the University of Alcala.

**Table 1.** Experiments carried out in the University of Alcala

| Subject | # Sessions | # Trials | Left Hand | Right Hand |
|---------|-----------|----------|-----------|------------|
| $S_{01}$ | 3 | 180 | 90 | 90 |
| $S_{02}$ | 3 | 180 | 90 | 90 |
| $S_{03}$ | 4 | 240 | 120 | 120 |
| $S_{04}$ | 3 | 180 | 90 | 90 |

### 3.3 Feature Extraction: PSD

Due to the fact that EEG rhythms have been defined mainly in the frequency domain, the Power Spectrum Density (PSD) analysis of the signal is the non-parametric technique used for feature extraction. Other reasons that motivate this choice are the computational efficiency involved, the direct relation between PSD and power, power components can be interpreted in terms of cerebral rhythms and the estimations (via FFT) of spectral components are not biased as those estimated via AR models, as described in [5].

The PSD is estimated here via he Welch's Method, computed over sections of 1 s, averaging spectral estimates of 3 segments of 0.5 s each (64 samples, sampling rate of 128 Hz) with 50% of overlap between segments. The maximum size of each segment is important in order to consider the stationary behavior of the EEG signal [5,6]. A weighting *Hanning* window is applied to the signal due to its considerable attenuation in the side-lobes. The spectral components extracted from the signal and used as features spans from 8 Hz to 30 Hz, with a frequency resolution of 2 Hz. Thus, 12 components are generated, in connection to each channel. This feature extraction procedure is illustrated in Fig. 4.

### 3.4 Feature Extraction: AAR/RLS

The other technique used for feature extraction is based on Adaptive Autoregressive parameters (AAR), estimated via Recursive Least Squares (RLS) algorithm, as described in [3,7]. This procedure is performed according to

$$E_t = Y_t - \mathbf{a}_{t-1}^T \mathbf{Y}_{t-1} \tag{1}$$

$$\mathbf{r}_t = (1 - UC)^{-1} \mathbf{A}_{t-1} \mathbf{Y}_{t-1} \tag{2}$$

$$\mathbf{k}_t = \mathbf{r}_t / (\mathbf{Y}_{t-1}^T \mathbf{r}_t + 1) \tag{3}$$

$$\mathbf{a}_t = \mathbf{a}_{t-1} + \mathbf{k}_t E_t \tag{4}$$

$$\mathbf{A}_t = (1 - UC)^{-1} \mathbf{A}_{t-1} - \mathbf{k}_t \mathbf{r}_t^T, \tag{5}$$

where

$$\mathbf{a}_t = [a_{1,t} \ldots a_{p,t}]^T \tag{6}$$

$$\mathbf{Y}_{t-1} = [Y_{t-1} \ldots Y_{t-p}]^T. \tag{7}$$

The initial values adopted were $\mathbf{A}_0 = I$, $\mathbf{a}_0 = \mathbf{0}$ and $UC = 0.007$, and the model order was chosen as $p = 6$. Although the RLS algorithm has a higher computational

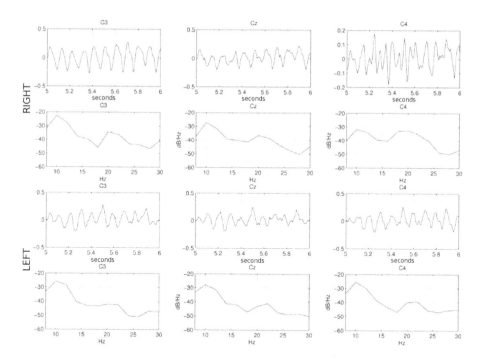

**Fig. 4.** Example of feature extraction using PSD components. Signals related to $C_3$, $C_z$ and $C_4$ (bipolar) during hands motor imagery. PSD is presented from 8 up to 30 Hz in dB/Hz.

complexity in comparison with the Least Mean Squares (LMS), it has some advantages: the faster convergence, the higher accuracy of the estimate and the fact that no matrix inversion is necessary. Fig. 5 shows the temporal evolution of six AAR parameters. In this case, the channel $C_3$ of the first trial included in the Graz dataset was considered.

### 3.5 Classifier: SVM

Although the concept of Support-Vector Machines (SVM) was introduced in COLT-92 (Fifth Annual Workshop on Computational Learning Theory) [8], its evaluation in BCIs is quite recent.

Briefly speaking, the main idea of a SVM is to find an optimal separating hyperplane for a given feature set. Given a training set of instance-label pairs $(\mathbf{x}_i, y_i), i = 1, \ldots, l$, where $\mathbf{x}_i \in R^n$ and $y \in \{1, -1\}^l$, the SVM requires the solution of the optimization problem

$$\min_{\mathbf{w},b,\xi} \frac{1}{2}\mathbf{w}^T\mathbf{w} + C\sum_{i=1}^{l} \xi_i, \tag{8}$$

subject to

$$y_i(\mathbf{w}^T\phi(\mathbf{x}_i) + b) \geq 1 - \xi_i \tag{9}$$

$$\xi_i \geq 0. \tag{10}$$

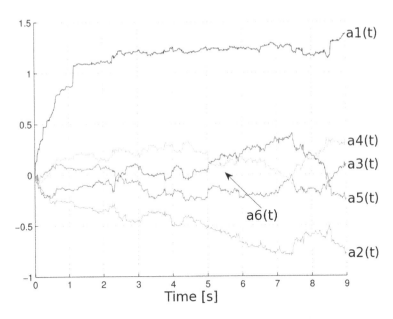

**Fig. 5.** Temporal evolution of six AAR parameters

Training vectors $\mathbf{x}_i$ are mapped into a higher dimensional space (maybe infinite) by the function $\phi$. The SVM finds a linear separating hyperplane with the maximal margin in this higher dimensional space. $C > 0$ is the penalty parameter of the error term. The function $K(\mathbf{x}_i, \mathbf{x}_j) \equiv \phi(\mathbf{x}_i)^T \phi(\mathbf{x}_j)$ is called *kernel*. The kernel function used in this paper is a Radial Basis Function (RBF) defined as

$$K(\mathbf{x}_i, \mathbf{x}_j) = exp(-\gamma \|\mathbf{x}_i - \mathbf{x}_j\|^2), \gamma > 0. \tag{11}$$

The choice of a SVM-based classifier and a RBF kernel function relies on previous works that considered this configuration [9,10,11]. Furthermore, a SVM classifier has improved the accuracy in 13% when compared to LDA (Linear Discriminant Analysis) and 16.3% when compared to NN (Neural Networks), using the same features [12].

The scripts developed during this work are based on the library *libsvm* [13].

## 4   Results

Taking into account the data distribution, 75% of each dataset was used for training and validation, while the other 25% were used for test. After evaluating two different techniques for feature extraction (based on PSD components and AAR parameters), the results are presented in the following four tables. The first one (Table 2) shows the classification accuracy obtained for each subject of the UAH dataset, when PSD-SVM is used. The gray cells represents the best classification accuracy found for each subject. The higher values are related to the central period of the experiment (4-6s) and, except by the subject $S_{01}$, these values are obtained with only two channels. The classification accuracy (ACC) is obtained from the confusion matrix $H$ as follows

$$ACC = p_0 = \frac{1}{N} \sum_i H_{ii}, \tag{12}$$

where $N = \sum_i \sum_j H_{ij}$ is the total number of samples and $H_{ii}$ are the elements of the confusion matrix $H$ on the main diagonal [14]. The classification accuracy of each class is calculated using the amount of true-positives (TP) and false-negatives (FN) of each class as follows:

$$ACC_{class} = \frac{TP_{class}}{TP_{class} + FN_{class}}. \tag{13}$$

**Table 2.** PSD-SVM (UAH dataset)

| Subject | 2 channels ($C_3$ $C_4$) | | | | | 3 channels ($C_3$ $C_z$ $C_4$) | | | | |
|---|---|---|---|---|---|---|---|---|---|---|
| | 3-5s | 4-6s | 5-7s | 6-8s | 7-9s | 3-5s | 4-6s | 5-7s | 6-8s | 7-9s |
| $S_{01}$ | 71.1 | 71.1 | 73.3 | 73.3 | 66.7 | 73.3 | **80.0** | 66.7 | 75.6 | 64.4 |
| $S_{02}$ | 68.7 | **80.0** | 73.3 | 73.3 | 71.1 | 75.6 | 75.6 | 73.3 | 73.3 | 66.7 |
| $S_{03}$ | 68.3 | **76.7** | 66.7 | 66.7 | 66.7 | 68.3 | 70.0 | 66.7 | 65.0 | 70.0 |
| $S_{04}$ | 75.6 | **91.1** | 82.2 | 84.4 | 84.4 | 73.3 | 86.7 | 82.2 | 84.4 | 75.6 |
| $\mu$ | 70,9 | 79,7 | 73,9 | 74,4 | 72,2 | 72,6 | 78,1 | 72,2 | 74,6 | 69,2 |

Table 3 contains the results for (AAR/RLS-SVM), the other explored configuration. Four subjects of the UAH dataset are evaluated using different CTs and channel configuration. Once more, the gray cells represents the best classification values for each subject during the test. Equal values are all highlighted (gray cells) to show in which situations they can appear. As in PSD case, the higher classification rates are related to the middle of the experiment (Table 2 and Table 3 (4-6s)). The best results can also be reached with only 2 channels, taking into account that all the high values obtained with 3 channels appear on the left side of the Table 3 (2 channels).

**Table 3.** AAR/RLS-SVM (UAH dataset)

| Subject | 2 channels ($C_3$ $C_4$) | | | | | | 3 channels ($C_3$ $C_z$ $C_4$) | | | | | |
|---|---|---|---|---|---|---|---|---|---|---|---|---|
| | 3s | 4s | 5s | 6s | 7s | 8s | 3s | 4s | 5s | 6s | 7s | 8s |
| $S_{01}$ | 66.7 | **73.3** | 66.7 | 68.9 | 62.2 | **73.3** | 66.7 | **73.3** | 64.4 | 66.7 | 71.1 | 66.7 |
| $S_{02}$ | 66.7 | 68.9 | 66.7 | **80.0** | 57.8 | 57.8 | 68.9 | 68.9 | 64.4 | 73.3 | 64.4 | 64.4 |
| $S_{03}$ | 60.0 | 66.7 | **73.3** | 61.7 | 66.7 | **73.3** | 58.3 | 61.7 | 68.3 | 66.7 | 65.0 | 70.0 |
| $S_{04}$ | 66.7 | 71.1 | **86.7** | 82.2 | 75.6 | 73.3 | 66.7 | 73.3 | **86.7** | 77.8 | 71.1 | 75.5 |
| $\mu$ | 65,0 | 70,0 | 73,4 | 73,2 | 65,6 | 69,4 | 65,2 | 69,3 | 71,0 | 71,1 | 67,9 | 69,2 |

Thus, the best results with the UAH dataset can be found using PSD-SVM, 2 channels ($C_3$ $C_4$) and in the middle of the experiment. A summary of the results is presented in Table 4.

As the next step of the proposed methodology, this configuration was applied to the Graz dataset, in order to evaluate it. The results obtained for this configuration and the other are shown in Tables 5 and 6.

**Table 4.** Best Results (UAH dataset)

| Subject | Accuracy | Left Hand | Right Hand | Configuration |
|---------|----------|-----------|------------|---------------|
| $S_{01}$ | 80.0 | 82,6 | 77,3 | PSD-SVM,$C_3$ $C_z$ $C_4$,4-6s |
| $S_{02}$ | 80.0 | 78.3 | 81,8 | PSD-SVM,$C_3$ $C_4$,4-6s |
| $S_{03}$ | 76.7 | 80.0 | 73.3 | PSD-SVM,$C_3$ $C_4$,4-6s |
| $S_{04}$ | 91.1 | 95,7 | 86,4 | PSD-SVM,$C_3$ $C_4$,4-6s |

**Table 5.** PSD-SVM (Graz dataset)

| | 2 channels ($C_3$ $C_4$) | | | | | 3 channels ($C_3$ $C_z$ $C_4$) | | | | |
|---|---|---|---|---|---|---|---|---|---|---|
| Subject | 3-5s | 4-6s | 5-7s | 6-8s | 7-9s | 3-5s | 4-6s | 5-7s | 6-8s | 7-9s |
| $S_{Graz}$ | 88.6 | **97.1** | 85.7 | 74.3 | 77.1 | 85.7 | 94.3 | 85.7 | 80.0 | 71.4 |

**Table 6.** AAR/RLS-SVM (Graz dataset)

| | 2 channels ($C_3$ $C_4$) | | | | | | 3 channels ($C_3$ $C_z$ $C_4$) | | | | | |
|---|---|---|---|---|---|---|---|---|---|---|---|---|
| Subject | 3s | 4s | 5s | 6s | 7s | 8s | 3s | 4s | 5s | 6s | 7s | 8s |
| $S_{Graz}$ | 65.7 | 74.3 | **91.4** | **91.4** | 82.9 | 80.0 | 74.3 | 68.6 | **91.4** | **91.4** | 80.0 | 80.0 |

## 5  Conclusions

This paper evaluates the use of two set of features (PSD components and AAR/ RLS parameters of an EEG signal) as inputs for a SVM classifier, in order to distinguish between two mental tasks related to hands motor imagery.

The approach based on PSD (Welch's Method) components and a SVM (RBF kernel) generated the best results. The highest classification rates are related to the middle of the experiment, usually between seconds 4 and 6. It can be explained taking into account that the subject needs some time to setup him/herself (the cue, an arrow, is presented to the subject at instant $t = 3\ s$) to the end of the trial (the trial finishes at $t = 9\ s$).

The best results can be accomplished using only 2 channels, four electrodes placed around positions $[C_3\ C_4]$ of the 10-20 International System.

After evaluating the system with the UAH dataset, the algorithms were applied to the Graz dataset and the best classification rate (accuracy) was 97.1% (100% to mental task 1 and 94.1% to mental task 2).

The replacement of the method currently used to select a symbol in a PDA, which requires a brief eye-closing, by another based on motor imagery, such as the one here discussed, is the next step of our research. In other words, the idea is that motor imagery of a hand (the one with higher accuracy) it is enough to the user without eyes control to select desired symbols in the PDA that will be translated into commands to the robotic wheelchair or into some communication outputs, also available in this system.

**Acknowledgements.** The authors thank CAPES, a foundation of the Brazilian Ministry of Education (Project 150/07), and FACITEC/PMV, a fund of the Vitoria City Hall to support scientific and technological development (Process 061/2007), for their financial support to this research.

# References

1. Sajda, P., Muller, K.R., Shenoy, K.: Brain-computer interfaces [from the guest editors]. IEEE Signal Processing Magazine 25(1), 16–17 (2008)
2. Pons, J.L.: Wearable Robots: Biomechatronic Exoskeletons, 1st edn. Wiley, Madrid (2008)
3. Schlögl, A., Neuper, C., Pfurtscheller, G.: Subject specific EEG patterns during motor imaginary [sic.: for imaginary read imagery]. In: 19th Annual International Conference of the IEEE Engineering in Medicine and Biology Society, vol. 4, pp. 1530–1532 (1997)
4. Schlögl, A.: Data set: BCI-experiment (2003), http://ida.first.fraunhofer.de/projects/bci/competition_ii
5. Mouriño, J.: EEG-based Analysis for the Design of Adaptive Brain Interfaces. PhD thesis, Universitat Politècnica de Catalunya, Barcelona (2003)
6. McEwen, J.A., Anderson, G.B.: Modeling the stationarity and gaussianity of spontaneous electroencephalographic activity (5), 361–369 (1975)
7. Haykin, S.: Adaptive Filter Theory, 4th edn. Prentice-Hall, Englewood Cliffs (2001)
8. Boser, B.E., Guyon, I.M., Vapnik, V.N.: A training algorithm for optimal margin classifiers. In: COLT 1992: The Fifth Annual Workshop on Computational Learning Theory, New York, pp. 144–152 (1992)
9. Shoker, L., Sanei, S., Sumich, A.: Distinguishing between left and right finger movement from EEG using SVM. In: 27th Annual International Conference of the Engineering in Medicine and Biology Society, pp. 5420–5423 (2005)
10. Guler, I., Ubeyli, E.: Multiclass support vector machines for EEG-signals classification. IEEE Transactions on Information Technology in Biomedicine 11(2), 117–126 (2007)
11. Khachab, M., Kaakour, S., Mokbel, C.: Brain imaging and support vector machines for brain computer interface. In: 4th IEEE International Symposium on Biomedical Imaging: From Nano to Macro, pp. 1032–1035 (2007)
12. Nicolaou, N., Georgeou, J., Polycarpou, M.: Autoregressive features for a thought-to-speech converter. In: International Conference on Biomedical Electronics and Devices, Funchal, pp. 11–16 (2008)
13. Chang, C.C., Lin, C.J.: LIBSVM: a library for support vector machines (2001) Software available, http://www.csie.ntu.edu.tw/~cjlin/libsvm
14. Pfurtscheller, G., Brunner, C., Schlögl, A., da Silva, F.H.L.: Mu rhythm (de)synchronization and EEG single-trial classification of different motor imagery tasks. Neuroimage 31(1), 153–159 (2006)

# Wavelet-Based and Morphological Analysis of the Global Flash Multifocal ERG for Open Angle Glaucoma Characterization

J.M. Miguel-Jiménez[1], S. Ortega[1], I. Artacho[1], L. Boquete[1],
J.M. Rodríguez-Ascariz[1], P. De La Villa[2], and R. Blanco[3]

[1] Department of Electronics, University of Alcalá, 28701 Alcalá de Henares, Spain
[2] Department of Physiology, University of Alcalá, 28701 Alcalá de Henares, Spain
[3] Department of Surgery, University of Alcalá, 28701 Alcalá de Henares, Spain
jmanuel@depeca.uah.es, boquete@depeca.uah.es, roman@ski.org

**Abstract.** This article presents one of the alternative methods developed for the early detection of ocular glaucoma based on the characterisation of mfERG (multifocal electroretinography) readings. The digital signal processing technique is based on Wavelets, hitherto unused in this field, for detection of advanced-stage glaucoma and the study of signal morphology by means of identity patterns for detection of glaucoma in earlier stages. Future research possibilities are also mentioned, such as the study of orientation in the development of the disease.

**Keywords:** Wavelet Transforms, Glaucoma, m-sequence, Multifocal Electroretinogram, Morphological Analysis.

## 1 Introduction

Glaucoma is currently deemed to be a high-risk eye disease since a large percentage of the population suffer from its effects. The method proposed herein has been developed for study and analysis of OAG (open angle glaucoma), the commonest form in today's society.

The sheer complexity of the disease and its occultation make early and reliable detection essential. The traditional techniques for clinical analysis of the retina are based on indirect methods (measurement of the intraocular pressure, visual inspection of the eyeground, campimetric tests, etc). Their main drawback is that they do not give objective information on the functioning of the retinal photoreceptors [1], essential elements in the perception of light energy. A new technique has recently been developed for obtaining this retina-functioning information in a quick and reproducible way; this technique is known as the multifocal electroretinogram (mfERG). The mfERG enables a functional exploration to be made of the light sensitivity of the retinal cells and also the spatial distribution of this sensitivity [2]. The mfERG basically involves recording the variations in retinal potential evoked by a light stimulus and then mapping out the results in a 2D or 3D diagram showing those regions that respond to the visual stimuli [3][4].

A. Fred, J. Filipe, and H. Gamboa (Eds.): BIOSTEC 2009, CCIS 52, pp. 74–84, 2010.

The mfERG technique allows simultaneous recording of local responses from many different regions of the retina, building up a map of its sensitivities. As in the conventional electroretinogram (ERG), also called the full-field electroretinogram, the potential is measured as the sum of the electric activity of the retina cells. In the full-field ERG, however, the signal recorded comes from the whole retina surface, so it is hard to detect smaller one-off defects that do not affect the whole retina. The mfERG, by contrast, gives detailed topographical information of each zone and can therefore detect small-area local lesions in the retina and even in its central region (fovea) [5].

From the technical point of view, equipment is needed for capturing the visually evoked potentials at retina level (presented as a set of hexagons of varying sizes and intensities). Due to the low amplitude of the signals generated (down to nanovolt level), the technique calls for suitable hardware equipment (recording electrodes, instrumentation amplifiers, digitalisation, etc) and also signal processing algorithms (filtering, averaging or smoothing procedures, rejection of artefacts, etc) to ensure that the results are clinically useful [6].

This paper gives a description of the recording and arrangement of the signals we have used in our research, the signal analysis by the Wavelet transform for recording possible advanced-stage glaucoma markers, the detection of smaller lesions by means of morphological analysis of the signal; it also mentions possible future research lines.

## 2  Methods

### 2.1  Obtaining the Signals

A total of 50 patients with diagnosis of advanced open angle glaucoma (OAG) as well as an identical number of healthy subjects were included in our mfERG record database, used for obtaining markers by means of the wavelet transform. Moreover, to study the efficiency of our morphological analysis, a second database was drawn up formed by 15 patients diagnosed with early-stage Glaucoma plus an identical number of healthy controls.

The signal recording system was the VERIS 5.1 multifocal recording system (Electro-Diagnostic Imaging, San Mateo, USA). The stimulus consisted of an m-sequence applied to a group of 103 hexagons, as shown in figure 1, displayed on a 21-inch monitor and covering a 45° arc of the retina. The local luminance of each hexagon was 200 cd/m$^2$ in the on phase and less than 1.5 cd/m$^2$ in the off phase, determined by the pseudorandom sequence.

The monitor frequency was 75 Hz and the m-sequence was modified so that each step was followed by 4 frames in the following order: flash-dark-flash-dark, as shown in figure 2. In the flash frames all the hexagons were illuminated with a maximum luminance of 200 cd/m$^2$, with a minimum luminance of less than 1.5 cd/m$^2$ in the dark frames. The background luminance of the rest of the monitor surface surrounding the hexagons was held steady at 100 cd/m$^2$. This stimulation protocol is especially adapted for obtaining responses from the retinal ganglion cells and their axons [7]. It

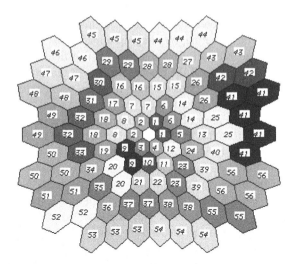

**Fig. 1.** Geometry of the multifocal stimulus and regrouping of the hexagons

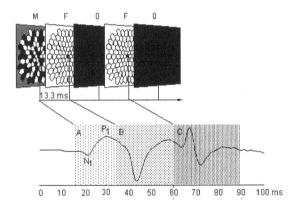

**Fig. 2.** Modification of the m-sequence

is based on the effect of the focal responses (M) on the following global stimulus (F), which amplifies the signals coming from the ganglion cells.

Basically, the protocol (M-F-O-F-O) consists of five steps. In the first step (M) each hexagon follows a luminous stimulation (200 cd/m$^2$) determined by a pseudorandom binary m-sequence. In the second step the whole area is illuminated (200 cd/m$^2$) (F), followed by a dark sequence (O) (<1.5 cd/m$^2$), followed by another global flash (200 cd/m$^2$) (F) and then darkness again (O) (<1.5 cd/m$^2$). This stimulation will give us an acceptable signal-to-noise ratio and also ensures a reasonably short recording time (9 minutes).

The stimulus was displayed through pharmacologically dilated pupils (minimum diameter of 7 millimetres) using a Burian-Allen bipolar contact lens (Hansen ophthalmics, Iowa City, IA). Contact lens adaptation was facilitated by a drop of topical anaesthetic (0.5% Proparacaine). The residual spherical refractive error was corrected by the VERIS™ autorefractor, mounted on the stimulation monitor. The alignment of the patient's pupil with the monitor optic and the fixation stability are controlled by an attached infrared camera. Each monocular recording lasts about 9 minutes (exponent of the stimulation m-sequence = 13). To make the process more comfortable for the patient, the recording process was divided into eighteen 30-second segments. Segments contaminated with ocular movements were discarded and recorded anew. The signals are amplified with a Grass Neurodata Model 15ST amplification system (Grass Telefactor, NH), with a 50,000 gain, filters with 10-300 Hz bandwidth and a sampling interval of 0.83 milliseconds (1200 Hz).

Each participant was given a complete ophthalmic exam, including general anamnesis, best-corrected visual acuity, slit lamp biomicroscopy, intraocular-pressure measurement using the Goldmann applanation tonometer, gonioscopy, dilated fundoscopic examination (90D lens), stereo retinographs and a 24-2 SITA Humphrey automated perimetry (Swedish Interactive Threshold Algorithm. Carl Zeiss Meditec Inc.). A diagnosis of open angle glaucoma was established where there were at least two consecutive abnormal visual fields in the Humphrey campimetry, (threshold test 24-2), defined by: 1) a pattern standard deviation (PSD) and/or corrected pattern standard deviation (CPSD) below the 95% confidence interval; or 2) a Glaucoma Hemifield Test outside the normal limits. We define as abnormal an altitudinal hemifield in the Humphrey visual field analysis giving three or more contiguous sectors below the 95% confidence interval, with at least one of them below the 99% confidence interval. The visual field was dismissed as unreliable if the rate of false positives, false negatives or fixation losses was higher than 33%. A control database was also established on the basis of normal eye records established within the longitudinal prospective study. All these normal eye records had an intraocular pressure of 21 mmHg or less (with no previous history of ocular hypertension). An ophthalmic examination of the optic papilla was also conducted to check that it fell within the normal structural parameters.

The signals obtained from the 103 hexagons were regrouped and averaged to build up a new 56-sector map as shown in figure 1. The purpose of this regrouping was to simplify the analysis and to improve the signal-to-noise ratio. A 56-sector topography was therefore chosen, similar to that studied in automated campimetry, the clinical "gold-standard" for evaluating the visual field. It should also be noted here that sector 41 is the average of a greater number of hexagons, since it is the area containing the blind spot and, as such, more difficult to analyse.

Two mfERG record databases were built up, one containing healthy or control individuals and the other glaucoma-affected individuals for study by means of the Discrete Wavelet Transform (DWT). Two other specific databases were also created to be studied by means of an alternative technique, Morphological Analysis, all made up by a complete 56-sector map as shown in figure 1.

Not all the sectors making up the map to be analysed by the Wavelet Transform belonged to a single patient; the map groups together 56 clearly glaucoma-identified sectors from among the fifty patients diagnosed with the same symptom. Following a similar procedure, a sector map comprising the control database was built up, this time on the basis of healthy individuals.

As regards the databases used for the morphological analysis, these were made up by two 15-record collections from the 56 sectors: the first coming from 15 patients affected with early-stage OAG and showing between 3 and 12 diseased sectors, and the other built up from the 15 healthy control subjects.

## 2.2  Study of Severe Lesions by Wavelet Analysis

DWT was better than morphological analysis as a mfERG-record analysis tool for detecting severe retina lesions. Conversely, morphological analysis was much more efficient for detecting early-stage glaucoma by extracting certain markers present in the records.

The great drawback of the Fourier transform-based analysis is that the time information is forfeited when the signal is transformed into the frequency domain. The drawback is particularly telling when the signal to be analysed is transitory in nature or of finite duration, as in the case of mfERG signals, whose frequency content changes over time. The discrete wavelet transform (DWT) surmounts this drawback by analysing the signal in different frequencies with different resolutions, using regions with windowing of different sizes and obtaining a two-dimensional time-frequency function as a result. Wavelet analysis uses finite-length, oscillating, zero-mean wave forms, which tend to be irregular and asymmetrical. These are the windowing functions called mother wavelets. In principle there may be an infinite number of possible waves that are eligible for use as wavelets, but in practice a more limited number of wavelets are used, of well-known characteristics, efficacy and implementation: Haar, Daubechies, Coiflets, Mexican Hat, Symlets, Morlet, Meyer, etc. In the study we are dealing with here a great number of them were explored; it was with the Bior3.1 wavelet that the best subjective results were obtained for visual identification of certain markers that help us to differentiate normal mfERG signals from those belonging to subjects with advanced glaucoma [8].

The signal to be analysed is decomposed on the basis of shifted and dilated versions of the mother wavelet or analysing wavelet that we have decided to use; this is all done by means of the correlation between the signal to be decomposed and the abovementioned versions of the mother wavelet. Mathematically, the discrete wavelet transform (DWT) is defined as:

$$C(j,k) = \sum_{n \in z} f(n) 2^{-j/2} \psi(2^{-j}n - k) \tag{1}$$

where the resulting $C(j,k)$ is a series of coefficients indicating the correlation between the function $f(n)$ to be decomposed and the wavelet $\psi a,b(t)$ dilated to a scale $a=2^j$ and with a shifting $b=k2^j$, with $j,k \in Z$. The resulting $C(j,k)$ includes time and frequency information of the function $f(n)$, according to the values of $j$ and $k$, respectively. In practice we obtain two sets of time-function signals, one of them made up by the signals $A_1$ to $A_n$ which represent successive approximations of increasing smoothness or declining frequency of the signal $f(n)$, and the other by $D_1$ to $D_n$ which represent the successive details, also of falling frequency.

The signals were analysed by applying up to 5 levels of wavelet decomposition to each one of the different sectors and for two different time windows: one from 10 to 190 ms and another from 60 to 90 ms. The first contains the global response to the multifocal stimulus used here and the second contains the most important information on the induced response generated by this type of stimulus. Several superimposed records were obtained from different sectors to obtain an overview of the markers that might differentiate normal signals from abnormal signals.

## 2.3 Study of Slight Lesions by Morphological Analysis

The mfERG readings from patients with early-stage glaucoma, with slight lesions or isolated sectors developing the disease, do not show a uniform pattern over the healthy or diseased retina sectors. This makes the analysis thereof more critical. To detect lesions of this type a morphological signal study was conducted in the IC time interval (induced component) falling between P1 and P2, as shown in figure 3.

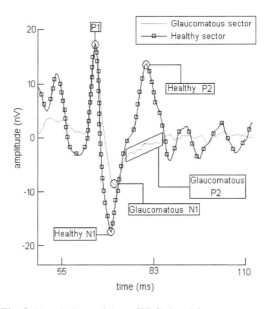

**Fig. 3.** Morphology of the mfERG signal from one sector

Although the claim cannot be made across the board for all cases, there is usually a series of morphological characteristics held in common in the records of healthy sectors, differentiating them from the diseased ones. These are called identity patterns [9]. The identity pattern of the healthy sectors shows little variation and contains a quick signal response in and near the induced component, thus building up more energy at mid frequencies. This conduct reflects the behaviour of the healthy retina cells, which tend to respond quickly and efficiently to the mfERG stimulus. The behaviour of a glaucomatous sector, on the other hand, shows much more high frequency oscillatory potentials in the IC interval, with a more blurred definition of signal peaks and troughs and a long drawn-out response. Given the signal characteristics in said interval, our morphological analysis studies the behaviour of the following signal parameters:

- Localisation of points P1, N1 and P2.
- Distance between P1 and P2.

- Sample width at N1.
- Slope in the interval N1 - P2.
- Signal oscillations in the interval N1 - P2.

The waveform of the mfERG reading changes from one sector to another, depending on the retina position of each one. To allow for this effect the analyses have been carried out under different performance parameters, depending on the sector's position in the retina. Results show that the individualised study of each sector zone gives our method an enhanced spatial resolution.

## 3   Results

In the DWT analysis, several superimposed records were obtained from different sectors to obtain an overview of the markers that might differentiate normal signals from abnormal signals.

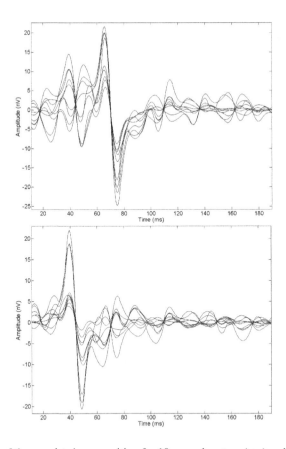

**Fig. 4.** Detail $D_4$ of the wavelet decomposition for 10 normal sectors (top) and 10 glaucomatous (bottom)

The top graph of figure 4 shows superimposed the $D_4$ details of the Wavelet decomposition between 10 and 190 ms from ten different sectors corresponding to various healthy individuals. The bottom graph of the same figure shows a similar representation for ten glaucomatous sectors and with an identical topographical position to the former. One of the most obvious features here is that the signals corresponding to healthy individuals show their greatest negative edge at about 70 ms, while signals in the hexagons affected by glaucoma tend to bottom out at about 45 ms. The efficiency of this marker was quantified against a time window running from 25 to 90 ms, looking for the greatest negative edge. When this edge came in the first half of the window the signal was classified as glaucomatous, while if it came in the second half it was classified as healthy.

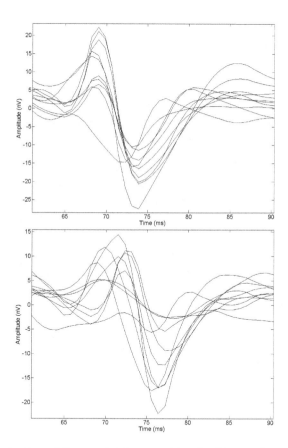

**Fig. 5.** $A_2$ approximation of the wavelet decomposition for 10 normal sectors (top) and 10 glaucomatous (bottom)

Figure 5 (top) shows superimposed the $A_2$ approximations corresponding to the wavelet decomposition between 60 and 90 ms of ten different hexagons belonging to different healthy individuals. The lower part of this figure shows a similar representation for ten hexagons affected with glaucoma and with the same topographical position as those above. In this case a trough appears at about 73 ms for healthy signals, coming slightly later for abnormal subjects. Since there might be more troughs, the efficiency of this second marker is quantified against a time window running from 65 to 87 ms., seeking this trough. When the trough comes in the first half of the window the signal was classified as healthy, while if it came in the second half it was classified as glaucomatous.

Table 1 shows the results, using both markers separately, for true and false healthy and glaucomatous out of a set of 56 sectors belonging to different healthy individuals and 56 with glaucoma.

**Table 1.** Results obtained using DWT markers separately (M=Marker, TH= True Healthy, FG=False Glaucomatous, TG= True Glaucomatous, FH=False Healthy)

| M | TH | FG | TG | FH |
|---|----|----|----|----|
| D4 | 55 | 1 | 48 | 8 |
| A2 | 54 | 2 | 51 | 5 |

The morphological analysis of slight lesions shows that the duration of the N1 interval is less in healthy than in glaucomatous sectors, the time-lag of P2 behind P1 is less in healthy than in glaucomatous sectors, the amplitude of P2 has to be positive, the glaucomatous signal shows greater sensitivity in P2 than in N1 and in P1 (accepting a 2% variation).

The disease also shows a change in the deterioration of healthy sectors according to whether the lesion is slight or severe (see Figure 6 from left to right). This evolution can be seen in P2, changing from a healthy sector morphology with a sharp P2 peak rising quickly from N1, to a flat morphology with high frequency alterations in P2 (slight case) and lastly to an even flatter P2 morphology (severe case). The study's statistical results are shown in table 2.

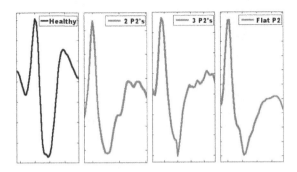

**Fig. 6.** P2 wave morphology trend

**Table 2.** Results of the morphological study TH= True Healthy, FG=False Glaucomatous, TG= True Glaucomatous, FH=False Healthy)

| TH | FG | TG | FH |
|------|------|------|------|
| 80 % | 20 % | 90 % | 10 % |

## 4  Conclusions

The morphology of the signals recorded in each hexagon varies according to the position that this hexagon occupies in the retina and the type of stimulus used. It is also known that the optic nerve head component (ONHC) is the main cause of the asymmetries in the records [9][10], whereby said component arrives in each hexagon with a different time-lag depending on the distance between the hexagon and the optic nerve. This will enhance or cancel out some components as a result of the different retina levels below the hexagon under study. Loss of the ONHC has already been mooted as an early indicator of glaucoma [11][12], so there is obviously a need for adjustment of the various time windows and types of markers used in this study, according to the position of the hexagon in the retina map, to optimise and fine tune the results obtained herein.

A more in-depth investigation needs to be carried out to adjust the parameters obtained herein by means of DWT analysis, to find out best values in terms of the retinal quadrants and rings to which the sector under study belongs, in view of the abovementioned hexagon dependency.

The type of markers used herein and the tool used to obtain them, i.e., the Wavelet transform, make it impossible a priori to establish any association with a specific physiological origin, since there are no precedents to go on. It does not fall within the remit of this study to establish a physiological cause-effect relationship for the marker but rather to search for technical tools to help experts to diagnose glaucoma in humans in its early stages of development.

It is obvious that a joint and complementary use of all the techniques studied herein would be the best way to improve OAG diagnosis. In this way the sectors detected as healthy in the Wavelet study would be introduced into the signal morphology analysis to check whether there might be any slight lesions that Wavelet analysis was incapable of picking up.

**Acknowledgements.** This work was supported by grants from Comunidad de Madrid-Universidad de Alcalá (ref. nº CCG06-UAH/BIO-0711) and Ministerio de Educación y Ciencia (ref. nº SAF2004-5870-C02-01) awarded to Pedro de la Villa.

## References

1. Catalá Mora, J., Castany Aregall, M., Berniell Trota, J.A., Arias Barquet, L., Roca Linares, G., Jürgens Mestre, I.: Electrorretinograma Multifocal y Degeneración Macular Asociada a la Edad. Archivos de la Sociedad Española de Oftalmología 80(7) (2005)

2. Miguel, J.M., Blanco, R., Boquete, L., Rodríguez, J.M., De la Villa, P.: Electroretinography Multifocal. Técnicas y Aplicaciones. In: CISTI 2007, Porto, Portugal (2007)
3. Sutter, E., Tran, D.: The field topography of ERG components in man–I. The photopic luminance response. Vision Res. 32, 433–446 (1992)
4. Sutter, E.: Imaging visual function with the multifocal m-sequence technique. Vision Res. 41, 1241–1255 (2001)
5. Hood, D.C., Odel, J.G., Chen, C.S., Winn, B.J.: The Multifocal Electroretinogram. J. Neuro-Ophthalmol. 23(3), 225–235 (2003)
6. Marmor, M.F., Hood, D.C., Keating, D., Kondo, M., Seeliger, M.W., Miyake, Y.: Guidelines for basic multifocal electroretinography (mfERG). Documenta Ophthalmologica 106, 105–115 (2003)
7. Hagan, R.P., Fisher, A.C., Brown, M.C.: Examination of short binary sequences for mfERG recording. Doc. Ophthalmol. 113, 21–27 (2006)
8. Miguel, J.M., Blanco, R., Boquete, L., Rodríguez, J.M., De la Villa, P.: Multifocal Electroretinography. Glaucoma Diagnosis by Means of TheWavelet Transform. In: IEEE CCECE (2008) ISBN: 978-1-9244-1643-1
9. Fortune, B., Bearse Jr., M.A., Cioffi, G.A., Johnson, C.A.: Selective Loss of an Oscillatory Component from Temporal Retinal Multifocal ERG Responses in Glaucoma. IOVS 43(8), 2638–2647 (2002)
10. Zhou, W., Rangaswamy, N., Ktonas, P., Frishman, L.J.: Oscillatory potentials of the slow-sequence multifocal ERG in primates extracted using the Matching Pursuit method. Vision Research 47, 2021–2036 (2007)
11. Rangaswamy, N.V., Zhou, W., Harwerth, R.S., Frishman, L.J.: Effect of Experimental Glaucoma in Primates on Oscillatory Potentials of the Slow-Sequence mfERG. IOVS 47(2), 753–767 (2006)
12. Hood, D.C.: Assessing Retinal Function with the Multifocal Technique. Progress in Retinal and Eye Research 19(5), 607–646 (2000)

# Biotin-Streptavidin Sensitive BioFETs and Their Properties

Thomas Windbacher, Viktor Sverdlov, and Siegfried Selberherr

Institute for Microelectronics, TU Wien, Gußhausstraße 27–29/E360, A-1040, Wien, Austria
{Windbacher,Sverdlov,Selberherr}@iue.tuwien.ac.at

**Abstract.** In this work the properties of a biotin-streptavidin BioFET have been studied numerically with homogenized boundary interface conditions as the link between the oxide of the FET and the analyte which contains the bio-sample. The biotin-streptavidin reaction pair is used in purification and detection of various biomolecules; the strong streptavidin-biotin bond can also be used to attach biomolecules to one another or onto a solid support. Thus this reaction pair in combination with a FET as the transducer is a powerful setup enabling the detection of a wide variety of molecules with many advantages that stem from the FET, like no labeling, no need of expensive read-out devices, the possibility to put the signal amplification and analysis on the same chip, and outdoor usage without the necessity of a lab.

## 1 Introduction

Today's technology for detecting tumor markers, antigen-antibody complexes, and pathogens is time-consuming, complex, and expensive [1, 2]. For instance, a typical procedure to detect a given DNA complex is to increase the concentration by Reverse Transcription (RT) or Polymerase Chain Reaction (PCR), followed by a process step that will add a label to the DNA enabling detection by light or radiation. After all these steps the sample is applied to a microarray. The microarray consists of an array of spots, and every single spot is able to detect a different type of molecule. After the reaction has taken place the array is read by an expensive microarray reader.

Replacing the above sensing mechanism by electrical detection has several benefits. First, the optical microarray reader becomes superfluous. Detection by FET (field-effect transistor) makes the integration of amplifying and analyzing circuits on the same chip possible, thus saving also equipment. The advanced development of semiconductor process technology allows mass production of such devices, decreasing the price dramatically. Various kinds of reaction pairs are possible and have been studied, like detection of DNA [3,4,5,6,7], cancer markers [8], proteins, e.g. biotin-streptavidin [9,10,11,12], albumin [13], and transferrin [14]. In principle, every molecule which is charged in the solute and which can be bound to the surface layer can be detected by a BioFET. The field of applications is very wide and spans from DNA sequencing and point of care applications to controlling environmental pollution and the spread of diseases. The BioFET can be easily integrated into the chip environment. By putting a microfluidic channel above the functionalized gate of the BioFET the chip can be turned into a mini-laboratory - the lab-on-chip. This enables better control of the environmental parameters (e.g. local pH or detecting the amount of a special protein) and gives the possibility

A. Fred, J. Filipe, and H. Gamboa (Eds.): BIOSTEC 2009, CCIS 52, pp. 85–95, 2010.

of local measurement (e.g. how a cell reacts to a stimulus), thus providing a complete lab-on-a-chip.

## 2 Method

A BioFET consists of several parts: a semiconductor transducer, a dielectric layer, a biofunctionalized surface, and the analyte (Figure 1). The semiconductor transducer is a conventional FET. The dielectric layer is the gate oxide, and the biofunctionalized surface contains immobilized biomolecule receptors attached, so it is able to bind the desired molecule. The analyte is in an aqueous solution. If a target molecule binds to a receptor, the local charge density at the surface changes and thus the potential in the semiconductor and so the conductivity of the channel of the field-effect transducer is changed.

The binding of the target with the receptor happens at the Angstrom length scale, while the semiconductor device is in the micrometer length scale. Thus a proper way of describing the semiconductor-solution interface is crucial.

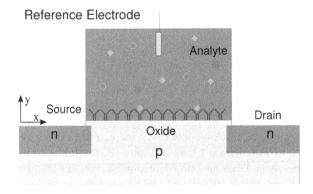

**Fig. 1.** Schematic diagram of a BioFET.

Transport in a FET with a gate length of $1\mu$m, can be modeled via the drift-diffusion approach [15, 16]. The aqueous solution is described by the Poisson-Boltzmann equation.

$$\epsilon_0 \nabla \cdot (\epsilon_{\text{Ana}} \nabla \psi(x,y)) = -\sum_{\sigma \in S} \sigma\, q\, c_\sigma^\infty\, e^{-\sigma \frac{q}{k_{\text{B}} T}(\psi(x,y)-\psi_\mu)} \tag{1}$$

$k_{\text{B}}$ denotes Boltzmann's constant, $T$ the temperature in Kelvin, and $\sigma$ the indices out of the set $S$ containing the valences of the ions in the electrolyte. $\epsilon_0$ describes the permittivity of vacuum, and $q$ the elementary charge. $\psi_\mu$ is the chemical potential. $c_\sigma^\infty$ is the bulk ion concentration in equilibrium, while $\epsilon_{\text{Ana}} \approx 80$ is the relative permittivity of water.

The sum describes the carrier densities arising from the Boltzmann model. Assuming sodium-chloride as salt, which is a 1 : 1 salt, the expression given in (1) can be reduced to

$$\epsilon_0 \nabla \cdot (\epsilon_{\text{Ana}} \nabla \psi(x,y)) = 2q\, c_\sigma^\infty\, \sinh(\frac{q}{k_{\text{B}} T}(\psi(x,y) - \psi_\mu)) \ . \tag{2}$$

The charge on the surface due to chemical reaction of the $H^+$ and $OH^-$ has been modeled at $pH = 7$ with the site-binding model [2]:

$$Q_{Ox} = q\,N_S\ \frac{\dfrac{[H^+]_b}{K_a}\,e^{-\frac{q}{k_BT}\Psi(x,y)} - \dfrac{K_b}{[H^+]_b}\,e^{\frac{q}{k_BT}\Psi(x,y)}}{1 + \dfrac{[H^+]_b}{K_a}\,e^{-\frac{q}{k_BT}\Psi(x,y)} + \dfrac{K_b}{[H^+]_b}\,e^{\frac{q}{k_BT}\Psi(x,y)}}\ . \tag{3}$$

$N_S$ denotes the surface binding site density, while $K_a$ and $K_b$ are the equilibrium constants for charging the surface positively and negatively, respectively. $[H^+]_b$ describes the positive hydrogen ion concentration of the bulk and is corrected to the activity of the hydrogen concentration by the $e^{\frac{q}{k_BT}\Psi(x,y)}$ terms.

The biomolecules are modeled in a physics-based bottom-up approach. By calculating the charge and dipole moment for a single molecule (see for example Figure 2, [17]), a mean charge density and a mean dipole moment density of the boundary layer is obtained. This bridges the gap between the Angstrom length scale of the biomolecules and the micrometer dimensions of the FET [18, 19, 20, 21].

**Fig. 2.** Biotin-streptavidin complex [22] on the oxide surface. Two iso-surfaces for plus and minus $0.1\ \frac{k_BT}{qA^2}$ are shown.

The link between the gate oxide and the aqueous solution is realized by two interface conditions, [18, 19, 23],

$$\epsilon_0\epsilon_{Oxid}\,\partial_y\psi(0-,x) - \epsilon_0\epsilon_{Ana}\,\partial_y\psi(0+,x) = -C(x), \tag{4}$$

$$\psi(0-,x) - \psi(0+,x) = -\frac{D_y(x)}{\epsilon_{Ana}\epsilon_0}\ . \tag{5}$$

The x-axis is parallel oriented to the oxide surface, while the y-axis points into the liquid. $\psi(0-)$ describes the potential in the oxide, while $\psi(0+)$ relates to the potential in the solute. The first equation describes the jump in the field, while the second introduces a dipole moment which causes a shift of the potential taken into account by adjusting the potential in the analyte (Figure 3).

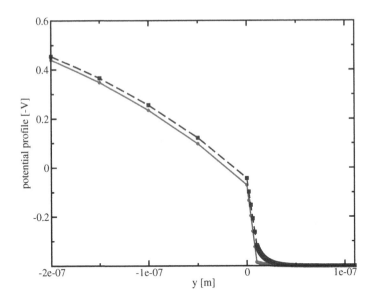

**Fig. 3.** Potential profile at the interface (from left to right: semiconductor, oxide, solute). The dashed line shows the profile, when biotin and streptavidin are attached to the surface (for 10 nm density), while the full line shows the potential with water and salt only.

## 3   Simulation

Three different types of dielectric were simulated. $SiO_2$ as a reference, $Al_2O_3$, and $Ta_2O_5$ as possible high-k materials, with relative permitivies of 3.9, 10, and 25, respectively. As solute 1 mMol sodium-chloride at pH $= 7$ was considered. The parameters for the site-binding model can be found in Table 1 [24]. For each dielectric the unprepared state (just water and salt), the prepared state (water, salt, and biotin), and the bound state, when the chemical reaction has taken place (water, salt, and biotin-streptavidin), were calculated for two different mean distances between molecules ($\lambda = 10$ nm, $\lambda = 15$ nm). The data used for calculating charge and dipole moment of biotin and streptavidin are obtained from http://www.pdb.org/ ( Figure 2, Figure 11, [25] ). The potential distribution across the device is shown in Figure 4 and output curves were calculated for every parameter combination mentioned above, assuming a 100% binding efficency. The potential of the reference electrode is set to 0.4 V so that the FET will be in moderate inversion as proposed by [26].

**Table 1.** The parameters needed for the site-binding model using different dielectric

| Oxide | $pK_a$ | $pK_b$ | $N_S$ [cm$^{-2}$] | Reference |
|-------|--------|--------|-------------------|-----------|
| $SiO_2$ | $-2$ | 6 | $5 \cdot 10^{14}$ | [27] |
| $Al_2O_3$ | 6 | 10 | $8 \cdot 10^{14}$ | [27] |
| $Ta_2O_5$ | 2 | 4 | $10 \cdot 10^{14}$ | [28] |

**Fig. 4.** Potential profile for $Ta_2O_5$ water, salt, and biotin-streptavidin at $\lambda = 10$ nm average distance. Blue denotes $-1$ V while red stands for 1 V.

## 4   Results

Figures 5, 6, and 7 show a decrease in the output current for biotin attached to the surface in comparison to the unprepared surface. This downward shift for the bound state in comparison to the unbound state is due to the increase of negative charges at the interface, which is also confirmed by the difference between the curves for $\lambda = 10$ nm and $\lambda = 15$ nm, since for 10 nm the molecules are located denser than for 15 nm.

As can be seen in the Figures 5, 6, and 7, the bigger the $\epsilon_r$ of the dielectric the bigger is the output current. Thus high-k materials deliver stronger output signals. According to [29] however, higher $\epsilon_r$ dielectric constants may lead to higher trap densities and thus

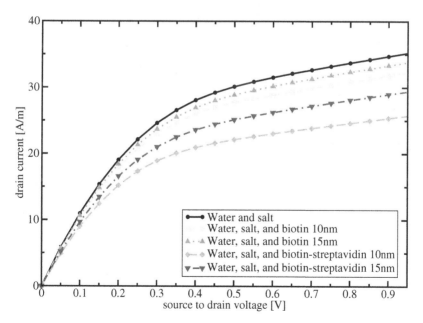

**Fig. 5.** Output curve for $SiO_2$ for unprepared, prepared but unbound, and bound state at $\lambda = 10$ nm and $\lambda = 15$ nm, respectively.

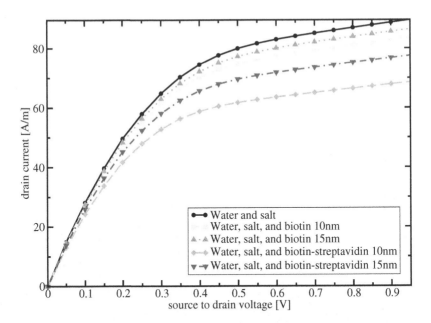

**Fig. 6.** Output curve for $Al_2O_3$ for unprepared, prepared but unbound, and bound state at $\lambda = 10\,nm$ and $\lambda = 15\,nm$, respectively.

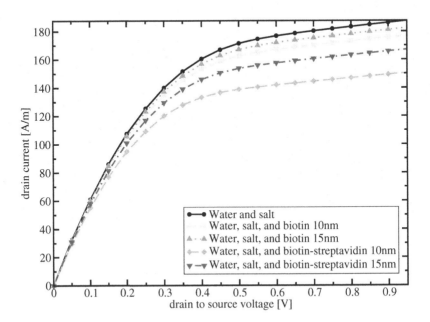

**Fig. 7.** Output curve for $Ta_2O_5$ for unprepared, prepared but unbound, and bound state at $\lambda = 10\,nm$ and $\lambda = 15\,nm$, respectively.

to a decreased signal-to-noise ratio. Therefore, a trade-off between bigger output signal and signal-to-noise ratio has to be met.

Figure 8 shows the output curves as a function of dielectric and molecule orientation (0° means perpendicular to the surface and 90° means lying flatly on the surface) leading to the lowest output curves for 0° followed by 90° and the curves without dipole moment for each group. Figures 9 and 10 show the small signal or differential resistance as a function of dielectric and molecule orientation, displaying smaller values for higher relative permittivity $\epsilon_r$. A slightly larger differential resistance is observed for perpendicular molecule orientation, in agreement with the previous results shown in Figures 5, 6, and 7. This is expected, because biomolecules are inhomogeneously charged. Therefore, they possess a dipole moment which enters into the boundary conditions (5) and there should be a difference in the output curves of the BioFET for different orientation angles in relation to the surface.

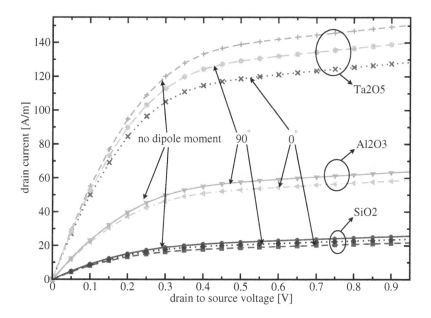

**Fig. 8.** Output curves for $SiO_2$, $Al_2O_3$, and $Ta_2O_5$ for calculation without dipole moment, angle 0° (perpendicular to surface), and angle 90° (parallel to surface).

In the biochemical community there is an ongoing discussion, if the orientation of the biomolecule is relevant for sensing. Several papers have shown contradictory results [30,31,32,33,34]. All these papers are based on optical detection. Although more study is needed, we mention that for optical detection it is more important to choose the linking molecule in a way that the reaction is not hindered by steric effects (receptors block each other) or the binding sites are blocked or even broken by the crosslinker. In the case of a BioFET, however, a field-effect as working principle is used. Thus it is important to have a linker which is as short as possible, to be close to the surface.

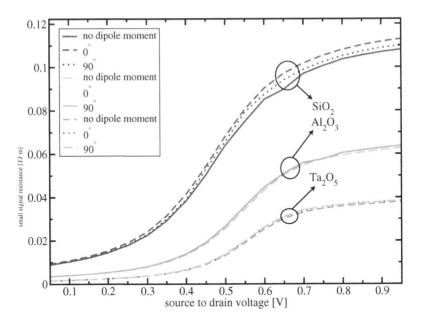

**Fig. 9.** Small signal resistance for $SiO_2$, $Al_2O_3$, and $Ta_2O_5$ calculated without dipole moment, angle $0°$ (perpendicular to surface), and angle $90°$ (parallel to surface) at biotin only.

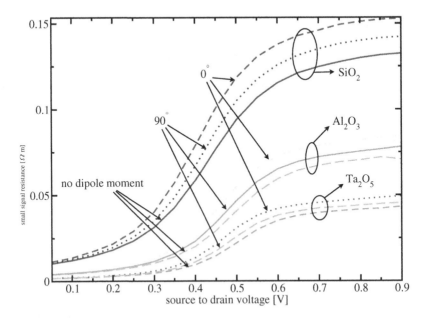

**Fig. 10.** Small signal resistance for $SiO_2$, $Al_2O_3$, and $Ta_2O_5$ calculated without dipole moment, angle $0°$ (perpendicular to surface), and angle $90°$ (parallel to surface) at bound state (biotin-streptavidin) for $10\,nm$ average distance.

To increase the signal-to-noise ratio, the linker should have as little charge as possible. For example, in order to detect streptavidin, biotin is used as a binding agent. A biotin molecule is attached to the surface with a neutral linker. Streptavidin then binds to biotin thus forming a bound state. The charge difference between the unbound state of a biotin molecule alone, which is negatively charged with a single elementary charge and the bound state of biotin-streptavidin, which is negatively charged with five elementary charges, is large enough for detection. We also note that due to the tetrameric nature of streptavidin it has four sites to bind biotin as shown in Figure 11. Therefore, the linker binding biotin to the surface should be short enough in order to prevent binding several biotin molecules to a single molecule of streptavidin. If there is the freedom to choose attaching biotin or streptavidin via a linker to the surface, it will be better to use biotin. This will lead to a better signal to noise ratio because of the bigger change in surface charge, when the streptavidin charged with minus four elementary charges binds to biotin charged with minus one elementary charge. Additionally it offers the possibility to bind further proteins with biotin attached.

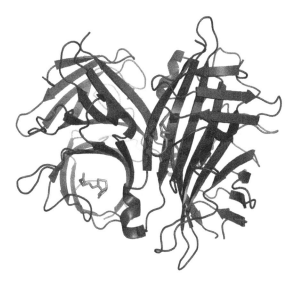

**Fig. 11.** Scheme of the tetrameric protein streptavidin and biotin.

## 5   Conclusions

The presented model shows a strong dependence on surface charges and indicates a detectable shift in the threshold voltage depending on their orientation related to the surface. The bound state (streptavidin-biotin) negatively charged with five elementary charges compared to the unbound state (biotin) negatively charged with one elementary charge leads to a reduced conductivity, when hybridization has taken place. Also the shift of the threshold voltage and output characteristics due to different molecule orientations (0°...perpendicular to surface, 90°...lying flat on surface) can be seen. This demonstrates the usefulness of the simulation method for the design of efficient BioFETs.

**Acknowledgements.** This work was supported by the Austrian Science Fund FWF, Project P19997-N14.

# References

1. Pirrung, M.C.: How to make a DNA chip. Angew. Chem. Int. Ed. 41, 1276–1289 (2002)
2. Shinwari, M.W., Deen, M.J., Landheer, D.: Study of the electrolyte-insulator-semiconductor field-effect transistor (EISFET) with applications in biosensor design. Microelectronics Reliability 47(12), 2025–2057 (2007)
3. Fritz, J., Cooper, E.B., Gaudet, S., Soger, P.K., Manalis, S.R.: Electronic detection of DNA by its intrinsic molecular charge. PNAS 99, 1412–1416 (2002)
4. Hahm, J., Lieber, C.M.: Direct ultrasensitive electrical detection of DNA and DNA sequence variations using nanowire nanosensors. Nano Letters 4(1), 51–54 (2004)
5. Gao, Z., Agarwal, A., Trigg, A., Singh, N., Fang, C., Tung, C., Fan, Y., Buddharaju, K., Kong, J.: Silicon nanowire arrays for label-free detection of DNA. Analytical Chemistry 79(9), 3291–3297 (2007)
6. Stern, E., Vacic, A., Reed, M.A.: Semiconducting nanowire field-effect transistor biomolecular sensors. IEEE Transactions on Electron Devices 55(11), 3119–3130 (2008)
7. Kim, S.N., Rusling, J.F., Papadimitrakopoulos, F.: Carbon nanotubes for electronic and electrochemical detection of biomolecules. Advanced Materials 19(20), 3214–3228 (2007)
8. Zheng, G., Patolsky, F., Cui, Y., Wang, W.U., Lieber, C.M.: Multiplexed electrical detection of cancer markers with nanowire sensor arrays. Nature Biotechnology 23(10), 1294–1301 (2005)
9. Im, H., Huang, X., Gu, B., Choi, Y.: A dielectric-modulated field-effect transistor for biosensing. Nature Nanotechnology 2(7), 430–434 (2007)
10. Cui, Y., Wei, Q., Park, H., Lieber, C.M.: Nanowire nanosensors for highly sensitive and selective detection of biological and chemical species. Science 293(5533), 1289–1292 (2001)
11. Gupta, S., Elias, M., Wen, X., Shapiro, J., Brillson, L.: Detection of clinical relevant levels of protein analyte under physiologic buffer using planar field effect transistors. Biosensors and Bioelectronics 24, 505–511 (2008)
12. Stern, E., Klemic, J., Routenberg, D., Wyrembak, P., Turner-Evans, D., Hamilton, A., LaVan, D., Fahmy, T., Reed, M.: Lable-free immunodetection with CMOS-compatible semiconducting nanowires. Nature Letters 445(1), 519–522 (2007)
13. Park, K.M., Lee, S.K., Sohn, Y.S., Choi, S.Y.: BioFET sensor for detection of albumin in urine. Electronic Letters 44(3) (January 2008)
14. Girard, A., Bendria, F., Sagazan, O.D., Harnois, M., Bihan, F.L., Salaün, A., Mohammed-Brahim, T., Brissot, P., Loréal, O.: Transferrin electronic detector for iron disease diagnostics. IEEE Sensors, 474–477 (October 2006)
15. Tang, T.W., Ieong, M.K.: Discretization of flux densities in device simulations using optimum artificial diffusivity. IEEE Transactions on Computer-Aided Design of Integrated Circuits and Systems 14(11), 1309–1315 (1995)
16. Selberherr, S.: Analysis and Simulation of Semiconductor Devices. Springer, Heidelberg (1984)
17. Poghossian, A., Cherstvy, A., Ingebrandt, S., Offenhäusser, A., Schöning, M.J.: Possibilities and limitations of label-free detection of DNA hybridization with field-effect-based devices. Sensors and Actuators, B: Chemical 111-112, 470–480 (2005)
18. Heitzinger, C., Kennell, R., Klimeck, G., Mauser, N., McLennan, M., Ringhofer, C.: Modeling and simulation of field-effect biosensors (BioFETs) and their deployment on the nanoHUB. J. Phys.: Conf. Ser. 107, 1–12 (2008)

19. Ringhofer, C., Heitzinger, C.: Multi-scale modeling and simulation of field-effect biosensors. ECS Transactions 14(1), 11–19 (2008)
20. Windbacher, T., Sverdlov, V., Selberherr, S., Heitzinger, C., Mauser, N., Ringhofer, C.: Simulation of field-effect biosensors (BioFETs). In: Proc. Simulation of Semiconductor Processes and Devices (SISPAD 2008), Hakone, Japan, September 2008, pp. 1–4 (2008)
21. Windbacher, T., Sverdlov, V., Selberherr, S., Heitzinger, C., Mauser, N., Ringhofer, C.: Simulation of field-effect biosensors (BioFETs) for biotin-streptavidin complexes. In: 29th International Conference on the Physics of Smeiconductors (ICPS 2008), Rio de Janeiro, Brasil (2008)
22. Protein data bank: (A resource for studying biological macromolecules), http://www.pdb.org/
23. Heitzinger, C., Klimeck, G.: Computational aspects of the three-dimensional feature-scale simulation of silicon-nanowire field-effect sensors for DNA detection. Journal of Computational Electronics 6, 387–390 (2007)
24. Landheer, D., Aers, G., McKinnon, W., Deen, M., Ranuárez, J.: Model for the field effect from layers of biological macromolecules on the gates of meta-oxide-semiconductor transistors. Journal of Applied Physics 98(4), 044701–1 –044701–15 (2005)
25. Stenkamp, R.E., Trong, I.L., Klumb, L., Stayton, P.S., Freitag, S.: Structural studies of the streptavidin binding loop. Protein Science 6(6), 1157–1166 (1997)
26. Deen, M.J., Shinwari, M.W., Ranuárez, J.C., Landheer, D.: Noise considerations in field-effect biosensors. Journal of Applied Physics 100(7), 074703–1 –074703–8 (2006)
27. Bousse, L.: The Chemical Sensitivity of Electrolyte/Insulator/Silicon Structures. Phd, Dissertation, Twente University of Technology, Enschede (1982)
28. Bousse, L., Mostarshed, S., Van Der Shoot, B., De Rooij, N.F., Gimmel, P., Gopel, W.: Zeta potential measurements of $Ta_2O_5$ and $SiO_2$ thin films. Journal of Colloid and Interface Science 147(1), 22–32 (1991)
29. Deen, M.J.: Highly sensitive, low-cost integrated biosensors. In: SBCCI 2007: 20th Symposium on Integrated Circuits and System Design, p. 1 (2007)
30. Oh, S.W., Moon, J.D., Lim, H.J., Park, S.Y., Kim, T., Park, J., Han, M.H., Snyder, M., Choi, E.Y.: Calixarene derivative as a tool for highly sensitive detection and oriented immobilization of proteins in a microarray format through noncovalent molecular interaction. FASEB Journal 19(10), 1335–1337 (2005)
31. Wacker, R., Schröder, H., Niemeyer, C.M.: Performance of antibody microarrays fabricated by either DNA-directed immobilization, direct spotting, or streptavidin-biotin attachment: A comparative study. Analytical Biochemistry 330(2), 281–287 (2004)
32. Kusnezow, W., Jacob, A., Walijew, A., Diehl, F., Hoheisel, J.D.: Antibody microarrays: An evaluation of production parameters. Proteomics 3(3), 254–264 (2003)
33. Peluso, P., Wilson, D.S., Do, D., Tran, H., Venkatasubbaiah, M., Quincy, D., Heidecker, B., Poindexter, K., Tolani, N., Phelan, M., Witte, K., Jung, L.S., Wagner, P., Nock, S.: Optimizing antibody immobilization strategies for the construction of protein microarrays. Analytical Biochemistry 312(2), 113–124 (2003)
34. Turková, J.: Oriented immobilization of biologically active proteins as a tool for revealing protein interactions and function. Journal of Chromatography B: Biomedical Sciences and Applications 722(1-2), 11–31 (1999)

# Improving Patient Safety with X-Ray and Anesthesia Machine Ventilator Synchronization: A Medical Device Interoperability Case Study*

David Arney[1], Julian M. Goldman[2], Susan F. Whitehead[2], and Insup Lee[1]

[1] University of Pennsylvania, Philadelphia, PA U.S.A.
arney@cis.upenn.edu, lee@cis.upenn.edu
[2] MD PnP Program, CIMIT, Cambridge, MA U.S.A.
jgoldman@mdpnp.org, swhitehead@partners.org

**Abstract.** When a x-ray image is needed during surgery, clinicians may stop the anesthesia machine ventilator while the exposure is made. If the ventilator is not restarted promptly, the patient may experience severe complications. This paper explores the interconnection of a ventilator and simulated x-ray into a prototype plug-and-play medical device system. This work assists ongoing interoperability framework development standards efforts to develop functional and non-functional requirements and illustrates the potential patient safety benefits of interoperable medical device systems by implementing a solution to a clinical use case requiring interoperability.

**Keywords:** MDPnP interoperability, Plug-and-play, Interoperable interconnected medical devices, X-ray ventilator, Formal methods, Verification model, Checking, apnea, Patient safety.

## 1 Introduction

Medical devices are a key element in the modern health care environment. They assist medical staff by automatically measuring physiologic parameters such as blood pressure, oxygen level, and heart rate, or actively influence these parameters by means of infusion pumps for analgesia and insulin or ventilators for breathing support. Almost all modern medical care rely on electronic medical devices.

Despite the pervasive use of medical devices throughout modern health care, most devices work on their own and in isolation. In contrast, interoperable devices would allow connections for sharing patient data, device status, and enabling external control, even between devices from different manufacturers. Such interoperability would lead to clear benefits for the care provider and the patient such as more accurate assessment of the patient's health and error-resilient systems through safety interlocks, closed-loop control, and automatic hot swappable backups.

To realize these benefits, the MD PnP program at the Center for Integration of Medicine & Innovative Technology at the Massachusetts General Hospital (CIMIT.org)

---

* This research was supported in part by NSF CNS-0509327, NSF-CNS-0610297, NSF CNS-0720703, and NSF CNS-0834524.

A. Fred, J. Filipe, and H. Gamboa (Eds.): BIOSTEC 2009, CCIS 52, pp. 96–109, 2010.

has been developing techniques and standards to facilitate medical device interoperability via MD PnP (Medical Device Plug-and-Play), similar to the plug-and-play of PC devices.

This paper describes a prototype MD PnP case study that was conducted for two purposes: (1) for the MD PnP program to extrapolate functional and non-functional requirements for the interoperability standards in progress, and (2) to develop a demo interoperable medical device system which would illustrate the benefits of the work by implementing a solution to a clinical use case requiring interoperability.

The rest of the paper is organized as follows. Section 2 describes the clinical use case which motivated this case study. Our problem statement and challenges are in Section 3. Section 4 describes the details of our system implementation, and Section 5 tells how we modeled and verified the system and generated code from the model. Finally, our conclusions are in Section 6.

## 2    Clinical Use Case

This project was driven by a specific clinical use case. This use case was documented by the Anesthesia Patient Safety Foundation to illustrate a potential safety problem with the way x-ray images are usually taken during surgery.

> A 32-year-old woman had a laparoscopic cholecystectomy [gall bladder removal] performed under general anesthesia. At the surgeons request, a plane film x-ray was shot during a cholangiogram [bile duct image]. The anesthesiologist stopped the ventilator for the film. The x-ray technician was unable to remove the film because of its position beneath the table. The anesthesiologist attempted to help her, but found it difficult because the gears on the table had jammed. Finally, the x-ray was removed, and the surgical procedure recommenced. At some point, the anesthesiologist glanced at the EKG and noticed severe bradycardia. He realized he had never restarted the ventilator. This patient ultimately expired. [1]

It is common practice to stop the anesthesia machine ventilator for a short time during surgery when this type of x-ray is performed. This ensures that the patient's chest and abdomen are not moving when the exposure is made, thus providing a sharper image. This does not harm the patient provided that the ventilator is restarted promptly. Difficulties arise only if the ventilator is not restarted for some reason. This kind of problem can be mitigated by using interconnected devices. If the anesthesia machine ventilator can synchronize with the x-ray, then it is no longer necessary to manually stop the ventilator to make the exposure.

Synchronization between a camera and external devices like a flash is not new. Typically, the camera sends a trigger signal to the flash at the right time. Similarly, the ventilator could synchronize with the x-ray machine. Since ventilators are not currently built to send synchronized signals to x-ray machines, we designed our system to have a third device which sits between the ventilator and x-ray, reads status messages from the ventilator, and makes the decision about when to trigger the x-ray. This third component is called the supervisor and is described in detail in Section 4. Systems which synchronize x-rays and ventilators have been built in the past, see for instance [2], but these systems must be built one at a time for specific devices and are limited to experimental use. Ventilators and x-ray machines are manufactured by many companies. Cross-manufacturer interoperability would allow synchronized systems to be built from any combination of devices that support the necessary functionality. The aim of the MD PnP program is to develop techniques and standards that facilitate medical device interoperability in order to allow such systems to be easily assembled and used clinically.

## 3   Problem Statement and Challenges

Our goal was to explore the safety and engineering issues involved in building a system that would allow the x-ray machine to take a clear image of the patient without the need to turn off the ventilator. Furthermore, we wanted to build a system which would illustrate the benefits of interoperability in the medical domain. Interoperable medical devices are devices which are capable of connecting to each other to share data or to allow external control. Such devices must have an external interface, and the design of these interfaces is the subject of several ongoing standards processes such as ISO/IEC 11073, Health Level 7 (HL7), and others. The use case we addressed specifically requires interoperability supporting external control. The implementation we developed is not intended to be used clinically. This project is essentially a research platform for understanding the core issues involved in interfacing these devices for this particular use case.

Most medical devices currently manufactured are not designed to be interoperable. The challenges we faced in building this system are generally faced by anyone trying to connect medical devices and are a major reason such interconnection is not more common. Medical devices generally have proprietary interfaces which are only documented in technical manuals or other material not openly available. We were fortunate to have the cooperation of Dräger, the manufacturer of the ventilator we used. The interface of the ventilator was designed to be used for diagnosis of machine faults and to send data to the electronic medical record, not as a source of real-time status information. Thus, it runs at a relatively slow rate, and the low maximum sample rate (5 - 10 samples per second) was the limiting factor in designing our control algorithm.

A further challenge in interconnecting medical systems is proving the safety of the resultant system. Safety is defined as freedom from unnecessary risk, where risks are unmitigated hazards. FDA provides guidance on risk minimization for medical devices. [3] The risk assessment process starts with a hazards analysis or failure modes and effects analysis (FMEA). These documents gather potential hazards and their mitigations, that are used in writing requirements and safety properties. The risk analysis process and how we used hazards to derive safety properties with which to verify the system is described in Section 5.

Our development process started with informal system requirements which were used to build a state machine model of the desired system behavior. We checked this model for safety properties using model checking software and then generated code from the model to produce the supervisor.

## 4 System Description

Figure 1 shows the overall architecture of our approach. This architecture follows closely that of the ICE (integrating the clinical environment) standard[4]. The major components are a set of medical devices, a network controller, a supervisor, the patient, and a caregiver. Medical devices connect to each other and the supervisor through the network controller. The devices' connections to the network controller may go through physical adapters and data format converters if their connectors and formats are not directly compatible. The network controller may also connect to an external network such as a hospital information system. The supervisor runs the control software for the system. Supervisor software for our system is the subject of Section 4.2. The supervisor hosts the user interface for the caregiver and may also contain a data logger, which records network activity and information from the devices.

MD PnP requires three phases of operations each with its own safety, security, and functional requirements. The first phase is device discovery and connection establishment, when devices are first connected to a MD PnP network. When a new device is connected to the network, the device's capabilities need to be communicated to the rest of the system. The second phase is normal operation of the plug-and-play system. During this phase, the devices transmit data they produce and receive commands or data from other parts of the system. The final phase is disconnection. When devices are removed from the system, the supervisor must decide how to respond. If the device was necessary for continued operation of the supervisor program, then the supervisor might notify the user and shut down. If the device was not necessary, the supervisor might be able to continue operating in a limited manner.

Our system implementation, which follows the conceptual architecture, is shown in Figure 2. The devices we used were a Dräger anesthesia machine ventilator and a simulated x-ray machine. The role of network controller and adapter is filled by the LiveData RTI software program. LiveData Inc. is a company which produces software

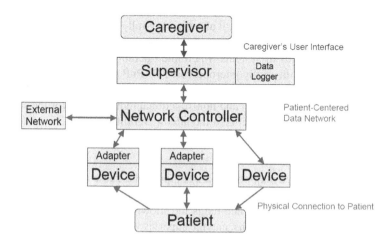

**Fig. 1.** Conceptual Architecture Overview

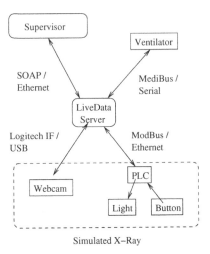

**Fig. 2.** Overview of the System

to integrate medical devices for common display data. We worked with LiveData to connect the ventilator and simulated x-ray. Their software translates the proprietary medical device formats and makes the data available through a single interface.

The supervisor program runs on the same computer as the LiveData RTI software. Finally, the patient was represented with a physical lung simulator consisting of a bellows and spring. While a simple lung simulator does not capture all the nuances of a real patient, it is sufficient for this application. Lung movement is the factor we can control in taking a clear x-ray, and a supervisor which can synchronize with a simulated lung can be expected to do the same with a real patient.

Our demo is not a full MD PnP system. It is an interconnected medical device system rather than an interoperable system. An interconnected system is one in which devices are functionally connected through an interface. It differs from an interoperable system in that the devices are hard-coded. The system is built around specific devices and will not operate with other, similar devices. It also does not fully implement the three phases described above. The demo is designed to illustrate the possibilities of interconnected systems and show the kinds of systems which interoperability would permit. It is not possible at this time to build a fully interoperable MD PnP system, since the standards are still under development. We believe that limited systems such as this demo still have value in identifying functional and non-functional requirements for the standards in progress and illustrating the benefits of the interoperability work.

### 4.1   Hardware

The system consists of three major hardware components. These are an anesthesia machine ventilator, an x-ray machine, and a supervisor computer. The ventilator breathes for the patient by pumping gas into their lungs and allowing them to exhale on their own. The x-ray machine takes radiographs, and the supervisor coordinates the actions of the other components. Each of these devices has its own physical interface and communication protocol, all of which are different. Medical devices are not generally developed with the intention of interconnection. Any external interfaces which are present are usually used for logging status information or debugging. There is presently little incentive for manufacturers to follow standards in building these interfaces, and few standards for them to follow. Thus, a wide range of interfaces are found on various devices.

A fourth component is the LiveData server, which translates formats between the other devices. The LiveData server communicates with the anesthesia machine ventilator using Dräger's MediBus protocol over a 9600 baud serial line, with the x-ray machine using the Modbus protocol over ethernet, and with the supervisor using SOAP on HTTP on TCP/IP over ethernet.

The x-ray machine is simulated using a PLC controller, a webcam, a small red light, and a pushbutton. The PLC allows the light to be turned on and off and the pushbutton's status to be read over ethernet. The webcam is a standard USB webcam which is controlled using proprietary software.

The supervisor computer, LiveData server, and PLC are connected with a standard ethernet switch. In our demo, the supervisor software and LiveData server were usually run on the same computer.

## 4.2   Software

The system's software is divided between the supervisor and the LiveData server. The supervisor controls the other devices in the system - it correlates information from the other devices and makes the decision when to trigger the x-ray. Supervisors in general are responsible for implementing the parts of the system which are specific to a particular clinical scenario. The supervisor checks to see whether the required devices are present, collects data from the various connected devices, and sends commands to the devices according to the particular scenario. For this demo, the supervisor gathers data from the ventilator and sends the signal to trigger an x-ray exposure.

The LiveData server receives SOAP requests from the supervisor and translates them into requests to individual devices in their proprietary formats, then takes the replies from the devices and formats them as SOAP responses.

The development of the supervisor software is described in more detail in Section 5.

## 4.3   SOAP

The SOAP interface is used for communication between the supervisor and the LiveData server. A typical SOAP transaction goes as follows:

1. the user requests the list of variables from the LiveData server
2. they use the generic *get* or *set* methods to do operations on those variables
3. the server receives the command, translates it to the Dräger MediBus protocol used for the ventilator, and sends the command along
4. the ventilator sends its response to the server, which translates it and passes the response to the user
5. the response is returned in an XML wrapper which contains typing information for the returned values

SOAP is a standard protocol for web services. We used it here primarily because it was supported by the LiveData program. It worked well enough for this application, which had relatively slow data rates and small amounts of data being passed, but it does have appreciable overhead. All queries and responses are passed as XML messages which need to be generated on the sending side and parsed when received.

Another issue was the latency of the SOAP server. Some of the latency is inherent in encoding and decoding XML messages and in passing the messages through a translator instead of directly sending them from one device to another. Additional latency in our demo system resulted from a commonly used congestion control method. When many small packets are sent over a TCP/IP network, the data being sent is a small portion of the transmitted packet - most of the packet is taken up with headers. This can lead to the network becoming congested when small packets are sent quickly.

Nagle's algorithm [5] is used to concatenate these small packets together to reduce overall network overhead at the expense of delayed message delivery. The data in many small packets can be bundled into one large packet, reducing the overhead but increasing the transmission time of the early packets. In this context, we had plenty of bandwidth and were much more concerned about the timely delivery of messages, so turning off this feature greatly reduced the latency of the SOAP server.

### 4.4 Synchronization Algorithms

The supervisor uses information from the ventilator to decide when to trigger the x-ray. The synchronization algorithm defines exactly how this decision is made. Figure 3 shows the respiratory cycle graphed as pressure over time. The pressure increases until the end of inspiration (at time *Tinsp* after start of breath), at which point it drops off quickly through expiration. There is usually a pause between the end of exhalation and the start of the next breath. For this case study, we want to support taking an x-ray when the 'lung' was not moving significantly. This occurs when the patient is relatively still at the peak of inspiration or between the end of expiration and the start of the next breath. An exposure is possible if the time the patient is still exceeds the time needed for the exposure plus the latency between triggering the x-ray and the actual exposure.

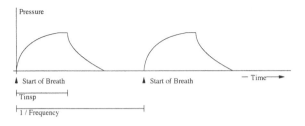

**Fig. 3.** Respiratory Cycle

**Synchronization Method 1: Dead Reckoning.** The first method used to determine when to trigger the x-ray is simple dead reckoning using the time of last breath, time of inspiration, and frequency. The variables used for this method are shown in Figure 4. All times are in seconds.

| name | description |
|------|-------------|
| $T_{now}$ | current time |
| $T_{lb}$ | time of last breath |
| $T_{nb}$ | time of next breath |
| $T_\delta$ | a small offset time to accommodate jitter |
| $T_{trigger}$ | time to send trigger signal to the X-ray |
| $T_{exp}$ | time of X-ray exposure |
| $freq$ | frequency, breaths / minute |
| $flow$ | instantaneous flow rate |

**Fig. 4.** Variables for dead reckoning

If we know the time of the start of the last breath and the frequency of breathing, then it is trivial to calculate the time of the start of the next breath.

$$T_{nb} = T_{lb} + 60/freq \tag{1}$$

There is probably time to trigger the x-ray just before start of the next breath, as long as the patient has finished exhaling before the start of the next inhalation.

$$T_{trig} = T_{nb} - T_{exp} - T_{delta} \tag{2}$$

We can check whether the patient has actually finished exhaling by sampling the instantaneous flow rate just before the start of the next breath. If it is close to zero, then the patient is not inhaling or exhaling and is still enough to allow taking the x-ray.

1. Get values of the variables $T_{now}$, $T_{lb}$, $freq$
2. Calculate $T_{trig}$
3. Sleep for $T_{trig} - T_{now}$ seconds
4. Wake up and sample $flow$
5. If $flow = 0$, trigger X-ray
   else, start over

This method of synchronization makes many assumptions. The most critical assumption is that the respiratory frequency is not going to change between the last breath and the next one. If it does, or if the system setup changes in other ways, this method of synchronization will not work. The check of instantaneous flow rate should prevent the system from triggering the x-ray when the patient is moving, but the system may not be able to take an image in situations where a different synchronization method would allow an exposure.

**Synchronization Method 2: Dynamic.** Another way to calculate the trigger time is to sample the real-time flow rate rapidly enough to build a picture of the flow graph. We experimented with two techniques for doing this. The variables used in the following descriptions are listed in Figure 5.

We originally envisioned sampling at a high enough rate to be able to integrate the total flow volume by multiplying the sampled flow rate by the time interval of the samples. This would allow the supervisor to trigger the x-ray at the right time no matter

| name | description |
|------|-------------|
| $flow$ | instantaneous flow rate |
| $T_{flow}$ | time of last $flow$ sample |
| $S_{current}$ | value of current flow sample |
| $T_{current}$ | time of current flow sample |
| $S_{last}$ | value of last flow sample |
| $T_{last}$ | time of last flow sample |
| $slope$ | calculated slope value |
| $Threshold$ | slope threshold |

**Fig. 5.** Variables for dynamic synchronization

what changes were made to the ventilator's programming or how the patient reacted. However, the ventilator was not able to provide samples at a high enough rate to enable this method to be used. The SOAP server and interface introduced additional latency and jitter into the samples, which further reduced their usefulness for this purpose.

Our second idea was to use the slope of the flow signal to find when inspiration is about to end. This meant taking two or more samples, calculating the rate of change of the flow rate between them, and triggering when this rate of change was low enough. The problem we ran into here is that the flow graph tails off very rapidly, making it unlikely that we would get even a single pair of samples in the short time when the breath is about to end. The low sample rate made this problem worse.

1. prime $S_{last}$, $T_{last}$, $S_{current}$, $T_{current}$ with two consecutive samples
2. $S_{last} = S_{current}$
3. $T_{last} = T_{current}$
4. $S_{current} = flow$
5. $T_{current} = T_{flow}$
6. $slope = S_{current} - S_{last}/T_{current} - T_{last}$
7. if $slope < Threshold$ and $flow$ is near 0, trigger x-ray
   else loop back to 2.

In the end, we found that dynamic synchronization is possible only at relatively low respiratory rates - under 8 to 10 breaths per minute. The dead reckoning method functions at much higher rates, up to approximately 25 to 30 bpm depending on the other ventilator settings. The supervisor program for our demo checks the respiratory rate and chooses whether to use the dynamic or dead reckoning method accordingly.

### 4.5   Alarms

The system should not trigger the x-ray if the ventilator has active alarms. The ventilator will take care of displaying the alarm condition to the caregiver and sounding alarms, so the supervisor just has to detect that the ventilator has active alarms and not trigger the x-ray on that respiratory cycle. It does this by getting a summary of all active alarms and warnings from the ventilator. If the list of active alarms is not empty, then the supervisor will not trigger the x-ray.

This technique is easy to implement and covers the most common situation where the alarm sounds sometime before the supervisor decides to trigger the x-ray. This is sufficient for the demo, but an implementation with a real x-ray machine and a real patient would have to take into account factors such as the alarm being raised after the supervisor checks the alarm status but before the exposure is made.

In the case where this happens, many conditions which would cause a ventilator alarm will not affect the synchronization algorithm. These include alarms like low gas levels, overpressure, some sensor failures, etc. Any alarm that does not indicate an unexpected change in ventilator settings will not stop the supervisor from being able to synchronize. Alarms for major mechanical malfunctions are very rare, but would indicate conditions where we would not want an exposure to be made - though any failure which stopped the ventilator from operating would mean that the ventilator was not causing the patient's chest to move. The problem with taking an exposure during an alarm is not that the image would be blurred, but that the safety of any caregivers responding to the alarm could be compromised. Caregivers are also protected by the use of a 'dead man switch' that the x-ray technician holds during the exposure. If the switch is released, the x-ray will not be taken. The time interval where there was an active alarm and the exposure was being made would be a fraction of a second, but this should be taken into account in the risk management process. Any system using a real x-ray machine would also need to take into account alarms from the x-ray, and any system using medical devices which are capable of pushing alarms rather than having them polled (as we did with this ventilator) would also need to consider possible race conditions between the alarm handling and synchronization parts of the supervisor.

## 5   Modeling and Verification

The software for the supervisor is the key element of the system. The supervisor is the new piece which facilitates communication between the other devices. As was described in Section 4.2, the supervisor's role in this demo is to gather data from the ventilator, decide when to trigger the x-ray, and send the signal to the x-ray machine at the correct time. The supervisor interacts with the caregiver to get input such as whether to make the exposure during inspiration or expiration and to provide the caregiver with status information and, ultimately, with the x-ray image.

The functioning of the supervisor program is critical to the safety of the system, so we devoted a significant amount of time and effort to ensuring its correctness.

The supervisor software development process started with gathering informal requirements. These requirements were collected during discussions with caregivers and biomedical engineers and included functional requirements such as "when the exposure is made, the red light on the x-ray box should light up" and safety requirements like "the caregiver's x-ray trigger button must be held down for the x-ray exposure to be made". These requirements were refined and expanded upon throughout the development process. For instance, when we started development we did not know that we would need a dead-reckoning synchronization algorithm in addition to the dynamic method and thus did not include any requirements about when the supervisor should use one or the other of these techniques.

A state machine model of the supervisor was built and then verified to meet essential safety properties. We used the model to generate Java code which then ran the demo. This development process is described in more detail in the following sections.

We began by modeling the supervisor program as an extended finite state machine (EFSM). This format was chosen because it is expressive enough to capture the behavior of the program and tools are available to automatically translate the EFSM specification into the input languages of a number of tools.

**Verification.** Once the system was modeled as a state machine, we used a tool to translate it into the input format for the model checker UPPAAL and then manually added timing information. The model checker was used to simulate the system, to test the system for general properties like deadlock, and to test more specific properties. These activities suggested changes to the EFSM specification and UPPAAl model, and the process went though several iterations. Eventually, we produced a model that satisfied all the safety requirements.

We have continued to refine the system model, and have created a model which closely resembles the clinical system. It consists of nine communicating state machines. The caregiver, xray, ventilator, and patient are each modeled in a single machine, and the supervisor is decomposed into five. There is a top level supervisor machine that interacts with the caregiver and four machines that implement the dynamic and dead reckoning algorithms at peak of inspiration and peak of expiration.

Figure 6 shows the UPPAAL state machine for the dynamic algorithm for synchronizing at peak of inspiration. The algorithm waits for the start_I_dyn signal from the top level supervisor, resets its local clock, and then enters a loop where it gets samples of

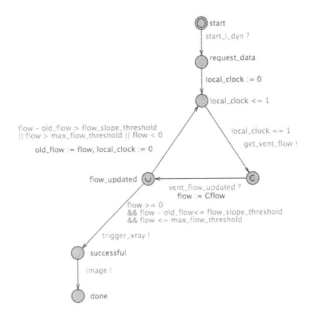

**Fig. 6.** UPPAAL model for dynamic inspiration

the ventilator flow rate once per clock tick. It compares each pair of successive samples to check if the difference between then (and hence the slope of the glow signal is less than the 'flow slope threshold'. If it is, and if the flow value is also less than the 'max_flow_threshold', then it sends a trigger signal to the x-ray. After the x-ray synchronizes with its trigger signal, the algorithm sends an 'image' signal to tell the top level algorithm that it is done and the image should be ready.

Safety requirements for the system were gathered by talking with clinicians and working though an informal hazard analysis process. For a device intended for use with patients, this process would be much more thorough.

The primary hazard introduced by this system is triggering the x-ray at the wrong time. This could potentially endanger the x-ray technician or other clinicians. Triggering the x-ray when the patient is moving will result in a blurred x-ray and the need to take another exposure, meaning additional radiation exposure for the patient.

Another hazard is that an image might not be taken even though it is possible. This is less significant, since the system will inform the clinician that the exposure was not possible and try again on the next breath. The exposure is delayed slightly, but this is a small cost compared to that of a failed exposure.

The EFSM model of the system was checked for structural properties like deadlock (that the system can't get 'stuck') and for specific safety properties. These focused on when the x-ray is triggered, since this is the single safety-critical action the system takes. We checked that the trigger signal was sent only at the correct time (as described in the algorithms in 4.4 and 4.4) and that the system would not trigger unless the flow rate reported by the ventilator was near zero.

$$AG \ xray = exposing \ implies \ T_{now} = T_{nb} - Texp - T_\delta \qquad (3)$$

Formula 3 is used for checking the system when it is being used to make an exposure at the peak of expiration (the lung is empty) in dead reckoning mode. This specification is in linear temporal logic (LTL) and it says that whenever the x-ray machine is in a state where it is exposing (AG xray = exposing) the current time must be the time of the next breath minus the exposure time minus a small offset ($T_{now} = T_{nb} - Texp - T_\delta$). This means that if there is any possible way that the EFSM could have the x-ray in the state 'exposing' when it is not that time, the model checker will show it as a counterexample. Similar formulas are used for checking exposure times for inspiration.

$$AG \ xray = exposing \ implies \ flow <= flow\_threshold \qquad (4)$$

Formula 4 states that when the x-ray is exposing, the instantaneous flow rate must be less than the flow threshold. This threshold is defined to be low enough that the lung will not be moving enough to blur the image, but also high enough to allow an exposure when there are very small movements.

## 6  Code Generation and Implementation

The final EFSM specification was used to automatically generate Java code which was used in the demo implementation. The demo includes a handwritten GUI frontend which is the user interface and the supervisor application, which is largely generated code. The generated code interacts with some handwritten functions which perform low-level actions. For instance, the model simply uses values like $flow$, while the generated code replaces references to such variables with calls to handwritten library functions which actually provide the values.

The demo implementation starts with a screen describing the clinical use case. This is followed by giving the user a choice of taking an image at the peak of inspiration (when the lungs are full) or the peak of expiration (when the lungs are empty). The user is asked to confirm their choice and taken to a screen describing the image-taking process. The user is asked to play the role of an x-ray technician and to pick up a physical button which they will hold while the exposure is made. In a non-synchronized x-ray, this button would trigger the x-ray directly. In our system, the button is held down to give the system permission to make the exposure. The clinician holds the button for several seconds while the system waits for the lung to reach the proper phase of respiration and the system checks to make sure the button is held before taking an image. If the clinician decides that it is not safe to make an exposure (e.g., if someone walks into the room), they can simply release the button and no exposure will occur. This allows us to keep a human in the loop as an additional safety precaution. Assuming the button is held down, when the lung reaches the proper phase the exposure is made and the webcam image is displayed on the screen.

The system consisted of many components from a variety of sources, written in several languages. The main difficulty in implementing the demo was integrating these diverse components into a single, functional system. As was described in Section 4, the system was tied together using LiveData and SOAP. While there were significant disadvantages to this approach (especially in terms of latency), we were successful in making a working demo. This demo was shown at the CIMIT Innovation Congress and as a Scientific Exhibit at the American Society of Anesthesiologists annual meeting and presented at the High Confidence Medical Devices, Software and Systems and Medical Device Plug-and-Play Interoperability workshop [6].

## 7    Conclusions

We successfully built a system which was able to synchronize the ventilator with a simulated x-ray machine, demonstrating that the approach is feasible. In the process, we learned lessons for building more general systems. These include the importance of recognizing the limitations of device interfaces in the supervisor algorithm design and the need to have supervisors which can respond to the changing settings of the devices. We had two synchronization algorithms, one which was more accurate but only usable at low breath rates and a less accurate but faster algorithm for high breath rates. We used formal methods in the development of the supervisor and have presented a methodology for ensuring that the integrated device systems meet their specified safety properties.

This work started with an unfortunate use case, resulting from the lack of a respiratory pause feature on the ventilator and the ventilator's inability to synchronize with the x-ray machine. The exposure that our demos brought to this problem has led to a proposed change to the international anesthesia workstation standard. Hopefully in the future such changes and the introduction of safe, inter-connected systems will help to improve patient safety.

**Acknowledgements.** We would like to thank the following people who were involved in creating the x-ray ventilator synchronization demo. Without their contributions, this work would not have been possible.

- Steve Boutrus, Tufts Medical School
- Philippe Cortes, Compiegne Univ. of Technology, France
- Jennifer Jackson, BWH Biomedical Engineering
- Shankar Krishnan, MGH Biomedical Engineering
- Ersel Llukacej, LiveData, Inc.
- Heidi Perry, Draper Laboratory
- Tracy Rausch, DocBox, Inc.
- Jeff Robbins, LiveData Inc.
- Rick Schrenker, Biomedical Engineering
- Dan Traviglia, Draper Laboratory
- Sandy Weininger, U.S. Food and Drug Administration

# References

1. Lofsky, A.S.: Turn Your Alarms On! APSF Newsletter 19(4), 41–60 (2004)
2. Langevin, P.B., Hellein, V., Harms, S.M., Tharp, W.K., Cheung-Seekit, C., Lampotang, S.: Synchronization of Radiograph Film Exposure with the Inspiratory Pause. Am. J. Respir. Crit. Care Med. 160(6), 2067–2071 (1999)
3. U.S. Food and Drug Administration Center for Biologics Evaluation and Research (CBER): Guidance for Industry Development and Use of Risk Minimization Action Plans. Technical report, Office of Training and Communication, Division of Drug Information (2005)
4. ASTM F29 WK19878: New Specification for Essential Safety Requirements for Equipment Comprising the Patient-Centric Integrated Clinical Environment (ICE) Part 1: General Requirements and Conceptual Model (2008)
5. Nagle, J.: Request for Comments: 896, Congestion Control in IP/TCP Internetworks. Technical report, Ford Aerospace and Communications Corporation (1984)
6. Arney, D., Goldman, J., Lee, I., Llukacej, E., Whitehead, S.: Use Case Demonstration: X-Ray / Ventilator. In: High Confidence Medical Devices, Software, and Systems and Medical Device Plug-and-Play Interoperability, June 2007, p. 160 (2007)

# A Ceramic Microfluidic Device for Monitoring Complex Biochemical Reactive Systems

Walter Smetana[1], Bruno Balluch[1,2], Ibrahim Atassi[1,2], Philipp Kügler[4],
Erwin Gaubitzer[2,3], Michael Edetsberger[2,3], and Gottfried Köhler[3]

[1] Institute of Sensor and Actuator Systems, Vienna University of Technology
Gusshausstrasse 27-29, 1040 Wien, Austria
[2] OnkoTec GmbH, Vestenötting 1, 3830 Waidhofen/Thaya, Austria
[3] Max. F. Perutz Laboratories, Department of Biomolecular Structural Chemistry
University of Vienna, Campus Vienna Biocenter 5/1, 1030 Wien, Austria
[4] Industrial Mathematics Institute, University of Linz
Altenbergerstrasse 69, 4040 Linz, Austria

**Abstract.** A 3-dimensional mesofluidic biological monitoring module has been successfully designed and fabricated using a low-temperature co-fired ceramic (LTCC) technology. This mesofluidic device consists of a network of micro-channels, a spherical mixing cavity and measuring ports. A selection of appropriate commercially available ceramic tapes has been chosen with regard to their biocompatibility performance. Specific processing procedures required for the realization of such a complex structure are demonstrated. Three dimensional numerical flow simulations have been conducted to characterize the concentration profiles of liquids at a specific measuring port and verified by experiment.

**Keywords:** LTCC-technology, microfluidic, Finite Element Analysis.

## 1 Introduction

Microfluidic and mesofluidic analytical systems are becoming increasingly popular in chemical and biomedical applications due to the need of small volume reagents, small wastes and short reaction times. Microfluidic devices may be classified into functionally limited labs-on-chip (LOC) and micro total analysis systems (μTAS). Most LOCs are single function and single layer devices such as mixers, separation channels, etc. Micro total analysis systems, on the other hand, are more complicated and are capable of performing many functions such as mixing, reaction, separation, etc. on a single module. Generally these devices handle nanoliters of reaction volumes and often accomplish their specific tasks in milliseconds of reaction times. Most of these systems are based on silicon, glass, poly-methylmethacrylate (PMMA) or polydimethyl-siloxane (PDMS) substrates [1]. A range of rapid-prototyping methods using laser-techniques for the fabrication of microfluidic devices are reported in literature. Micro-stereo-lithography, selective laser sintering, laser writing method and microcladding techniques are applied for generating complex 3-dimensional (3D)

A. Fred, J. Filipe, and H. Gamboa (Eds.): BIOSTEC 2009, CCIS 52, pp. 110–123, 2010.

microparts of polymer, metal and metal-matrix composite [2], [3]. Micropatterned ceramic components may be fabricated in a rapid prototying process combining stereolithography for the supply of master models with low pressure moulding [4]. But true 3D microfluidic structures cannot be easily implemented using these techniques due to process limitations or material properties. The use of LTCC-technology for microfluidic devices enables the realization of multiple 3D microchannels, a feature not easily attainable in other MEMS technologies [5]. Therefore, based on this experience LTCC-technology has been considered as an adequate approach to realize a compact temperature controlled monitoring module for biological reactions with low sample consumption which may be considered as a µTAS device.

The process technique for making a 3D-architecture with LTCCs is rather simple and standardized. Device production in LTCC-technology covers the machining, punching or laser drilling of vias and channels on individual layers. The individual layers are stacked and laminated in a heated platen press or in a heated isostatic press. Subsequently the laminated stack is exposed to the firing cycle which is a somewhat critical process where heating rate, dwell time at burnout temperature and total firing cycle time have to be matched to the thickness of the ceramic stack.

The field of nonlinear chemical kinetics has been investigated about half a century. Still only a few complex chemical reactions have been described by means of an experimentally backed system of coupled chemical equations. Moreover, in biological relevant nonlinear systems most of the reactants, e.g. proteins etc., are expensive to prepare and generally only available in limited quantities. Therefore, it is essential to use tiny reaction volumes for continuous flow experiments, as realized in this reaction cell. This hereby presented contribution outlines the procedures of fabrication and characterization of a reaction module designed to follow quantitatively complex biochemical regulatory reaction networks in vitro as a basis for advanced mathematical fitting. This device consists of a mixing module which provides fast mixing of reactants and four sensor ports for simultaneous measurements. The functionality was proved using the well known oscillatory behaviour of the chlorite-iodide reaction [6].

## 2  Design

Figure 1 shows the schematic of the monitoring module also used as model for establishing finite element analyses (FEA).

The module comprises a spherical reactor cell where continuous mixing of the reagent fluids is provided. Besides this mixing chamber the module is equipped with pH-, oxygen-, temperature- and iodide sensitive sensors for reaction monitoring as well as with SMA-connectors for glass fibres required for absorption or fluorescence spectroscopic analyses (figure 2). Pumping a thermal fluid through embedded ducts provides temperature control. A network of microchannels with cross section dimensions varying from 200 µm x 200 µm up to 2 mm x 2 mm are connecting the different measuring sections within the ceramic module. Special attention has to be spent on the selection of appropriate ceramic tape material with regard to biocompatibility and suitability to build up a complex 3-dimensional structure which contains a large number of cavities and channels.

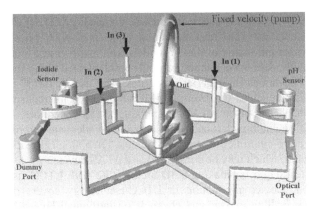

**Fig. 1.** Scheme of the monitoring module with reactor cell and channel system

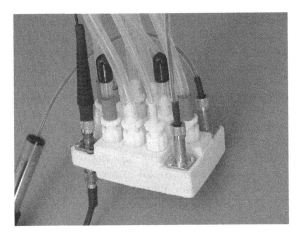

**Fig. 2.** Biological monitoring module (completely assembled with sensor-, inlet-, outlet- ports and sockets for optical fibers)

## 3   Device Fabrication

### 3.1  Biocompatibility Testing

Three different lead-free tapes have been considered for this application like the ESL 41020, Ferro A6 and CeramTec GC-tape. The tapes should be biocompatible but nevertheless cells should not adhere and proliferate to a large extent on their surfaces. The latter requirements are critical with regard to clogging of channels by cell agglomerations. In order to compare the biocompatibility, proliferation, viability and adherence of cells on LTCC–tapes, HeLa (human, cervix epithelial) cells have been grown on sintered LTCC-tapes for evaluation purpose using standard test procedures.

### 3.1.1 Proliferation Test

As a first biocompatibility testing the ability of HeLa cells to proliferate on different LTCC-tapes in contrast to glass and standard plastic surfaces was observed. The influence on cell proliferation was tested with the Bromodeoxyuridine (BrdU) assay [7] (Calbiochem, USA). BrdU incorporation is detected immunochemically with unlabelled primary antibodies and HRPO labelled secondary antibodies.

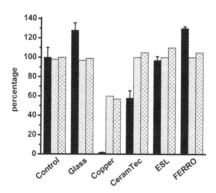

**Fig. 3.** Biocompatibility testing for HeLa cells on fired LTCC- samples (CeramTec GC, ESL 41020, FERRO A6), glass-cover slides (Assistent, Germany), Copper (Cu) and standard plastic dishes (control-sample); *Black filling: proliferation rates compared to the control group, hatched filling: percentage of viable cells compared to the control group, checked filling: percentage of total cells compared to the control group.*

HeLa cells were exposed in a 96 well plate (1 x $10^5$ cells/well) with the different test disks (4 mm x 4 mm) for 23 hours in DMEM (Dulbecco's mod. Eagle-Medium) containing 4.500 mg Glucose, 4.5 mM L-Glutamine, 44 mM Na-bicarbonate, 100 units/ml Penicillin, 100 µg/ml Streptomycin, 0.9 mM Na-pyruvate and 10 % fetal calf serum in a 6 volume % $CO_2$ humified atmosphere at 37 °C (standard conditions). During the final 3 hours of incubation BrdU is added. The BrdU concentration is measured using a multi-plate reader (BIORAD) at $\lambda_{abs}$= 455nm and $\lambda_{reference}$= 655 nm.

The results of the proliferation test are shown in figure 3 (bars with black filling). With exception of the CeramTec GC-tape and the Cu-disks, the detected proliferation-rates of all other test samples are equal or even higher (glass support) compared to the growing rate of the control group on plastic support (100 ± 8.7 %). Also the deviation is comparable to the control group. In contrast CeramTec GC-tapes show a 30-40 % reduction in proliferation. Nevertheless this reduced proliferation is low in comparison with cells incubated with Cu-disks, which under the same conditions show a proliferation rate of 2-3 % compared to control group.

### 3.1.2 Viability Testing

Next to the determination of proliferation rates it is necessary to analyze the ability of cells to adhere to surfaces and to determine the percentage of viable cells. For this test series HeLa cells were incubated in standard culture dishes with a diameter of 10 cm (about 5 x $10^5$ cells/dish) for 23 hours in DMEM under standard conditions. The fired

LTCC-tapes to be tested, glass cover slides and Cu-plates were used with lateral dimensions of about 2.5 cm x 4 cm. Cells were incubated in the presence of these materials and on standard culture surfaces as control.

The number of cells and their viability was evaluated with propidium-iodide [8] using a Nucleocounter (Chemometec, Denmark). The results of the viability tests are presented in figure 3.

No significant reduction in total number of cells (bars with checked filling) has been detected for all LTTC-samples and glass as their percentage of total cells, in relation to the control group, are comparable within the error limits caused by inaccuracies of seeding the cells. Only samples grown on Cu show about 40 % less cells than the other samples.

Additionally not any significant reduction in percentage of viable cells (bars with hatched filling) was shown for cells grown on LTCC-tapes or glass support. Again, only samples grown on Cu-plates show a diminution of about 40 %.

### 3.1.3  Adhesion Testing

The side walls of channels and cavities are formed by the laser machined edges of tapes. To observe cell adhesion, which may rely on tape material but also on surface finishing a test method has been designed which enables to examine the potential adhesion of cells on the laser machined edges of the tapes. To evaluate the influence of the ceramic tapes on the adhesion of cells on glass surfaces, CeramTec and ESL ceramic tapes with laser-micromachined tapering channels with decreasing channel width (see figure 4-A) were mounted on standard glass cover slides (Assistant, Germany) with a super adhesive (Loctite). HeLa cells were seeded on the cover slides ($10^5$ cells) and incubated over night in DMEM medium under standard conditions. The growth behaviour of the cells was evaluated using a light transmission microscope.

Exemplary the results for the CeramTec and the ESL tapes are shown in figure 4. It could be demonstrated that the initial circular opening and channel 1 machined in the CeramTec GC tape and the ESL tape show only a poor influence on the adherence of HeLa cells (figure 4-C, D; cell density in circular opening: 250 – 300 cells/mm$^2$, channel 1: 200 – 220 cells/mm$^2$) but already at channels 2 and 3 (figure 4-E) with a width of about 0.6 mm a significant decrease in adherence can be observed for CeramTec GC-tapes (cell density in channel 2: 45 cells/mm$^2$, channel 3: 19 cells/mm$^2$) as no significant number of cells is observed at this area. In contrast the fired ESL-tape shows quite a different cell adherence performance. Also in channels 3-4 (figure 4-F) a rather dense cell population of cells (cell density in channel 3 and 4: ca. 100 cells/mm$^2$) and even aggregation near to the edges are observed. Only for smaller channels with width of about 0.2 mm a decrease of adherence can be observed (data not shown). The mounting glue has not any effect on the adherence behavior of HeLa cells to the glass surface (figure 4-B).

### 3.1.4  Summary of Biocompatibility Testing

It can be concluded that all three LTCC-materials tested do not affect the viability of the cells, as the total number of cells counted after the same period was found identical within an error interval of ± 10 % for all LTCC-materials, standard plastic as well as glass. An increased cell death compared to the control group could not be found for any of the standard materials. Only copper showed a significant influence

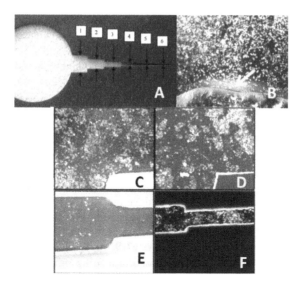

**Fig. 4.** A: Test sample of a ceramic tape coupon carrying a continuous row of channel segments with decreasing width (central circular opening is 10 mm in diameter, the channel width is 2.122, 1.175, 0.672, 0.394, 0.228, 0.168 mm and 2.208, 1.224, 0.653, 0.396, 0.102, 0.054 mm for channels 1 to 6 of the ESL (D, F) and CeramTec (C, E) tape, respectively. C, E: HeLa cells grown on glass cover slides with a mounted fired CeramTec-tape at channels 0-1 (C) and 2-3 (E). D-F: HeLa cells grown on glass cover slides with a mounted fired ESL-tape at channels 0-1 (D) and 3-4 (F). B: HeLa cells grown on glass cover slides surrounded by mounting glue (arrow). Micrographs were performed with a Zeiss Axiovert S100TV at 10x magnification and a Digital Camera (Nikon DMX1200).

on viability, cell adherence and proliferation. If only the proliferation characteristic is considered, a distinction in tape performance can be made. Whereas the ESL-tape shows a similar behavior with respect to the control assay of a standard plastic support, FERRO-tape reacts similar to a glass support and shows an increased proliferation. Only the fired CeramTec GC-tape exhibits a significant decrease in proliferation (figure 3). These results are also validated by the adherence test. When cell growth on a glass support was restricted by sidewalls of a narrow channel configuration, the cells reacted differently to ESL- and CeramTec GC-tapes: Whereas on ESL-tapes the cells grow even in channels of very small width, it was not the case for channels machined into CeramTec-tapes. For these samples the area near to the glass – ceramic interface was widely free of cells and an efficient growth of cells was only detectable in rather wide channels. The reasons of the difference in cell adhesion on the channel walls are still under investigation. It might be related to the difference in surface energy as result of laser machining. The quality of the cutting edges and the surface of channel walls are defined by the absorption of laser energy in the tape material which may vary in dependence on the composition of the selected LTCC-material.

The results show, that on the one hand side fired ceramic tapes are generally biocompatible and do not restrict the viability of cells, and on the other hand side cells adherence can be specifically influenced by the material selected. CeramTec GC-tape should be preferred, when adherence to the walls of the chamber or channels should be restricted as e.g. in microscopic observation chambers where single cells are preferentially observed in the center of an optical window and a cell agglomeration along the side walls or growth into micro fluidic flow channels should be avoided.

## 3.2  Processing Procedures

The CeramTec GC-tape has been proved as a suitable candidate for the realization of the monitoring module not only with regard to the results of the biocompatibility tests but also due to its performance characteristic during processing. Since the module comprises a complex network of channels and cavities a tape material is required which does not tend to sag during firing. Beyond the green CeramTec GC-tape with a thickness of 325 μm provides an adequate ruggedness for handling.

The three dimensional LTCC-structure is realized by forming a stack of adequately laser machined (diode pumped Nd:YAG-laser (Rofin Sinar, Germany) equipped with an acoustic optical switch, operating power: 12 W at TEM00-mode) single layers of tapes which are laminated and finally exposed to a firing process. For the realization of the complex module a single tape shows a rather delicate structure since it is penetrated by large-area laser machined perforations. A standard LTCC-processing technology cannot be easily applied for the realization of the module since it comprises 133 ceramic layers, which deviates from the number of tapes usually applied for conventional applications.

Special attention has to be paid on the lamination of the ceramic tapes since the module contains a large number of cavities and channels (e.g. the reactor cell has a cavity volume of 1 cm$^3$). The finished sheets are collated in a mould (figure 5) and aligned by registration pins providing that successive layers are rotated by 90° to compensate for the texture (preferential orientation) induced by the fabrication of green tape. The lamination of the stack of tapes has been carried out in a heated platen press (Wabash).

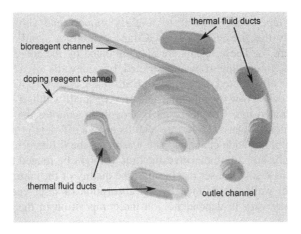

**Fig. 5.** Inside view of reactor cell with various channels and ducts (collated stack of tapes before lamination)

A range of experimental work has been conducted [9] in order to optimize lamination parameters which should contribute to provide the shape integrity of channel and cavity structures. Sagging and delamination are typical for relatively wide channels (width equal to 500 µm or more) whilst contraction is characteristic for narrow channels (width equal to 200 µm and less). The application of sacrificial material for filling cavities and channel structures is a valuable approach to avoid sagging as well as the risk of delamination. The selection of an appropriate sacrificial material is absolutely essential for this complex microstructure device. It has been found out that the main task of the sacrificial material is to provide a uniform pressure distribution within the LTCC-tape stack during lamination. An additional supporting function of cavity structures during firing is not required since the considered LTCC-material shows in all the phases of the firing process an adequate strength and stability. This substance should evaporate during the burnout phase of the sintering process without damaging the structure of the module. So the approach was to find a material, which is accommodated to the shrinkage performance of the tape. An obstruction of material's shrinkage should be avoided. Some authors recommend carbon black as sacrificial volume material (SVM) which may be applied as tape or paste [10]. This material decomposes and exhausts at rather high temperature when sintering of tape already starts. Different polymer materials have been tested as potential candidates acting as SVM since they decompose and burn out at a temperature $< 400$ °C (before tape-shrinkage is starting).

Best results have been attained with PMMA chosen as SVM. It is also part of the organic constituents of the considered tape and exhausts free of residues at the burnout temperature of tape. The organic sacrificial material has been used in powder form. Channels and cavities are filled by a vacuum sucking technique. It has been found out that a uniform pressure of only 30 bar (sample temperature: 70 °C, lamination time: 3 minutes) has to be applied onto the ceramic layer stack which enables to maintain the rectangular cross-section of the buried channels while still avoiding delamination of the ceramic batch. The applied pressure was reasonably lower than recommended for conventional applications.

Thermal gravimetric analyses have been conducted to optimize the sintering profile (especially to provide a complete and slow burn out of the organics before sintering) in order to avoid crack formation or delamination. The binder decomposition process was studied by means of Thermogravimetric Analysis (TGA) coupled with a Mass-Spectrometer (MS) to detect simultaneously the evolved gases. Additionally, the binder burnout was analyzed by Dynamic Scanning Calorimetry (DSC). TGA revealed multistage decomposition behaviour of the binder system. The binder degradation is predominately governed by exothermic reactions in the temperature range up to 400 °C and the course of the heat flow may be correlated with the evolution of carbon dioxide. The heat flow of GC-tape shows two exothermic peaks: one at 250 °C and a major one at 364 °C. For the CeramTec GC-tape the first exothermic peak corresponds to the maximum degradation rate in the TGA as can be seen in the insert of figure 6.

The results of this study yield a useful approach to establish the firing profile of the considered LTCC with regard to heating rate and intermediate dwell time for the preheat phase of firing process. Based on the results of TGA it becomes evident that for the burnout stage of the CeramTec GC-tape firing process an appropriate dwell time depending on the mass of LTCC-module has to be established. It has to be provided that the PMMA starts to pyrolyze slowly which enables to exhaust the gaseous decomposition products via the channels. In contrast a rapid decomposition of the organic filler induces a sudden intensified production of gaseous burnout products which cannot escape adequately via the channels and finally results in a destruction of the LTCC-module. The burnout phase of the firing schedule has to be adapted to these requirements whereas an adequate dwell time (depending on the mass of the LTCC-module) at the critical degradation temperature of the filler is of great importance. Practical experiments have shown that a dwell time at 250 °C is ignorable if an adequate slow heat rate is selected. A sufficient long dwell time at 350 °C for the total binder burnout is obligatory to be provided.

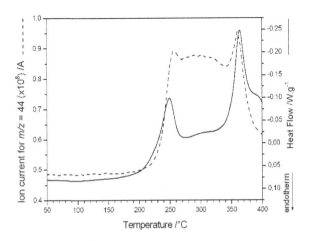

**Fig. 6.** Comparison of the heat flow signal with the ion current for carbon dioxide for GC-tape

Firing of the samples has been conducted in a box furnace of Heraeus (heating rate for temperature range 25 °C – 350 °C: 0.8 °C/min, holding time at 350 °C: 6 h, heating rate for temperature range 350 °C – 500 °C: 0.8 °C, holding time at 500 °C: 4 h, heating rate for temperature range 500 °C – 920 °C: 1.8 °C/min, holding time at peak temperature of 920 °C: 1 h, cooling rate: 2 °C/min). Another parameter, which has to be considered during the firing cycle, was the temperature distribution within the sample. A temperature gradient > 6 °C within the ceramic module during the total firing cycle has to be avoided with regard to potential crack formation and fracture of the module.

# 4  Device Characterization

## 4.1  Mixing Performance

The influence of arrangement of inlet-channels along the perimeter of the reactor cell and the sink on the bottom of the cell on the flow performance of liquids within the reactor cell has been studied for different flow conditions by means of FEA using the

FLUENT program package. The characterization of the device is based on time - dependent flow simulations. The numerical model describes the flow of dyed and pure water entering through different inlets at varying mass flow rates and passing into the spherical cavity of the reactor cell and the channel system. The local distribution of fluid at the spectroscopic port is predicted and contrasted with the results attained by spectroscopic analyses of light absorption at the considered port.

Exemplarily the flow characteristic for a representative assumption has been derived by numerical simulation and validated by experiment. The flow condition described starts with filling the cavity by injection of pure water at the radial inlets 1 and 2 (figure 1 and 7). The mixing initiated as a step inflow of dyed water (0.5 % trypan blue) is imposed on the radial inlet 2, which means that a stream of constant mass flow–rate and dye–concentration is applied at the respective inlet (figure 7). In the meantime the average residence time of dye concentration is measured at the optical port (figure 1) for different mass flow-rate ratios.

The propagation of the streamlines depends strongly on the position of the inlet along the meridian of the spherical cavity as can be seen in figure 7 for the corresponding inlet arrangement. The inlet 1 ("In 1") is placed 0.45 mm higher than inlet 2 ("In 2"). Within a period of 2.5 s (constant mass flow rate of 2 ml/min at both inlets) the streamlines of the fluid entering inlet 1 are already diffusing into the interconnecting channels while those of the fluid starting from inlet 2 are still ending in the mixing chamber.

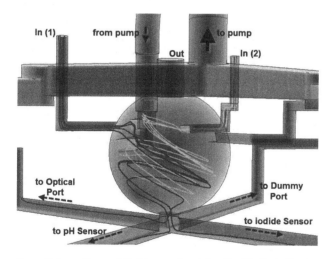

**Fig. 7.** Propagation of streamlines of fluids entering inlet "In 1" (dark lines) and inlet "In 2" (bright lines) with flow rates of 2 ml/min within a period of 2.5 s

The curves in figure 8 describe the increase of the dye volume in the optical port from initially pure water to the mixed state. For the supply of inlets with liquids two Flodos Stepdos 03 pumps were used and for providing the flow circulation a Flodos NF 5 diaphragm pump was applied. The flow rates considered in simulation and experiment are 2 ml/min at both inlets. At the optical port the absorbance was measured using a uv-vis array-spectrometer (EPP2000-50 μm Slit, StellarNet Inc.).

The absorbance–characteristic in a scaled presentation is shown in figure 8. The steady state mixed flow condition at the optical port corresponds to a dye concentration of 47 %. It can be noticed that the ideal and the computed curves are almost identical. However, the experimentally derived characteristic shows a slightly less rate of rise as well as oscillating peaks. The oscillating peaks may be attributed to the pulsating characteristic of the pumps in use. Furthermore it must be noted that the equilibrated state concentration condition is attained also within the same period if the injection flow rate at the inlet ports has been varied synchronously while keeping the total sum of mass flow rates constant. If all three inlet ports are used simultaneously for the supply of the module different mass flow rates have to be selected to provide the desired mean concentration profile at the optical port.

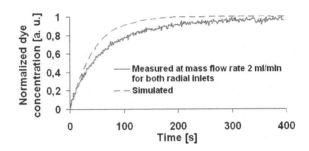

**Fig. 8.** Dye concentration vs. time characteristic at the optical port

## 4.2 Real-Time Analysis of a Complex Diffusion-Reaction Network

The classical iodide-chlorite reaction is one of the few nonlinear chemical reactions, which have been described by means of an experimentally backed system of ordinary differential equations and was, therefore, chosen to test the functionality of the microfluidic reaction module. This system involves only two inorganic ions and is characterized by an established set of reaction mechanisms according to the LLKE-model, which is named after its authors Lengyel, Li, Kustin and Epstein [6]. To demonstrate its complexity the network of chemical reactions (1) it is given below:

$$
\begin{array}{ll}
\text{M1} & ClO_2 + I^- \rightarrow ClO_2^- + \tfrac{1}{2}I_2 \\
\text{M2} & I_2 + H_2O \rightleftharpoons HOI + I^- + H^+ \\
\text{M3} & HClO_2 + I^- + H^+ \rightarrow HOI + HOCl \\
\text{M4} & HClO_2 + HOI \rightarrow HIO_2 + HOCl \\
\text{M5} & HClO_2 + HIO_2 \rightarrow IO_3^- + HOCl + H^+ \\
\text{M6} & HOCl + I^- \rightarrow HOI + Cl^- \\
\text{M7} & HOCl + HIO_2 \rightarrow IO_3^- + Cl^- + 2H^+ \\
\text{M8} & HIO_2 + I^- + H^+ \rightleftharpoons 2HOI \\
\text{M9} & 2HIO_2 \rightarrow IO_3^- + HOI + H^+ \\
\text{M10} & HIO_2 + H_2OI^+ \rightarrow IO_3^- + I^- + 3H^+ \\
\text{M11} & HOCl + Cl^- + H^+ \rightleftharpoons Cl_2 + H_2O \\
\text{M12} & Cl_2 + I_2 + 2H_2O \rightarrow 2HOI + 2Cl^- + 2H^+ \\
\text{M13} & Cl_2 + HOI + H_2O \rightarrow HIO_2 + 2Cl^- + 2H^+ \\
\text{M14} & HClO_2 \rightleftharpoons ClO_2^- + H^+ \\
\text{M15} & H_2OI^+ \rightleftharpoons HOI + H^+ \\
\text{M16} & I_2 + I^- \rightleftharpoons I_3^-
\end{array}
\tag{1}
$$

It involves an autocatalytic step and the coupling of the different reactions can, therefore, cause complex reaction patterns and unusual dynamic behavior under certain conditions. The reaction can be followed measuring the optical absorption due to $I_2$ formed in reaction M1. After fast mixing of the two inorganic ions chlorite $ClO^{2-}$ and iodide $I^-$, the color of the solution turns more and more yellow, corresponding to a monotonic increase of the iodine concentration. But then the solution suddenly becomes colorless again indicating a sudden drop of iodine concentration (see figure 9). This chemical phenomenon is referred to as clock-type behavior (also called the *iodine clock*), and the moment of concentration drop is called clock-time.

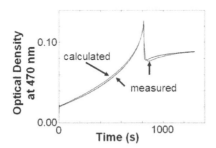

**Fig. 9.** Example of clock-time behavior of the iodide-chlorite reaction

Rate and equilibrium constants of these models are usually not directly accessible and have to be indirectly inferred from experimental observations of the system. Because of a limited amount of information and uncertainties in the data, the solutions of such parameter identification typically lack uniqueness and stability properties and hence cannot be found in a reliable way by the discrepancy between experimental observations and simulated model output. Regularization methods have to be used to overcome these difficulties. It has been shown previously [6] that a sparsity promoting regularization approach that eliminates unidentifiable model parameters (i.e., parameters of low or no sensitivity to the given data) allows to reduce the model to a core reaction mechanism with manageable interpretation while still remaining in accordance with the experimental observations.

Even more challenging is the treatment of the chlorite-iodide reaction in a continuous flow system as much higher sensitivity and time resolution is needed to detect concentration changes. In that case oscillatory behavior can be found under certain conditions. An intermediate of the chemical reaction leads also to production of the starting compound, which accelerates the reaction until the respective reaction partner is used up, but when the reaction is performed under continuous flow conditions oscillatory behavior results which can be observed continuously. Figure 10 shows the oscillations observed for the iodine absorption measured around 470 nm (i.e. in a range between 456 to 484 nm). This means, that the optical density of the product is measured as a mean value over 60 pixels of the CCD-array (Charge Coupled Device: camera). The thick line shows the final result when the time dependence is additionally averaged over 20 seconds. These measurements can now be used for mathematical modeling. Good agreement between the modeled and the experimental time dependence was achieved (figure 9) [6].

These results demonstrate that such complex chemical reactions can quantitatively be measured in the microfluidic reaction module, although the total reaction volume and also the optical path of the optical detection port are very small. Nevertheless oscillatory behaviour can also be found under such conditions. The device is sensitive enough to record small concentration changes giving rise to changes in the optical density below 0.001. In future experiments the microfluidic module will be used for measuring complex kinetics in biochemical systems. In that case, limited amounts of reactive compounds, e.g. of enzymes, are available and the use of small reaction volumes is an essential prerequisite to allow quantitative measurements.

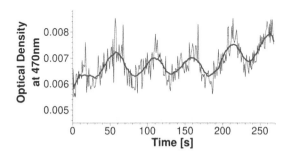

**Fig. 10.** Oscillations of the $I_2$ absorption observed in the module produced by the reaction of $I^-$ and $ClO_2^-$ in continuous flow

## 5   Conclusions

The influence of different ceramic materials on the viability and proliferation of HeLa cells has been tested as a first basic characterization of the biocompatibility. In respect to their use in micro fluidic devices the growth of cells in channels of different width has been visualized. The LTCC-technology has been proved as a very versatile method to build up complex three-dimensional multilevel channel structures including even large-volume cavities, which are suitable for biological and diagnostic applications.

**Acknowledgements.** The financial support of this work by the WWTF, project MA05 "Inverse methods in biology and chemistry" as well as by the FFG, project N209 "Nanoflu" is greatly acknowledged by the authors. The maintenance with ceramic tapes and many useful advices during the course of tape processing by Dr. C. P. Kluge (CeramTec AG) are especially appreciated.

## References

1. Anderson, J.R., Chiu, D.T., Jackman, J., Cherniavskaya, O., McDonald, J.C., Wu, H., Whitesides, S.H., Whitesides, G.M.: Fabrication of topologically complex three-dimensional microfluidic systems in PDMS by rapid prototyping. Anal. Chem. 72, 3158–3164 (2000)

2. Kathuria, Y.P.: An overview of 3D structuring in microdomain. J. Indian Inst. Sci. 81, 659–664 (2001)
3. Yu, H., Balogun, O., Li, B., Murray, T.W., Zhang, X.: Fabrication of three dimensional microstructures based on single-layered SU-8 for lab- on-chip applications. Sensors Actuat. A Phys. 127, 228–234 (2006)
4. Knitter, R., Bauer, W., Göhring, D.: Microfabrication of ceramics by rapid prototyping process chains. J. Mechan. Eng. Sci. 217, 41–51 (2003)
5. Golonka, L.J., Zawada, T., Radojewski, J., Roguszczak, H., Stefanow, M.: LTCC Microfluidic System. Int. J. Appl. Ceram. Technol. 3, 150–156 (2006)
6. Kügler, P., Gaubitzer, E., Müller, S.: Parameter identification for chemical reaction systems using the adjoint technique and sparsity enforcing stabilization - a case study for the Chloride Iodide reaction. J. Phys. Chem. A 113, 2775–2785 (2009)
7. Gire, V., Wynfford-Thomas, D.W.: Reinitiation of DNA Synthesis and Cell Division in Senescent Human Fibroblasts by Microinjection of Anti-p53 Antibodies. Mol. Cell. Biol. 18, 1611–1621 (1998)
8. Zamai, L., Canonico, B., Luchetti, F., Ferri, P., Melloni, E., Guidotti, L., Cappellini, A., Cutroneo, G., Vitale, M., Papa, S.: Supravial Exposure to Propidium Iodide Indentifies Apoptosies on Adherent Cells. Cytometry 44, 57–64 (2001)
9. Smetana, W., Balluch, B., Stangl, G., Lüftl, S., Seidler, S.: Processing procedures for the realization of fine structured channel arrays and bridging elements by LTCC-technology. Microelect. Rel. (in print)
10. Birol, H., Maeder, T., Jacq, C., Straessler, S., Ryser, P.: Fabrication of Low-Temperature Co-fired Ceramics micro-fluidic devices using sacrificial carbon layers. Int. J. Appl. Ceram. Technol. 2, 364–373 (2005)

# Knee Angle Estimation Algorithm for Myoelectric Control of Active Transfemoral Prostheses

Alberto López Delis[1,2], João Luiz Azevedo de Carvalho[1,3], Adson Ferreira da Rocha[1], Francisco Assis de Oliveira Nascimento[1], and Geovany Araújo Borges[1]

[1] Department of Electrical Engineering, University of Brasília, Brasília-DF, Brazil
`lopez_delis@yahoo.com, adson@unb.br, assis@unb.br`
`gaborges@ene.unb.br`
[2] Medical Biophysics Center, University of Oriente, Santiago de Cuba, Cuba
[3] UnB-Gama Faculty, University of Brasília, Gama-DF, Brazil
`joaoluiz@gmail.com`

**Abstract.** This paper presents a bioinstrumentation system for the acquisition and pre-processing of surface electromyographic (SEMG) signals, and a knee angle estimation algorithm for control of active transfemoral leg prostheses, using methods for feature extraction and classification of myoelectric signal patterns. The presented microcontrolled bioinstrumentation system is capable of recording up to four SEMG channels, and one electrogoniometer channel. The proposed neural myoelectric controller algorithm is capable of predicting the intended knee joint angle from the measured SEMG signals. The algorithm is designed in three stages: feature extraction, using auto-regressive model and amplitude histogram; feature projection, using self organizing maps; and pattern classification, using a Levenberg-Marquardt neural network. The use of SEMG signals and additional mechanical information such as that provided by the electrogoniometer may improve precision in the control of leg prostheses. Preliminary results are presented.

**Keywords:** Electromyographic signal, prosthesis control, microcontrolled bioinstrumentation, feature extraction, dimensionality reduction, neural networks.

## 1 Introduction

The use of microprocessors in myoelectric control has grown notably, benefitting from the functionality and low cost of these devices. Microprocessors provide the ability to employ advanced signal processing and artificial intelligence methods as part of a control system, while easily conforming to control options, and adjusting to the input characteristics. They also provide the ability to implement pattern-recognition-based control schemes, which increases the variety of control functions, and improves robustness.

Surface electromyographic (SEMG) signals provide a non-invasive tool for investigating the properties of skeletal muscles [1]. The bandwidth of the recorded potentials is relatively narrow (50–500 Hz), and their amplitude is low (50 μV – 5 mV) [2].

A. Fred, J. Filipe, and H. Gamboa (Eds.): BIOSTEC 2009, CCIS 52, pp. 124–135, 2010.

These signals have been used not only for monitoring muscle behavior during rehabilitation programs [3], but also for mechanical control of prostheses. In this context, it is important to be able to correctly predict which movement is intended by the user. The SEMG signal is very convenient for such application, because it is non-invasive, simple to use, and intrinsically related to the user's intention. However, there are other useful variables, especially those related to proprioception, for example: the angle of a joint, the position of the limb, and the force being exerted.

This project is motivated by the ongoing development of an active leg prosthesis prototype (Fig. 1). The prosthesis has three degrees of freedom: one for the knee (sagittal plane), and two movements for the foot (sagittal and frontal planes). The three degrees of freedom are associated with the angles $\theta_1$, $\theta_2$ and $\theta_3$, controlled by DC reduction motors. The prototype will be fixed to the patient's upper leg through a fixing capsule, where the SEMG sensors will be placed. The prosthesis will receive control commands through digital signal processing, feature extraction, and pattern classification. For the development of an active leg prosthesis that also possesses ankle and foot axes, it is necessary to use other sources of information besides the SEMG signal [4]. In this context, the use of myoelectric signals combined with other variables related to proprioception may improve the reliability in closed-loop control systems. The bioinstrumentation system should also be as immune to noise and interference as possible. This can be achieved by proper board and shielding design, as well as the use of filters whenever they are necessary.

Figure 2 presents the typical main components of a general myoelectric algorithm based on pattern recognition. The SEMG signals are acquired by surface electrodes placed on the skin over antagonistic leg muscles of the subject. The signals originating from the electrodes are pre-amplified to differentiate the small signals of interest, and then are amplified, filtered and digitized. Finally, the information is transferred to a myoelectric knee angle estimation algorithm.

In the design and implementation of myoelectric control algorithms [5], the system's precision is essential for an accurate accomplishment of the user's intention. Precision is an important factor on the development of multi-sensory controllers, and can be improved by extracting more information from the muscle's state, and using a classifier that is capable of processing this information. The myoelectric control algorithm should be capable of learning the muscular activation patterns that are used

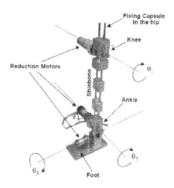

**Fig. 1.** Mechanical structure of the prosthesis prototype

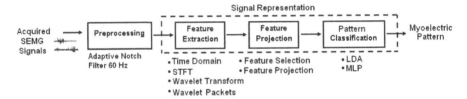

Fig. 2. Typical main components of a general myoelectric control algorithm based on pattern recognition

in natural form for typical movements. It also needs robustness against condition variations during the operation. The response time cannot create delays that are noticeable to the user.

This article presents a micro-controlled bioinstrumentation prototype system as part of the development of an active leg prosthesis structure that allows the acquisition and processing of electromyographic signals and other data related to articulate movement, specifically the angle of the knee joint. The information obtained is processed in order to obtain appropriate myoelectric patterns for prosthesis control. Preliminary results on the design of pattern recognition algorithms for the estimation of the knee joint angle are presented.

## 2  Methods

The front end stage of the designed bioinstrumentation system acquires up to four SEMG channels. The SEMG signals are measured on a pair of agonist and antagonist muscles of the leg (Figs. 3a and 3b). An electrogoniometer is used to measure the flexion and extension angles of the knee joint (Fig. 3c).

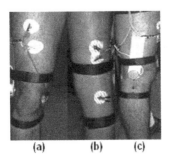

Fig. 3. Experimental setup. Surface electrodes are placed over a pair of agonist and antagonist muscle groups of the leg: (a) vastus intermedius, (b) semitendinosus. An electrogoniometer is used to measure the flexion and extension angles of the knee joint (c).

Differential amplifiers, used in the bipolar configuration, significantly reduce the common mode interference signals (CMRR > 110 dB). A band-pass filter in the 20–500 Hz frequency range is used. It is composed by a low-pass filter and high-pass filter with a programmable gain stage based on digital potentiometers, controlled by the microcontroller. These elements allow subject-based setting of the SEMG gain levels. To minimize power consumption and increase noise immunity, operational amplifiers with JFET inputs were used. To obtain adequate myoelectric amplitude, an overall gain of up to 20000 can be programmed at the front end [2].

A second block, micro-controlled and optically isolated from the front end (Fig. 4), centralizes all the functions associated with the analog/digital conversion process, implementing the digital gain control for the front end amplifiers and the synchronized sampling of SEMG signals. The microcontrollers are from the ARM SAM7S64 ATMEL family of high performance processors, based on 32-bit RISC architecture with an integrated group of peripherals that minimize the number of external components.

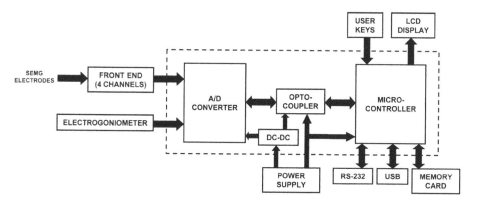

**Fig. 4.** Block diagram of the bioinstrumentation system

A 13-bit A/D converter with a serial peripheral interface (SPI) is used for signal sampling, and allows discriminating small amplitude levels. The electrogoniometer channel is coupled to the system, and generates an electric signal corresponding to the angular position ranging from 30 to 240 degrees. The sampling frequency of each channel is 1744.25 Hz. Figure 5 presents example data acquired during an experimental measurement.

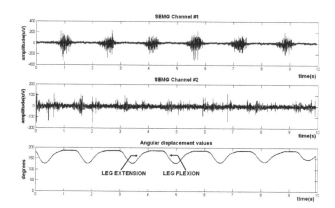

**Fig. 5.** Recorded SEMG signals (rectus femoris and opposite muscles) and measured knee joint angle, from a representative 10-second experiment

The microcontroller is linked through RS-485 protocol to the central processor of the prosthesis, which is responsible for coordinating the tasks in the control process. Besides the RS-485 protocol, which provides the interaction of the block with the central processor, RS-232C and USB interfaces are available for the communication with a PC when the system is configured in stand alone mode (Fig. 6). In this mode, the system allows the visualization of the state of the experiments during their realization using a LCD display. The instrumentation system is designed using low power consumption components, which increases the system's portability.

**Fig. 6.** Bioinstrumentation module (with electrogoniometer) configured in stand alone mode

### 2.1 Adaptive Filter Implementation

The power line interference usually has its first harmonics (60 Hz, 120 Hz, 180 Hz, and 240 Hz) in a portion of the spectrum with major SEMG energy concentration. The use of an analog notch filter may distort the signal; therefore it should only be used when really necessary. Generally, the best option is to use an adaptive notch filter. An embedded subroutine in the SAM7S64's core implements an adaptive notch filter in real time. This filter maintains a running estimate of the 60 Hz interference, and the current noise at time $t$ can be estimated from the previous two noise estimates [6], as shown in equations (1) and (2),

$$e(t) = N \cdot e(t - nT) - e(t - 2nT). \tag{1}$$

where $T$ is the sample period and $N = 2\cos(2\pi 60T)$. In the filter, the output is generated by subtracting the estimated noise, $e(t)$, from the input signal, $x(t)$. The following expression is used to implement the filter:

$$f(t) = [x(t) - e(t)] - [x(t - nT) - e(t - nT)]. \tag{2}$$

If $f(t) > 0$, then the estimate was too low, so we adjust the noise estimate upwards by incrementing by $d$:

$$e(nT + T) = e(nT + T) + d. \tag{3}$$

If $f(t) < 0$, the estimate was too high, so it is decremented by $d$:

$$e(nT + T) = e(nT + T) - d. \tag{4}$$

Using large vales of $d$, the filter adapts more rapidly, and exhibits a broad bandwidth. For small values of $d$, the filter adapts more slowly, and has a narrower bandwidth. The selection of the $d$ factor is empiric, based on test realizations, and its value is small compared to the dynamic range of the A/D converter [6]. Figure 7 shows the adaptive filtering of a SEMG signal measured on the rectus femoris muscle.

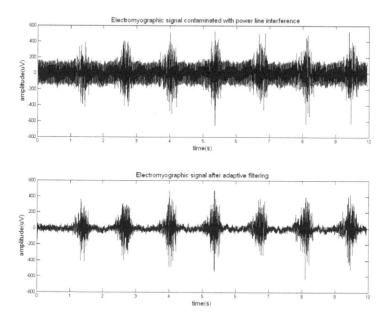

**Fig. 7.** Adaptive filtering performed on a SEMG signal contaminated with power-line interference

## 2.2  Myoelectric Knee Joint Angle Estimation Algorithm

Presenting the myoelectric signal directly to a classifier is impractical, because of the dimensionality and the random characteristics of the signal. It's necessary that the signal is represented as a vector with reduced dimensionality, i.e., a feature vector. The myoelectric knee angle estimation algorithm proposes the use of three stages for feature extraction and pattern classification. The first stage consists in the mixture of feature vectors from time domain and spectral analysis. A second stage will perform the reduction of the feature space, and the last stage has the goal of estimating the knee angle.

**Feature Vector Extraction.** Given the stochastic nature of the myoelectric signal, it can be considered as a time series, and modeled as a linear combination of their past and present values. The auto-regressive (AR) model is a convenient structure for model identification, especially when the computational speed and response time are important, as in the recognition and classification of myoelectric patterns. The auto-regressive coefficients provide information about the muscular contraction. The estimate of the coefficients is performed using a recursive least squares (RLS) technique, with a forgetting factor. This method gives more weight to the most recent samples at the moment of the iteration cycle. The parameters are calculated recursively [7], as presented in equations (5), (6) and (7):

$$\hat{\eta}_k = \hat{\eta}_{k-1} + L_k [y_k - \varphi_k^T \hat{\eta}_{k-1}] \tag{5}$$

$$P_k = \left[ P_{k-1} - \frac{P_{k-1} \varphi_k \varphi_k^T P_{k-1}}{\lambda_k \varphi_k^T P_{k-1} \varphi_k} \right] \frac{1}{\lambda_k} \tag{6}$$

$$L_k = \frac{P_{k-1} \varphi_k}{\lambda_k \varphi_k^T P_k \varphi_k}, \tag{7}$$

where $\hat{n}_k$ are the vector coefficients that are estimated at discrete time $k$; $\varphi_k$ are the regressive vectors, $P_k$ is the inverse correlation matrix and $L_k$ is the gain vector of the filter. The forgetting factor $\lambda_k$ controls the system response time. The coefficient estimated at instant $k$ can be interpreted as a characteristic of the SEMG signal within the time interval specified by the forgetting factor, and it is a way of determining the angular displacement that the patient wants to impose to the prosthesis [4]. The coefficients form a feature vector for the pattern classification process.

Recent research has demonstrated that a functional and efficient configuration consists of a mixture of time domain feature vectors with auto-regressive coefficients [8]. This configuration provides good classification precision, and is computationally efficient, which facilitates its implementation in embedded systems. It is also more robust to the displacement of the surface electrodes [8].

This work uses a combination of the auto-regressive model with the EMG histogram method. The EMG histogram is an extension of the zero crossing and the Willison amplitude [9]. Myoelectric signals reach relatively higher levels during the contraction process, compared to the base line amplitudes. Thus, vectors obtained from the histogram provide a measure of the frequency in which the signal reaches each level of amplitude, associated with different histogram bins. For the implementation of the histogram, the SEMG dynamic range was symmetrically subdivided into 9 bins. These bins represent intervals of amplitude in which the SEMG signal is grouped. The resulting feature vectors (auto-regressive coefficients and histogram bin counts) are concatenated, and then used as the input vector of the feature projection stage.

**Feature Projection.** A feature projection stage is used to reduce the dimension of the feature space of the input vectors, before the pattern classification process. This reduction is performed using an unsupervised Kohonen self-organizing map (SOM) neural network. The groups of vector coefficients obtained from each SEMG channel using the AR and histogram methods are transformed into two-dimensional vectors. With the reduction in input dimension, the SOM is able to reduce noise and absorb the large variations that appear in the original features. In addition, the SOM can shorten the training time of the supervised pattern-classification neural network. The unsupervised SOM can find the winning neuron on a 2-D map to represent the original pattern. To find the output neuron (winning node), the following steps are used, according to the learning rule of the Kohonen feature map [10].

**Step 1:** Choose random values for the initial weight vectors $W_j(0)$.

**Step 2:** Find the winning neuron $y_c$ at time step $t$ (similarity matching), by using the minimum-distance Euclidean criterion:

$$y_c = \mathrm{argmin} \|x(t) - W_j(t)\|, j = 1, 2, \ldots, t. \tag{8}$$

**Step 3:** Update the synaptic weight vectors of all neurons by using the following update rule:

$$W_j(t+1) = W_j(t) + \eta(t) h_{j,y_c}(t)[x(t) - W_j(t)]. \tag{9}$$

where $\eta(t)$ is the learning rate, and $h_{j,y_c}(t)$ is the neighbor function centered around the winner. $\eta(t)$ and $h_{j,y_c}(t)$ are varied dynamically during the learning stage, in order to obtain optimal results.

**Step 4:** Go back to step 2 until no changes in the feature map are observed.

The inputs of the Kohonen's SOM are features from each channel, and the output is the 2-D coordinate (on the $x$ and $y$ axes) on the 2-D topological net. A 2-D coordinate is a condensed feature for each channel (Fig. 8).

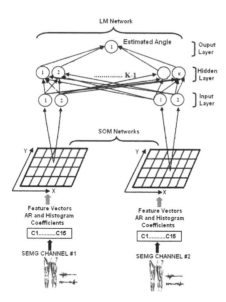

**Fig. 8.** Block diagram of the proposed myoelectric knee angle estimation algorithm

**Myoelectric Classification.** Multi-layer neural networks have been successfully applied to many difficult and nonlinear problems in diverse domains and there is considerable research on methods to accelerate the convergence time of the multi-layer feedforward neural network algorithm [11, 12]. The method used in this paper is the Levenberg-Marquardt (LM) algorithm [13], which consists in the use of the nonlinear least squares algorithm to the batch training of multi-layer perceptrons. The

LM algorithm can be considered a modification of the Gauss-Newton method. The key step in the LM algorithm is the computation of the Jacobian matrix. The LM algorithm is very efficient when training networks which have up to a few hundred weights. Although the computational requirements of the LM algorithm become much higher after each iteration, this is fully compensated by its higher efficiency. This is especially true when high precision is required [13]. Figure 8 presents the complete block diagram of the myoelectric control algorithm.

## 3   Results

As a prototype implementation, the training and testing processes were performed in off-line mode, and the algorithms described above were implemented in Matlab. At a later stage, the full validation of the algorithm will be the executed from an embedded system running on a Linux platform.

For this demonstration, SEMG measurements were captured from four healthy subjects using 10 mm Ag/AgCl surface electrodes with conductive gel, placed on a pair of antagonistic muscles, associated with the flexion and extension movements of the knee (Fig. 3). The electrodes were arranged in bipolar configuration. The distance between the centers of the electrodes was 3–5 cm, and the reference electrode was placed over the lateral condyle bone. Each subject was studied over the course of five days. Four 10-second measurements were performed on each day, with 5-minute rest periods between measurements. For each measurement, the subject was asked to walk in a particular direction at a constant pace. Some variability in pace was observed between measurements. The first and third measurements from each day were used for training, and the second and fourth measurements were used for testing. Thus, a total of 80 measurements were obtained, with half of them being used for training and the other half being used for testing.

For training purposes, it is essential to know information about the input and output, comparing the dimensional vectors obtained from the SOM network to the displacement angle measured with the electrogoniometer. The electrogoniometer was placed and strapped over the external side of the leg, so that it would measure the angular displacement of the knee in sagittal plane (Fig. 3c).

Figure 9 shows the estimated angle compared to the measured angle from the electrogoniometer for a representative set of signals. Although the estimated angle follows the measurement satisfactorily, the output of the LM network presents impulsive noise (Fig. 9a), which is reduced using a moving average recursive filter with a 50-sample window (Fig. 9b). This filter presents a delay of $(M-1)/2$ samples, where $M$ is the number of averaged samples [14]. The results obtained with 50 samples of average were satisfactory, decreasing the variance and conserving the waveform's shape.

A preliminary comparison was performed between the proposed algorithm and the methods proposed by Ferreira et al. [4]. The proposed algorithm is an alternative to the approach by Ferreira et al., which consists in using the AR model for feature extraction, and a LM multi-layer perceptron neural network for pattern classification.

This quantitative comparison was based on the classification error and on the correlation coefficient. The classification error was calculated as follows:

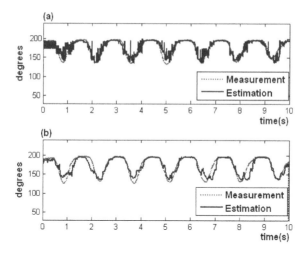

**Fig. 9.** Comparison of the estimated knee angle (dashed line) with the measured angle from the electrogoniometer (solid line): (a) before filtering; (b) after filtering

$$error = \sqrt{\frac{\sum\limits_{i=1}^{N} |x(i) - \hat{x}(i)|^2}{\sum\limits_{i=1}^{N} |x(i)|^2}} \times 100\ (\%), \tag{10}$$

where $x(i)$ and $\hat{x}(i)$ represent the angular values from the electrogoniometer sensor and the estimated angle values, respectively, and $N$ is the dimension of the vectors. The correlation coefficient $r$ was calculated as

$$r = \frac{\sum\limits_{i=0}^{N-1} (x_i - \bar{x})(y_i - \bar{y})}{\sqrt{\sum\limits_{i=0}^{N-1} (x_i - \bar{x})^2 \sum\limits_{i=0}^{N-1} (y_i - \bar{y})^2}} \tag{11}$$

where, $x_i$ and $y_i$ are the $i$-th measured and estimated joint angle value, respectively, $\bar{x}$ and $\bar{y}$ are the mean measured and estimated joint angular displacements, respectively, and $N$ is the dimension of the vectors. This coefficient provides a measurement of the degree of linear dependence between the two variables. The closer the coefficient is to either $-1$ or $+1$, the stronger the correlation between the variables.

Table 1 presents the average classification error rate and correlation coefficient, respectively, measured in each subject's group of test signals, with the associated standard deviation. The proposed method provides moderately lower classification error and slightly higher correlation than the method by Ferreira *et al.* for all four subjects.

**Table 1.** Comparison between the myoelectric knee joint angle estimation algorithms

|  | Classification error (%) | | Correlation coefficient | |
|---|---|---|---|---|
|  | Ferreira et al. | Proposed | Ferreira et al. | Proposed |
| Subject A | 8.02 ± 4.21 | 6.56 ± 1.85 | 0.75 ± 0.20 | 0.84 ± 0.07 |
| Subject B | 8.18 ± 4.70 | 5.33 ± 1.13 | 0.54 ± 0.27 | 0.61 ± 0.22 |
| Subject C | 6.54 ± 4.36 | 5.77 ± 3.64 | 0.59 ± 0.16 | 0.59 ± 0.90 |
| Subject D | 6.63 ± 3.06 | 5.23 ± 1.47 | 0.71 ± 0.17 | 0.72 ± 0.09 |

## 4  Conclusions

This paper presented the current state of development of a bioinstrumentation system for active control of leg prostheses. Features of the system and of the signal processing algorithm used in the myoelectric knee joint angle estimation algorithm were presented. The system allows the acquisition of SEMG signals with maximum amount of signal information and minimum amount of contamination from electrical noise. The results show that the system has great potential for future developments in leg prosthesis control. Preliminary analysis showed that the computational complexity of the proposed algorithm increases for each iteration during execution of the LM network. Future work will aim at optimizing the code for its execution in real time.

**Acknowledgements.** This work was partially supported by CAPES and CNPq.

## References

1. Sommerich, C.M., Joines, S.M., Hermans, V., Moon, S.D.: Use of Surface Electromyography to Estimate Neck Muscle Activity. J. Electromyography Kinesiol 6, 377–398 (2000)
2. De Luca, C.J.: Encyclopedia of Medical Devices and Instrumentation. John G. Webster (2006)
3. Monseni-Bendpei, M.A., Watson, M.J., Richardson, B.: Application of Surface Electromyography in the Assessment of Low Back Pain: A Literature Review. J. Phys. Ther. Rev. 2, 93–105 (2000)
4. Ferreira, R.U., da Rocha, A.F., Cascão Jr., C.A., Borges, G.A., Nascimento, F.A.O., Veneziano, W.H.: Reconhecimento de Padrões de Sinais de EMG para Controle de Prótese de Perna. In: XI Congresso Brasileiro de Biomecânica (2005)
5. Asghari, M.O., Hu, H.: Myoelectric Control System — A survey. J. Biomedical Signal Processing and Control 2, 275–294 (2007)
6. Hamilton, P.S.: A comparison of Adaptive and Nonadaptive Filters for Reduction of Power Line Interference in the ECG. J. IEEE Trans. Biomed. Eng. 43, 105–109 (1996)
7. Ljung, L.: Linear System Identification. Prentice-Hall, Inc., Englewood Cliffs (1987)
8. Hargrove, L., Englehart, K., Hudgins, B.: A Training Strategy to Reduce Classification Degradation Due to Electrode Displacements in Pattern Recognition Based Myoelectric Control. J. Biomedical Signal Processing and Control 3, 175–180 (2008)
9. Zardoshti-Kermani, M., Wheeler, B.C., Badie, K., Hashemi, R.M.: EMG Feature Evaluation for Movement Control of Upper Extremity Prostheses. J. IEEE Trans. on Rehabilitation Eng. 3, 324–333 (1995)

10. Haykin, S.: Neural Networks: A Comprehensive Foundation. Prentice Hall, New Jersey (1999)
11. Batiti, R.: First and Second Order Methods for Learning: Between Steepest Decent and Newton's Method. J. Neural Computation 4, 141–166 (1992)
12. Charalambous, C.: Conjugate Gradient Algorithm for Efficient Training of Artificial Neural Networks. IEE Circuit, Devices and System (1992)
13. Hagan, M.T., Menhaj, M.B.: Training Feedforward Networks with the Marquardt Algorithm. J. IEEE Trans. Neural Networks 5, 989–993 (1994)
14. Smith, S.W.: The Scientist and Engineer's Guide to Digital Signal Processing. California Technical Publishing, San Diego (1999)

# Micro Droplet Transfer between Superhydrophobic Surfaces via a High Adhesive Superhydrophobic Surface

Daisuke Ishii[1,3], Hiroshi Yabu[2,3], and Masatusgu Shimomura[1,3]

[1] WPI-AIMR, Tohoku University, 2-1-1 Katahira, Aoba-ku
980-8577 Sendai, Japan
[2] IMRAM, Tohoku University, 2-1-1 Katahira, Aoba-ku
980-8577 Sendai, Japan
[3] CREST, Japan Science and Technology Agency, Hon-cho 4-1-8
332-0012 Kawaguchi, Japan
dishii@tagen.tohoku.ac.jp

**Abstract.** Micro droplet handling is very important for micro and nano fluidic devices and an intelligent bio interface. Micro droplet transfer via a high adhesive superhydrophobic surface has been reported in recent years. We demonstrated water droplet adhesion controllable superhydrophobic metal–polymer surfaces. Moreover we achieved micro droplet transfer between superhydrophobic surfaces by using different droplet adhesion properties. Water micro droplets were transferred from a low-adhesive superhydrophobic surface to a midium-adhesive superhydrophobic surface via a high-adhesive superhydrophobic surface without any mass loss. After droplet transfer, water contact angle was about 150°. Droplet handlings on the adhesive superhydrophobic surfaces will be expected for fluidic bio devices with energy saving.

**Keywords:** Superhydrophobicity, Microfluidics, Water droplet, Adhesion, Lotus effect.

## 1 Introduction

Droplet manipulations mimicking behaviors on plant or insect surfaces such as lotus leaf effect are now interesting because simple surface structures provide amazing functionalities. Superhydrophobic surfaces which have the water contact angle lager than or near 150° are much paid attention, since its good water repellent property is used various applications in coating and electronic technologies [1]. Many researchers have been reported to obtain strong water repellent surface such as a hydrophobic fractal surface [2] and a nanopin array surface [3]. Recently several reports were published about water droplet adhesive superhydrophobic surfaces in mimicry of gecko's feet [4] and rose's petals [5]. These adhesion properties were caused by van der Waals' force on large real surface area against small apparent surface area. It was difficult to control the adhesion forces because the adhesion was caused by the surface structures.

A. Fred, J. Filipe, and H. Gamboa (Eds.): BIOSTEC 2009, CCIS 52, pp. 136–142, 2010.

Herein we prepared that a superhydrophobic metal–polymer (MP) surface with different droplet adhesion properties. The adhesive superhydrophobic surfaces were composed of hexagonally ordered polymer pillar arrays made from a self-organized honeycomb-patterned polystyrene film [6] and metal micro domes deposited by nickel electroless plating [7]. Moreover we demonstrated that water micro droplet transfer was achieved by using the MP surfaces possessing different water adhesion forces. Micro droplet handlings on superhydrophobic surfaces by control of their wettabilities are important for further understanding of superhydrophobicity and application in microfluidic bio devices.

## 2  Experimental

Polystyrene (PS; Mw = 280 000 g mol-1), poly(allylamine hydrochloride) (PAH; Mw = 14 000 g mol$^{-1}$), palladium(II) chloride (PdCl$_2$), nickel hypophosphite hexahydrate (Ni(H$_2$PO$_2$)$_2$·6H$_2$O), boric acid (H$_3$BO$_3$), sodium acetate (CH$_3$COONa), ammonium sulfate ((NH$_4$)$_2$SO$_4$) were of reagent grade and were purchased from Wako, Japan. Amphiphilic copolymer (CAP;  Mw = 270 000 g mol$^{-1}$) was synthesized according to our previous report [8]. Fig. 1 shows chemical structures used in this report.

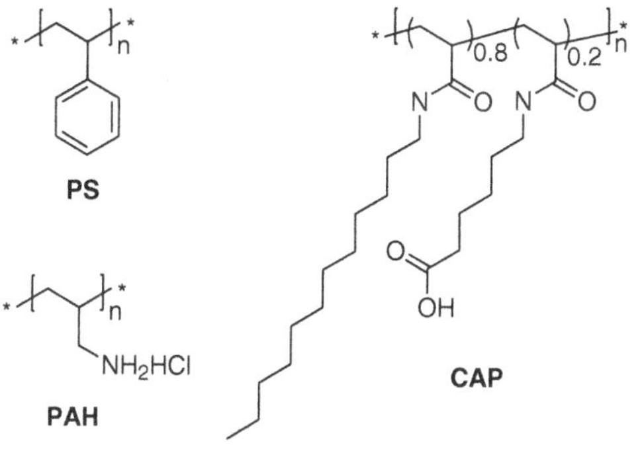

**Fig. 1.** Chemical structures used in this report

### 2.1  Preparation Method

The superhydrophobic metal–polymer surface (MP surface) was fabricated by electroless plating for honeycomb-patterned polymer films and peeling process (See Fig. 2). According to our previous report [8], the honeycomb films were prepared by casting a chloroform solution of 10:1 mixture of PS and CAP on a glass substrate with hexagonally condensed water droplet arrays. The honeycomb film cut to 1 × 1 cm$^2$ was soaked in a catalytic mixture solution of 6.0 ml containing 0.010 mol dm-3 PAH and 0.010 mol dm$^{-3}$ PdCl$_2$ at 25°C. The catalytic solution was gradually heated to 30°C, 45°C, and 60°C, respectively, and kept for 10 min under horizontal shaking at 10 rpm. Treated honeycomb films were immersed in a nickel plating bath [9] at 25°C containing 0.10 mol dm$^{-3}$ Ni(H$_2$PO$_2$)$_2$·6H$_2$O, 0.19 mol dm$^{-3}$ H$_3$BO$_3$, 0.030 mol dm$^{-3}$ CH$_3$COONa and 0.0098 mol dm$^{-3}$ (NH$_4$)$_2$SO$_4$ without any rinse and drying. Then the plating bath including the treated honeycomb film was heated to 70°C and kept for 2h with no stirring. After rinsing and drying, a nickel layer was covered on the honeycomb film. After electroless plating, metallic faces of the nickel-covered honeycomb films were adhered on an acryl substrate by an epoxy resin. After heating at 70°C for 2h, a lower half layer of the nickel-covered honeycomb film was peeled off from the acryl substrate.

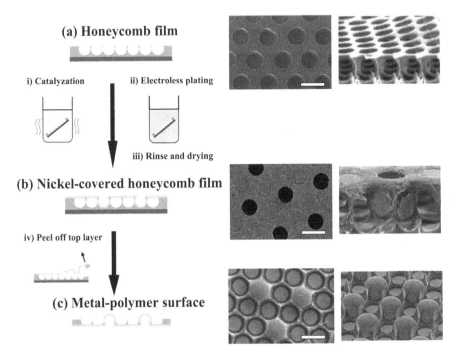

**(a) Honeycomb film**

i) Catalyzation          ii) Electroless plating

iii) Rinse and drying

**(b) Nickel-covered honeycomb film**

iv) Peel off top layer

**(c) Metal-polymer surface**

**Fig. 2.** Schematic illustrations of a preparation method of MP surfaces. SEM images of top (left) and tilt (right) views of (a) a honeycomb film, (b) a nickel-covered honeycomb film fabricated by immersion in the catalytic mixture solution at 45°C, and (c) a MP surface fabricated by peeling off a top layer of the nickel-covered honeycomb film shown in Fig. 2b. (Scale bar: 10 μm).

## 2.2  Physical Measurements

Surface structures of the MP surfaces were observed by a scanning electron microscope (SEM; Hitachi S-5200, Japan). A water contact angle (WCA) to 3 mg water droplet was measured by contact angle meter (Kyowa Interface Science DW-300, Japan). A sliding angle (SA) was measured to tilt the surfaces with a micro-droplet of 5 mg. Density of the metal dome which is defined by division of the number of metal domes by the number of honeycomb holes was calculated by means of low-magnified SEM images.

## 2.3  Droplet Transfer

Water Micro droplet transfer was demonstrated by using the MP surfaces with deferent adhesion properties. Fig. 3 shows a schematic illustration of water droplet transfer. The 5-mg water droplet was prepared on the MP surface fabricated by using the catalytic mixture solution at 30°C. Then the MP surface fabricated by using the catalytic mixture solution at 60°C was closed to the water droplet from above and touched a little bit. The upper MP surface was pulled up slowly from the lower MP surface. Then the upper MP surface catching the water droplet was closed and touched to the MP surface fabricated by using the catalytic mixture solution at 45°C. Finally the upper MP surface was pulled up again.

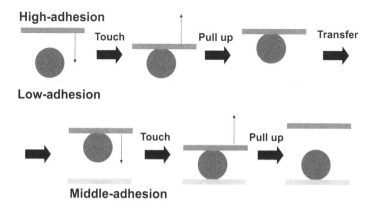

**Fig. 3.** Schematic model of a micro-droplet transfer by using the MP surfaces with different adhesion properties

# 3  Results and Discussion

## 3.1  Superhydrophobic Metal–Polymer Surface

SEM images of a honeycomb film, a nickel covered honeycomb film fabricated by immersion in the catalytic solution at 45°C, and a MP surface fabricated by immersion in the catalytic solution at 45°C are inserted in Fig. 2. An average diameter of a honeycomb hole was about 7 μm. A nickel-covered honeycomb film possessing some pores, which were distributed in the honeycomb holes, was obtained after electroless plating including immersion in the catalytic mixture solution. A tilted SEM image shown in a right column of Fig. 2b clears that the pores were openings of micro mono vessels. When the temperature of the catalytic mixture solution was changed low (30°C) and high (60°C), the number of the vessels was decreasing and increasing, respectively. Generally, a solution temperature influences surface wettability, because the surface tension of all solutions is represented by function of the temperature. This result indicates that the number of the vessels was dependent on wettability of the catalytic mixture solution to the honeycomb film. In the case of immersion in the catalytic solution at low temperature, wettability of the honeycomb film was low, so that the number of the vessels in the nickel-covered honeycomb film was a few. On the other hand, in the case of immersion in the one at high temperature, the number of the vessels was much because of good wettability to the honeycomb film. The number of the vessels of the nickel-covered honeycomb film was easily changed by the catalytic mixture solution temperature.

The MP surfaces after peeling off the top half of the nickel-covered honeycomb film were composed of superhydrophobic PS pillar arrays and hydrophilic nickel micro-domes as shown in Fig. 2c. The nickel micro-dome was reverse side of the micro mono vessel in the nickel-covered honeycomb film. This result anticipates that density of the nickel dome to the honeycomb hole is controlled indirectly by temperature of the catalytic mixture solution. Fig. 4 shows SEM images of the MP surfaces having different nickel dome density. The nickel dome density of the MP surface prepared by immersion in the catalytic mixture solution at 30°C, 45°C, and 60°C was about 3%, 15%, and 25%, respectively. The surface properties such as surface wettability and droplet adhesion properties were controlled easily, because hydrophilic-hydrophobic balance was varied by difference of the nickel dome density (See Table 1).

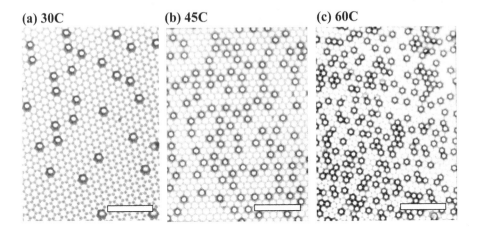

**Fig. 4.** SEM images of the superhydrophobic metal–polymer surfaces fabricated by using the catalytic mixture solution at (a) 30°C, (b) 45°C, and (c) 60°C. The black dots indicate the nickel domes. (Scale bar: 100 μm).

**Table 1.** Surface properties of the MP surfaces

| Sample | Density | WCA | SA | Adhesion |
|--------|---------|------|------|----------|
| 30°C | 3% | 155° | <5° | Low |
| 45°C | 15% | 150° | 30° | Middle |
| 60°C | 25% | 145° | N/A | High |

## 3.2  Micro Droplet Transfer

The water droplet adhesion properties were measured by water contact angles (WCAs) and sliding angles (SAs). The MP surface with nickel dome density of 3% possessed a WCA of 155° and a SA of less than 5°, and was abbreviated as a low-adhesion MP surface. The MP surface with dome density of 15% possessed a WCA of 150° and a SA of 30° (a middle-adhesion MP surface). The MP surface with dome density of 25% possessed a WCA of 145° and a not measured SA because the droplet adhered the surface when turned upside down (a high-adhesion MP surface). As the dome density was increasing, the WCA was decreasing and the SA was increasing, which means that the hydrophilic nickel domes gave the adhesion behaviors. This result made clear that the adhesion property was controlled by the quantity of the nickel dome easily changed by the catalytic mixture solution temperature for electroless plating.

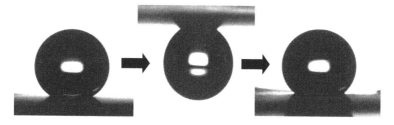

**Fig. 5.** Droplet transfer of 5.0-mg water droplet from the low-adhesion MP surface to the middle-adhesion MP surface via the high-adhesion MP surface

Water micro droplet transfer was attempted by using the MP surfaces with different adhesion properties as shown in Fig. 5. A water droplet of 5 mg on the low-adhesion MP surface was carried with the high-adhesion MP surface by means of pulling off after slight contact from above. However, in the case of the middle-adhesion MP surface, a water droplet did not remove from the low-adhesion MP surface. On the other hand, the high-adhesion MP surface was not caught a water droplet on the middle-adhesion MP surface from above. These results suggest that the adhesion force of the high-adhesion MP surface was stronger than that of the low-adhesion MP surface plus gravity on the water droplet, and weaker than that of the middle-adhesion MP surface plus gravity on the water droplet. By using this difference, the water droplet was transferred from the low-adhesion MP surface to the middle-adhesion MP surface via the high-adhesion MP surface. After transfer, the water droplet on the middle-adhesion MP surface was sliding when the surface was tilted at about 30°. A droplet transfer reported in the past is from a superhydrophobic surface to a hydrophilic surface via an adhesion superhydrophobic surface [4]. Therefore, after transfer, the water droplet is spreading, and is unable to be handled. The novel transfer method in this report remains a droplet shape after transfer, so that the droplet was handily manipulated again. These behaviors were useful to microfluidic devices, bio interfaces, and micro-reactors.

# 4  Conclusions

We could fabricate water repellency and adhesion properties of superhydrophobic metal-polymer surfaces by electroless plating for self-organized honeycomb films including immersion in a catalytic Pd salt and a cationic polymer mixture solution. It was found that a water contact angle and a water droplet adhesion property were changed by metal dome density that was easily controlled by the temperature of the catalytic mixture solution. Droplet transfer between superhydrophobic surfaces was demonstrated by means of using the metal-polymer surfaces with different adhesion properties.

# References

1. Zhang, X., Shi, F., Niu, J., Jiang, Y., Wang, Z.: Superhydrophobic surfaces: from structural control to functional application. J. Mater. Chem. 18, 621–633 (2008)
2. Onda, T., Shibuichi, S., Satoh, N., Tsujii, K.: Super-water-repellent fractal surface. Langmuir 12, 2125–2127 (1996)
3. Hosono, E., Fujihara, S., Honma, I., Zhou, H.: Superhydrophobic perpendicular nanopin film by the bottom-up process. J. Am. Chem. Soc. 127, 13458–13459 (2005)
4. Cho, W.K., Choi, I.S.: Fabrication of hairy polymeric films inspired by geckos: wetting and high adhesion properties. Adv. Funct. Mater. 18, 1089–1096 (2008)
5. Feng, L., Zhang, Y., Xi, J., Zhu, Y., Wang, N., Xia, F., Jiang, L.: Petal effect: a superhydrophobic state with high adhesive force. Langmuir 24, 4114–4119 (2008)
6. Yabu, H., Takebayashi, M., Tanaka, M., Shimomura, M.: Superhydrophobic and Lipophobic Properties of Self-Organized Honeycomb and Pincushion Structures. Langmuir 21, 3235–3237 (2005)
7. Ishii, D., Yabu, H., Shimomura, M.: Selective metal deposition in hydrophobic porous cavities of self-organized honeycomb-patterned polymer films by all-wet electroless plating. Col. Surf. A, 313–314, 590–594 (2008)
8. Karthaus, O., Maruyama, N., Cieren, X., Shimomura, M., Hasegawa, H., Hashimoto, T.: Water-assisted formation of micrometer-size honeycomb patterns of polymers. Langmuir 16, 6071–6076 (2000)
9. Ishii, D., Udatsu, M., Iyoda, T., Nagashima, T., Yamada, M., Nakagawa, M.: Electroless deposition mechanisms on fibrous hydrogen-bonded molecular aggregate to fabricate Ni-P hollow microfibers. Chem. Mater. 18, 2152–2158 (2006)

# Part II
# BIOSIGNALS

# Study on Biodegradation Process of Polyethylene Glycol with Exponential Glowth of Microbial Population

Masaji Watanabe[1,*] and Fusako Kawai[2]

[1] Graduate School of Environmental Science, Okayama University
1-1, Naka 3-chome, Tsushima, Kita-ku, Okayama 700-8530, Japan
[2] R & D center of Bio-based materials, Kyoto Institute of Technology
Hashigami-cho 1, Matsusaki, Sakyo-ku, Kyoto 606-8585, Japan
watanabe@ems.okayama-u.ac.jp

**Abstract.** Biodegradation of polyethylene glycol is studied mathematically. A mathematical model for depolymerization process of exogenous type is described. When a degradation rate is a product of a time factor and a molecular factor, a time dependent model can be transformed into a time independent model, and techniques developed in previous studies can be applied to the time independent model to determine the molecular factor. The time factor can be determined assuming the exponential growth of the microbial population. Those techniques are described, and numerical results are presented. A comparison between a numerical result and an experimental result shows that the mathematical method is appropriate for practical applications.

**Keywords:** Biodegradation, Polyethylene Glycol, Mathematical modeling, Numerical simulation.

## 1 Introduction

Biodegradation is an essential factor of the environmental protection against undesirable accumulation of xenobiotic polymers. It is particularly important for water soluble polymers, because they are not suitable for recycling nor incineration. It is also important for water-insoluble polymers, so-called plastics, because they are not completely recycled nor incinerated, and a significant portion of products remains in the environment after use. Microbial depolymerization processes are generally classified into either one of two types: exogenous type or endogenous type. In an exogenous depolymerization process, monomer units are separated from the terminals of molecules stepwise. The $\beta$-oxidation of polyethylene (PE) is an example of exogenous depolymerization process. Microbial depolymerization processes of PE are based on two primary factors : the gradual weight loss of large molecules due to the $\beta$-oxidation and the direct consumption or absorption of small molecules by cells. On the other hand, one of characteristics of endogenous depolymerization processes is the rapid breakdown of large molecules due to internal separations to yield small molecules. The enzymatic degradation of polyvinyl alcohol (PVA) is an example of endogenous depolymerization process. Mathematical models for those depolymerization processes have been proposed, and those models are analyzed to study the biodegradation of the xenobiotic polymers.

---

* This work was supported by KAKENHI (20540118).

A. Fred, J. Filipe, and H. Gamboa (Eds.): BIOSTEC 2009, CCIS 52, pp. 145–157, 2010.

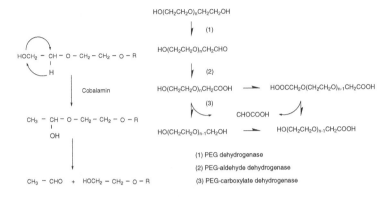

**Fig. 1.** Anaerobic metabolic pathway (left) and Aerobic metabolic pathway (right) of PEG

In this paper, the study of exogenous depolymerization processes is continued to cover the biodegradation of polyethylene glycol (PEG). PEG is one of polyethers which are represented by the expression $HO(R\text{-}O)_n H$, *e.g.*, PEG: $R = CH_2CH_2$, polypropylene glycol (PPG): $R = CH_3CHCH_2$, polytetramethylene glycol (PTMG): $R = (CH_2)_4$ [6]. Those polymers are utilized for constituents in a number of products including lubricants, antifreeze agents, inks, and cosmetics. They are either water soluble or oily liquid. Some portion of products are eventually discharged through sewage to be processed, while some others enter streams, rivers, and coastal areas. and therefore it is especially important to evaluate their biodegradability. PEG is produced more than any other polyethers, and the major part of production is consumed in production of nonionic surfactants. PEG is depolymerized by releasing $C_2$ compounds, either aerobically or anaerobically [7], [8], [12] (Figure 1).

High performance liquid chromatography (HPLC) patterns were introduced into analysis of an exogenous depolymerization model to set the weight distribution of PEG with respect to the molecular weight before and after cultivation of a microbial consortium E1 (Figure 2).

In the previous studies [13], the degradation rate was assumed to be independent of time. The time dependent degradation rate was considered in a recent study assuming a logistic growth in a microbial population [14], and using a cubic spline to take the change of microbial population into considerateion [15]. In this paper, the mathematical study of biodegradation of PEG is continued with the time dependent degradation rate incorporated into the exogenous depolymerization model. A change of variable reduces the model into the one for which the degradation rate is time independent. The techniques developed previously were applied to solve an inverse problem to determine the time independent degradation rate for which the solution of an initial value problem satisfies not only the initial weight distribution but also the weight distribution after cultivation. The time factor was determined by assuming the exponential growth of the microbial population. Once the degradation rate was found, the transition of the weight distribution was simulated by solving the initial value problem numerically.

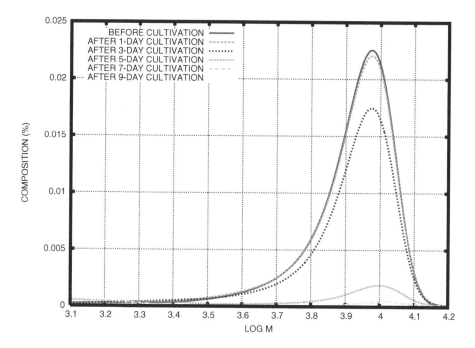

**Fig. 2.** Weight distribution of PEG before and after cultivation of a microbial consortium E1 for one day and three days

## 2   Model with Time Dependent Degradation Rate

The PE biodegradation model (1) is based on two essential factors: the gradual weight loss of large molecules due to terminal separations ($\beta$-oxidation) and the direct consumption of small molecules by cells [4], [5], [10].

$$\frac{dw}{dt}(t, M) = -\alpha(M)\, w(t, M) + \beta(M + L)\, \frac{M}{M + L} w(t, M + L). \qquad (1)$$

Here $t$ and $M$ represent the time and the molecular weight respectively. Let a $M$-molecule be a molecule with molecular weight $M$. Then $w(t, M)$ represents the total weight of $M$-molecules present at time $t$. Note that $w(t, M)$ is a function of time variable $t$, and that it also depends on the parameter $M$. The parameter $L$ represents the amount of the weight loss due to the $\beta$-oxidation. The variable $y$ denotes $w(t, M + L)$, and it is the total weight of $(M + L)$-molecules present at time $t$. The function $\alpha(M)$ denotes $\rho(M) + \beta(M)$, where the function $\rho(M)$ represents the direct consumption rate, and the function $\beta(M)$ represents the rate of the weight conversion from the class of $M$-molecules to the class of $(M - L)$-molecules due to the $\beta$-oxidation. The left-hand side of the equation (1) represents the rate of change in the total weight of $M$-molecules. The first term on the right-hand side of the equation (1) represents the amount lost by the direct consumption and the $\beta$-oxidation in the total weight of

$M$-molecules per unit time, and the second term represents the amount gained by the $\beta$-oxidation of $(M + L)$-molecules per unit time. The mathematical model (1) was originally developed for the PE biodegradation, but it can also be viewed as a general biodegradation model involving exogenous depolymerization processes. In the exogenous depolymeization of PEG, a PEG molecule is first oxidized at its terminal, and then an ether bond is split. It follows that $L = 44$ $(CH_2CH_2O)$ in the exogenous depolymerization of PEG. PEG molecules studied here are lagre molecules that can not be absorbed directly through membrene into cells. Then $\rho(M) = 0$, and $\alpha(M) = \beta(M)$.

The equation (1) is appropriate for the depolymerization processes over the period after the microbial population is fully developed. However the change of microbial population should be taken into consideration for the period in which it is still in a developing stage, and the degradation rate should be time dependent. Then the exogenous depolymerization model becomes

$$\frac{dw}{dt}(t, M) = -\beta(t, M)\,w(t, M) + \beta(t, M + L)\,\frac{M}{M + L}w(t, M + L). \quad (2)$$

to model the change of weight distribution of PEG. The solution $x = w(t, M)$ of (2) is associated with the initial condition:

$$w(0, M) = f(M), \quad (3)$$

where $f(M)$ is some prescribed function that represents the initial weight distribution. Given the the the degradation rate $\beta(t, M)$, the equation (2) and the initial condition (3) form an initial value problem to find the unknown function $w(t, M)$.

A time factor of the degradation rate such as microbial population, dissolved oxygen, or temperature should affect molecules regardless of their sizes. Then the dependence of the degradation rate on those time factors must be uniform over all the molecular weight classes, and the degradation rate should be a product of a time dependent part $\sigma(t)$ that represents the magnitude of degradability, and a molecular dependent part $\lambda(M)$ that represents the molecular dependence of degradability:

$$\beta(t, M) = \sigma(t)\,\lambda(M). \quad (4)$$

Let

$$\tau = \int_0^t \sigma(s)\,ds, \quad (5)$$

and

$$W(\tau, M) = w(t, M),$$
$$X = W(\tau, M),$$
$$Y = W(\tau, M + L).$$

Then

$$\frac{dX}{d\tau} = \frac{dx}{dt}\frac{dt}{d\tau} = \frac{1}{\sigma(t)}\frac{dx}{dt}.$$

It follows that

$$\frac{dX}{d\tau} = -\lambda(M)\,X + \lambda(M + L)\,\frac{M}{M + L}Y. \quad (6)$$

This equation governs the transition of weight distribution $W(t, M)$ which changes with the time independent or time averaged degradation rate $\lambda(M)$. Given the initial weight distribution $f(M)$, The solution of the initial value problem is the solution of the equation (6) subject to the initial condition

$$W(0, M) = f(M). \tag{7}$$

The solution of the inverse problem is the degradation rate $\lambda(M)$ for which the solution of the initial value problem (6), (7) also satisfies the final condition

$$W(T, M) = g(M). \tag{8}$$

When the solution $W(\tau, M)$ of the initial value problem (6), (7) satisfies this condition, the solution $w(t, M)$ of the intiail value problem (2), (3) satisfies the condition

$$w(T, M) = g(M), \tag{9}$$

where

$$T = \int_0^T \sigma(s)\, ds. \tag{10}$$

The inverse problem consisting of the equation (6) and the conditions (7) and (8) was solved numerically with techniques developed in previous studies. Figure 3 shows the graph of the function $\lambda(M)$ based on the weight distribution before and after cultivation for three days [13].

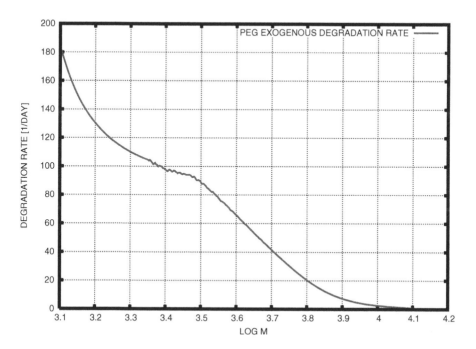

**Fig. 3.** Degradation rate based on the weight distribution of PEG before and after cultivation of a microbial consortium E1 for three days

## 3   Time Factor of Degradation Rate

A microbial population grows exponentially in a developing stage. Since the increase of biodegradability results from increase of microbial population, it is appropriate to assume that the time factor of the degradation rate $\sigma(t)$ is an exponential function of time:

$$\sigma(t) = e^{at+b}. \tag{11}$$

Then in view of the equation (5)

$$\tau = \int_0^t \sigma(s)\,ds = \int_0^t e^{as+b}\,ds = \frac{e^b}{a}\left(e^{at} - 1\right).$$

Suppose that the weight distribution is given at $t = T_1$ and $t = T_2$, where $0 < T_1 < T_2$, and let

$$\mathcal{T}_1 = \int_0^{T_1} \sigma(s)\,ds, \tag{12}$$

$$\mathcal{T}_2 = \int_0^{T_2} \sigma(s)\,ds. \tag{13}$$

It follows that

$$\sigma(t) = e^b e^{at} = \frac{a\mathcal{T}_1 e^{at}}{e^{aT_1} - 1} \tag{14}$$

and

$$\tau = \mathcal{T}_1 \frac{e^{at} - 1}{e^{aT_1} - 1}. \tag{15}$$

Now the equation (13) leads to

$$\mathcal{T}_2 = \mathcal{T}_1 \frac{e^{aT_2} - 1}{e^{aT_1} - 1},$$

which is equivalent to the equation

$$h(a) = 0, \tag{16}$$

where

$$h(a) = \frac{e^{aT_2} - 1}{e^{aT_1} - 1} - \frac{\mathcal{T}_2}{\mathcal{T}_1}.$$

Since

$$h'(a) = \frac{T_2 e^{aT_2}\left(e^{aT_1} - 1\right) - T_1 e^{aT_1}\left(e^{aT_2} - 1\right)}{\left(e^{aT_2} - 1\right)^2},$$

$h'(a) > 0$ if and only if

$$\frac{T_1 e^{aT_1}}{e^{aT_2} - 1} < \frac{T_2 e^{aT_2}}{e^{aT_2} - 1}.$$

For $x > 0$

$$q(x) = \frac{xe^{ax}}{e^{ax} - 1} = \frac{x}{1 - e^{-ax}}.$$

Then

$$q'(x) = \frac{1 - e^{-ax} - axe^{-ax}}{(1 - e^{-ax})^2} = \frac{e^{-ax}(e^{ax} - 1 - ax)}{(1 - e^{-ax})^2} > 0.$$

It follows that the $q(x)$ is a strictly increasing function, and it follows that $h'(a) > 0$. It is easily seen that

$$\lim_{a \to \infty} h(a) = \infty.$$

Suppose that

$$\frac{T_2}{T_1} < \frac{\mathcal{T}_2}{\mathcal{T}_1}. \qquad (17)$$

Then by L'Hospital's rule

$$\begin{aligned} \lim_{a \to 0+} h(a) &= \lim_{a \to 0+} \left\{ \frac{e^{aT_2} - 1}{e^{aT_1} - 1} - \frac{\mathcal{T}_2}{\mathcal{T}_1} \right\} \\ &= \lim_{a \to 0+} \frac{e^{aT_2} - 1}{e^{aT_1} - 1} - \frac{\mathcal{T}_2}{\mathcal{T}_1} \\ &= \lim_{a \to 0+} \frac{T_2 e^{aT_2}}{T_1 e^{aT_1}} - \frac{\mathcal{T}_2}{\mathcal{T}_1} \\ &= \frac{T_2}{T_1} - \frac{\mathcal{T}_2}{\mathcal{T}_1} < 0. \end{aligned} \qquad (18)$$

It follows that the condition (17) is a necessary and sufficient condition for the equation (16) to have a unique positive solution.

In order to determine $a$ and $b$, the values of $T_1$, $T_2$, $\mathcal{T}_1$, and $\mathcal{T}_2$ must be set. Let $T_1 = \mathcal{T}_1 = 3$. The initial value problem (6), (7) was solved numerically with the degradation whose graph is shown in Figure 3 to find the weight distribution at $\tau = 30$ (Figure 4). Figure 4 also shows the weight distribution after cultivation for five days.

Figure 4 shows that it is appropriate to set $T_2 = 5$ and $\mathcal{T}_2 = 30$. Figure 5 shows the graph of $h(a)$ with those values of parameters.

Figure 5 shows that there is a unique solution of the equation (16). It was solved numerically with the Newton's method, and a numerical solution, which was approximately equal to 1.136176 was found.

## 4    Simulation with Time Dependent Degradation Rate

Once the degradation rate $\sigma(t)\lambda(M)$ are given, the initial value problem (2) and (3) can be solved directly to see how the numerical results and the experimental results agree. Here the initial value problem was solved numerically with techniques base on previous results [5], [10], [11].

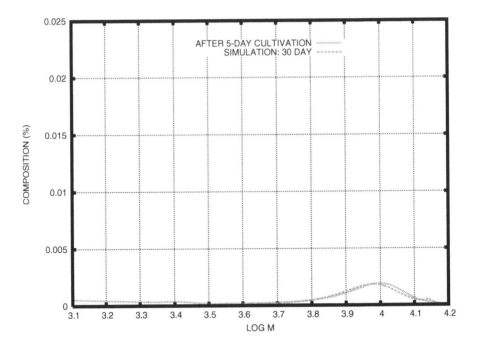

**Fig. 4.** Weight distribution of PEG after cultivation for 30 days according to the time independent model based on the initial value problem (6), (7), and the degradation rate shown in Figure 3. The experimental result obtained after cultivation for 5 days is also shown.

Choose a positive integer $N$ and set

$$\Delta M = \frac{b - a}{N},$$

$$M_i = a + i\Delta M, \quad i = 0, 1, 2, \cdots, N.$$

An approximate solution of the differential equation (1) at $M = M_i$ is denoted by $w_i = w_i(t)$ $(i = 0, 1, 2, \cdots, N)$. There is a non-negative integer $K$ and a constant $R$ such that $L = K\Delta M + R$, $0 \le R < \Delta M$, and that the inequalities

$$M_{i+K} \le M_i + L < M_{i+K+1}$$

hold. Then approximate values of $w(t, M_i + L)$ and $\beta(M_i + L)$ can be obtained by using the approximations

$$w(t, M_i + L) \approx \left(1 - \frac{R}{\Delta M}\right) w(t, M_{i+K}) + \frac{R}{\Delta M} w(t, M_{i+K+1}),$$

$$\lambda(M_i + L) \approx \left(1 - \frac{R}{\Delta M}\right) \lambda(M_{i+K}) + \frac{R}{\Delta M} \lambda(M_{i+K+1}).$$

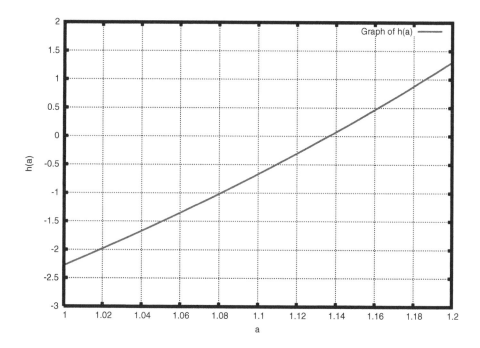

**Fig. 5.** Graph of $h\left(a\right)$ with $T_1 = T_1 = 3$, $T_2 = 5$ and $T_2 = 30$

Substituting these expressions in the differential equation (2) and setting $M = M_i$, we obtain the linear system:

$$\frac{dw_i}{dt} = \sigma\left(t\right)\left(-\alpha_i w_i + \beta_i w_{i+K} + \gamma_i w_{i+K+1}\right),$$

$$i = 0, 1, 2, \cdots, N.$$

(19)

The coefficients $\alpha_i$, $\beta_i$, and $\gamma_i$ are given by

$$\alpha_i = \lambda\left(M_i\right),$$

$$\beta_i = \phi_i \frac{M_i}{M_i + L}\left(1 - \frac{R}{\Delta M}\right),$$

$$\gamma_i = \phi_i \frac{M_i}{M_i + L} \cdot \frac{R}{\Delta M},$$

$$\phi_i = \left(1 - \frac{R}{\Delta M}\right)\lambda\left(M_{i+K}\right) + \frac{R}{\Delta M}\lambda\left(M_{i+K+1}\right).$$

Approximate values of the degradation rates $\lambda\left(M_i\right)$ can be obtained from the numerical solution of the inverse problem by the linear approximation.

For all sufficiently large $M$, the oxidation rate becomes 0. In particular, we may assume that the last two terms on the right-hand side of the equation (19) are absent when $i + K$ exceeds $N$, so that the system (19) becomes a closed system to be solved for unknown functions $w_i = w_i(t)$, $i = 0, 1, 2, \ldots, N$. In view of the condition (3), these functions are subject to the initial condition

$$w_i(0) = f_i = f(M_i). \tag{20}$$

Given the initial weight distribution shown in Figure 2, the degradation rate $\lambda(M)$ shown in Figure 3, and the function $\sigma(t)$ given by the equation (14) with the value of $a$ obtained numerically, the initial value problem (19) and (20) was solved numerically implementing the forth-order Adams-Bashforth-Moulton predictor-corrector in PECE mode in conjunction with the Runge-Kutta method to generate approximate solutions in the first three steps [9] by using $N = 10000$, and a time interval $\Delta t = 5/24000$. Figure 6 shows the transition of the weight distribution during cultivation of the microbial consortium E-1 for five days.

Figure 7 shows the numerical result and the experimental results for the weight distribution after one day cultivation of the microbial consortium E1.

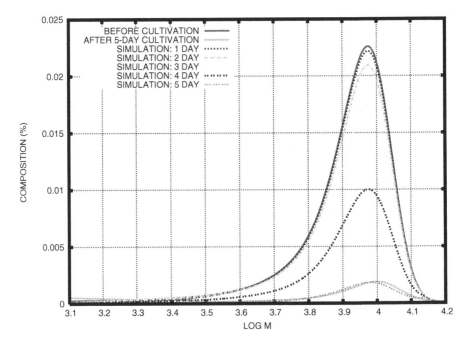

**Fig. 6.** The weight distribution of PEG before and after 5-day cultivation, and the transition of the weight distribution based on the initial value problem (2), (3) with $\sigma(t) = e^{at+b}$, $a \approx 1.136176$, $b = \ln\{aT_1/(e^{aT_1} - 1)\}$, $T_1 = T_1 = 3$

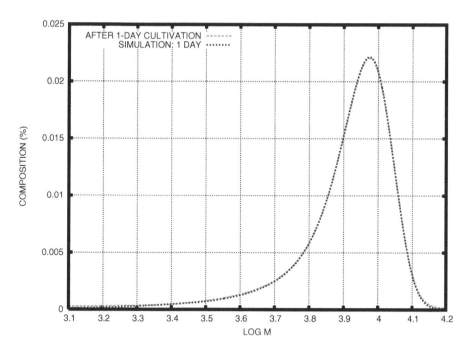

**Fig. 7.** The weight distribution of PEG after 1-day cultivation, and the weight distribution based on the initial value problem (2), (3) with $\sigma(t) = e^{at+b}$, $a \approx 1.136176$, $b = \ln\left\{aT_1/\left(e^{aT_1}-1\right)\right\}$, $T_1 = T_1 = 3, t = 1$

## 5   Discussion

Early studies of biodegradation of xenobiotic polymers are found in the second half of the 20th century. It was found that the linear paraffin molecules of molecular weight up to 500 were utilized by several microorganisms [1]. Oxidation of $n$-alkanes up to tetratetracontane ($C_{44}H_{90}$, mass of 618) in 20 days was reported [2]. Biodegradation of polyethylene was shown by measurement of $^{14}CO_2$ generation [3]. The weight distribution of polyethylene before and after cultivation of the fungus Aspergillus sp. AK-3 for 3 weeks was introduced into analysis based on the time dependent exogenous depolymerization model. The transition of weight distribution for 5 weeks was simulated with the degradation rate based on the initial weight distribution and the weight distribution after 3 weeks of cultivation. The numerical result was found to be acceptable in comparison with an experimental result [11]. The result shows that the microbial population was fully developed in 3 weeks, and that the biodegradation was with the constant rate.

The degradation rate changed over the cultivation period in the depolymerization processes of PEG. The development of microbial population accounts for the increase of degradability over the first five days of cultivation. In a depolymerization process where the microbial population becomes an essential factor, it is necessary to consider the dependence of the degradation rate on time. The numerical results based on the time dependent exogenous depolymerization model show reasonable agreement with the experimental results. Those results show that it is appropriate to assume that the degradation rate is a product of a time factor and a molecular factor. It has also been shown that the molecular factor can be determined by the weight distribution before and after cultivation experimentally. In the environment or sewer disposal, the time factor should also depends on other factors such as temperature or dissolved oxygen. Once those essentials are incorporated into the time dependent factor, the time dependent exogenous depolymerization model and the techniques based on the model should be applicable to assess biodegradability of xenobiotic polymers.

## Acknowledgement

The authors thank Ms. Y. Shimizu for her technical support.

## References

1. Potts, J.E., Clendinning, R.A., Ackart, W.B., Niegishi, W.D.: The biodegradability of synthetic polymers. Polym. Preprints 13, 629–634 (1972)
2. Haines, J.R., Alexander, M.: Microbial degradation of high-molecular-weight alkanes. Appl. Microbiol. 28, 1084–1085 (1974)
3. Albertsson, A.-C., Andersson, S.O., Karlsson, S.: The mechanism of bioidegradation of plyethylene. Polym. Degrad. Stab. 18, 73–87 (1987)
4. Kawai, F., Watanabe, M., Shibata, M., Yokoyama, S., Sudate, Y.: Experimental analysis and numerical simulation for biodegradability of polyethylene. Polym. Degrad. Stab. 76, 129–135 (2002)
5. Watanabe, M., Kawai, F., Shibata, M., Yokoyama, S., Sudate, Y.: Computational method for analysis of polyethylene biodegradation. Journal of Computational and Applied Mathematics 161, 133–144 (2003)
6. Kawai, F.: Biodegradability and chemical structure of polyethers. Kobunshi Ronbunshu 50(10), 775–780 (1993) (in Japanese)
7. Kawai, F.: Breakdown of plastics and polymers by microorganisms. Advances in Biochemical Engineering/Biotechnology 52, 151–194 (1995)
8. Kawai, F.: Microbial degradation of polyethers. Applied Microbiology and Biotechnology 58, 30–38 (2002)
9. Lambert, J.D.: Computational Methods in Ordinary Differential Equations. John Wiley & Sons, Chichester (1973)
10. Kawai, F., Watanabe, M., Shibata, M., Yokoyama, S., Sudate, Y., Hayashi, S.: Comparative study on biodegradability of polyethylene wax by bacteria and fungi. Polym. Degrad. Stab. 86, 105–114 (2004)
11. Watanabe, M., Kawai, F., Shibata, M., Yokoyama, S., Sudate, Y., Hayashi, S.: Analytical and computational techniques for exogenous depolymerization of xenobiotic polymers. Mathematical Biosciences 192, 19–37 (2004)
12. Kawai, F.: Xenobiotic polymers. In: Imanaka, T. (ed.) Great Development of Microorganisms, pp. 865–870. NTS. Inc., Tokyo (2002) (in Japanese)

13. Watanabe, M., Kawai, F.: Numerical simulation of microbial depolymerization process of exogenous type. In: May, R., Roberts, A.J. (eds.) Proc. of 12th Computational Techniques and Applications Conference, CTAC 2004, Melbourne, Australia (September 2004); ANZIAM J. 46(E), C1188–C1204 (2005),
    http://anziamj.austms.org.au/V46/CTAC2004/Wata
14. Watanabe, M., Kawai, F.: Mathematical study of the biodegradation of xenobiotic polymers with experimental data introduced into analysis. In: Stacey, A., Blyth, B., Shepherd, J., Roberts, A.J. (eds.) Proceedings of the 7th Biennial Engineering Mathematics and Applications Conference, EMAC 2005, Melbourne (2005); ANZIAM J. 47, C665–C681 (2007),
    http://anziamj.austms.org.au/V47EMAC2005/Watanabe
15. Watanabe, M., Kawai, F.: Numerical study of biodegradation of xenobiotic polymers based on exogenous depolymerization model with time dependent degradation rate. Journal of the Faculty of Environmental Science and Technology, Okayama University 12, 1–6 (2007)

# Variable Down-Selection for Brain-Computer Interfaces

Nuno S. Dias[1], Mst Kamrunnahar[2], Paulo M. Mendes[1],
Steven J. Schiff[2], and Jose H. Correia[1]

[1] Dept. of Industrial Electronics, University of Minho
Campus Azurem, 4800-058 Guimaraes, Portugal
{ndias,pmendes,higino.correia}@dei.uminho.pt
[2] Dept. of Engineering Sciences and Mechanics, The Pennsylvania State University
University Park, PA 16802, U.S.A.
muk11@psu.edu, sjs49@engr.psu.edu

**Abstract.** A new formulation of principal component analysis (PCA) that considers group structure in the data is proposed as a variable down-selection method. Optimization of electrode channels is a key problem in brain-computer interfaces (BCI). BCI experiments generate large feature spaces compared to the sample size due to time limitations in EEG sessions. It is essential to understand the importance of the features in terms of physical electrode channels in order to design a high performance yet realistic BCI. The proposed algorithm produces a ranked list of original variables (electrode channels or features), according to their ability to discriminate movement imagery tasks. A linear discrimination analysis (LDA) classifier is applied to the selected variable subset. Evaluation of the down-selection method using synthetic datasets selected more than 83% of relevant variables. Classification of imagery tasks using real BCI datasets resulted in less than 19% classification error. Across-Group Variance (AGV) showed the best classification performance with the largest dimensionality reduction in comparison with other algorithms in common use.

**Keywords:** BCI, EEG, Feature selection.

## 1 Introduction

Brain-Computer Interfaces (BCI) enable people to control a device with their brain signa [1]. BCIs are expected to be a very useful tool for impaired people both in invasive and non-invasive implementations. Non-invasive BCI operation commonly uses electroencephalogram (EEG) from human brain for the ease of applicability in laboratory set-ups as well as in clinical applications. BCI systems employing a pattern recognition approach tend to deal with high-dimensional variable sets, especially when no previous knowledge about EEG features is considered. The features commonly extracted from EEG are frequency band power, event-related desynchronization/synchronization (ERD/ERS) [2], movement-related potentials (MRP) [3] and event-related potentials (e.g. P300) [4], among others. The low ratio of the number of samples to the number of variables is described as the 'curse of dimensionality' [5]. In determining a suitable classifier for the BCI, the decision of

A. Fred, J. Filipe, and H. Gamboa (Eds.): BIOSTEC 2009, CCIS 52, pp. 158–172, 2010.

increasing the number of samples in order to compensate for high-dimensionality would imply longer experiments. However, long experiments should be avoided due to the progressive degradation of signal quality (i.e. electrolyte gel gets dry which increases electrode-skin interface impedance) and subject alertness. Thus, a variable down-selection method should be employed resulting in fewer variables to be considered in further processing. A variable subset calculation is feasible when few variables are relevant.

The variables in a dataset are frequently divided into irrelevant, weakly relevant and strongly relevant variables [6]. A good subset should include all the strongly relevant variables and some of the weakly relevant ones. The variable subset to choose should minimize the cross-validation error. Typically, the term 'feature' is used in the literature [7] instead of 'variable'. Nevertheless, we here use the latter to avoid confusion about the dimensions of the dataset (electrode channels) and the characteristic features (e.g. band power, ERP) extracted from EEG raw signals.

This work proposes a new variable down-selection method based on a different formulation of principal component analysis (PCA), introduced in [8] that accommodates the group structure of the dataset. In the PCA framework, data dimensionality reduction methods typically use a reduced set of principal components as a lower-dimensional representation of original variables, for discrimination purposes [8, 9]. However, if one wants to reduce the BCI computational complexity, the dimensionality reduction should take place on the original variable space. The proposed method was evaluated on both synthetic and real datasets. A synthetic dataset enabled us to evaluate this method in a controlled environment and simulate the 3 levels of variable relevance mentioned above. The real datasets were generated from EEG responses to 2 movement imagery tasks: left vs. right hand movement imageries. Event-related potentials (ERP) and power ratios were extracted from the EEG signals. Five subjects with no previous BCI experience participated in this experiment. The classification accuracy of the best subset in discriminating task performances, for each subject, was evaluated by cross-validation.

Additionally, the proposed algorithm was compared with 3 other methods that have been used in BCI research: RELIEF [10]; recursive feature elimination [11]; and genetic algorithm [12]. The data set IVa from the BCI competition III [13] was used for algorithm performance comparison.

## 2 Experimental Design

### 2.1 Synthetic Data

Among all variables $p$ in the synthetic dataset, $q$ relevant and $p-q$ irrelevant Gaussian distributed variables were generated. All the generated variables had the same standard deviation $\sigma$. In order to best simulate a typical multivariate dataset, the relevant features were generated in pairs with correlation between variables. In this way, the variables are more discriminative if considered together and the 3 levels of variable relevance are easy to be evaluated. The first variable in each pair is considered as the predominant variable (i.e. strongly relevant) since its mean has distance $d$ between groups. The distribution parameters were set similarly as shown in [14]. For each predominant variable, a highly correlated Gaussian variable (i.e. weakly relevant) with null distance between group means was generated. Pairs of correlated variables were generated until the required quantity of relevant variables is reached. The remaining $p$-$q$ variables (i.e. irrelevant) were generated from the same Gaussian distribution (i.e. no mean difference between groups) for both groups. Four different datasets were generated with 80 samples: $p=79$ and $q=6$; $p=79$ and $q=12$; $p=40$ and $q=6$; $p=40$ and $q=12$. The first 2 datasets were intended to simulate the high dimensional/low sample size problem. The last 2 represent a lower dimensional space. The standard deviation $\sigma$ was set to 2.5 in all datasets. The mean difference $d$ was set to be equal to $\sigma$ to simulate group overlapping. The values of both distribution parameters were set to best approximate the real variables extracted from the EEG data collected.

## 2.2 Electroencephalogram Data

Five healthy human subjects, 25 to 32 years old, three males and one female, were submitted to 1 session each of motor imagery. The experiments were conducted under Institutional Review Board (IRB) approval at Penn State University.

Each session had 4 runs of 40 trials each. Each subject was instructed to perform one of 4 tasks in each trial. The tasks were tongue, feet, left hand and right hand movement imageries. The following 2 imagery task discrimination cases were considered for variable subset selection (VSS) algorithm evaluation: tongue vs. feet; left hand vs. right hand.

Each trial started with the presentation of a cross centered on the screen, informing the subject to be prepared. Three seconds later, a cue was presented on top of the cross. The subject was instructed to perform the movement imagery during a 4 s period starting at the cue presentation. Then, both the cross and the arrow were removed from the screen, indicating the end of the trial. The inter-trial period was randomly set to be between 3-4.5 s long.

Data were acquired from 19 electrodes according to the standard 10-20 system (Fp1, Fp2, F7, F3, Fz, F4, F8, T7, C3, Cz, C4, T8, P7, P3, Pz, P4, P8, O1 and O2). All electrodes were acquired with respect to linked earlobes and were later referred to common average. Data were digitized at 256 Hz and passed through a 4th order 0.5-60 Hz band-pass filter. Each channel's raw EEG signal was epoched from the cue time point (0 s) to 4 s after the cue. The presence of artifacts in the epochs was checked through maximum allowed absolute value (50 μV) and maximum allowed absolute potential difference (20 μV) between two consecutive points. The epochs contaminated with artifacts (e.g. eye blinks, muscle artifacts) were excluded from further analyses.

### 2.2.1 Event-Related Potentials (ERP)

Event-related potentials (ERP) are slow, non-oscillatory EEG potential shifts in response to certain events (e.g. visual or auditory stimuli) [15]. ERPs may be defined as movement-related potentials when they are related to the initiation or imagination of a movement [3].

The epoch was subdivided in 1 s time windows with no overlap (4 time windows). Each epoch was low-pass filtered at 4 Hz with an 8th order Chebyshev type I filter. Then, the filtered 256 points time series was down sampled to be 10 points long. The

Matlab "decimate" function was used to accomplish both the filtering and down sampling. Only the $1^{st}$-$8^{th}$ data points of the resultant time series form the feature vector for each time window. The last 2 points of the time series were discarded because they seemed to be irrelevant on previous analyses. The feature matrix of each time window had 152 variables (8 features from each of the 19 electrodes).

### 2.2.2 Power Ratios

Alpha (8-14 Hz) and beta (16-24 Hz) EEG frequency bands include rhythms that are reactive to movement imagery [16]. Alpha is an idling rhythm which is also termed 'rolandic mu rhythm' when generated in a motor-related cortex area. Alpha amplitude decreases during the execution of, as well as imagined, limb movement at motor-related cortical locations. The beta rhythm generally increases in amplitude at limb movement initiation and termination at motor-related cortical locations [17].

The epochs used to extract the EEG power ratios were subdivided into 1 s time windows with no overlap. Five narrow frequency bands were defined: 8-12 Hz (i.e. low alpha band); 10-14 Hz (i.e. high alpha band); 16-20 Hz (i.e. low beta band); 18-22 Hz (i.e. mid beta band); and 20-24 Hz (i.e. high beta band). The sum of the power from all narrow frequency bands was used to normalize the power contained in each narrow band. The dot product of the frequency domain Fourier Transform with its complex conjugate gives the Power Spectral Density (PSD). Each band power was computed as the sum of all PSD components in corresponding frequency range. The power ratio feature matrices had 95 features (5 power ratio features × 19 electrodes).

## 3   Methods

The original feature matrix $Y$ has samples ($n$) in rows and features ($p$) in columns. Considering that just a few $p_{opt}$ features out of $p$ may be relevant for discrimination, data dimensionality should be reduced for robust and effective discrimination. We here introduce a PCA-based algorithm, as well as a cross-validation scheme that calculates the classification error, optimizes algorithm parameters and determines a classification model. The cross-validation error predicts the classifier's online performance.

### 3.1   Cross-Validation

The generalization error used to evaluate the down-selection algorithms' performance was calculated through a 10-fold double-loop cross-validation scheme. The optimal number of features to select ($p_{opt}$) is optimized in the inner cross-validation loop. The performance of the classifier previously trained with the selected features, is validated in the outer loop. Therefore, 10 validation error values are calculated. However, the 10-fold cross-validation results are considerably affected by variability. For this reason, the whole procedure was repeated 10 times. The median of the 100 validation error values was defined as the estimate of the classification online performance using the selected features.

A new feature subset was calculated at every validation fold. The number of times that each feature is selected for validation is indicative of its relevance for discrimination.

Likewise, channel discriminative ability is assessed by its frequency of selection. A channel is deemed selected when at least one of its particular features (either ERPs or power ratios) is selected.

### 3.1.1  Classification

A different approach of Fisher Discriminant Analysis that was robust on spatiotemporal EEG pattern discrimination was applied [18]. The canonical discrimination functions $Z_i$ are the result of a linear transformation of original data $Y$ according to equation (1). The discrimination coefficients of each $i^{th}$ canonical discrimination function are denoted by the columns of $b_i{}^T$:

$$z = Y b^T \tag{1}$$

The multivariate data vector $Y$ is transformed to the vector z which have mean $u$ and a normal $p$-variate distribution $f(z)$. Prior probabilities $\pi_j$ are calculated as the ratio of the samples within group $j$ to the total number of samples $n$. According to the Bayesian theory, the probability that the data came from group $j$ of 2 groups (in our case either left vs. right or tongue vs. feet), given a vector $z$, is calculated through $\pi_{jz}$ in equation (2):

$$\pi_{jz} = \frac{\pi_j f_j(z)}{\pi_1 f_1(z) + \pi_1 f_2(z)} . \, j=1,2 \tag{2}$$

The term $\pi_j f_j(z)$ was approximated by $\exp[q(z)]$ where $q(z) = u_j^T z - \frac{1}{2} u_j^T u_j + \ln\pi_j$ [19]. The mean of the canonical discrimination function for group $j$ is $u_j$. The highest $\pi_{jz}$ determined the predicted group membership for each sample. In order to robustly assess discrimination quality, the cross-validated prediction error rate was calculated.

### 3.2  Across-Group Variance

This proposed filter type method uses a different formulation of PCA [8] to select features while reducing data dimensionality. Initially, $Y$ is decomposed through singular value decomposition (SVD) into 3 matrices: $U_{n \times c}$ (component orthogonal matrix; $c$ is the number of principal components), $S_{c \times c}$ (singular value diagonal matrix), and $V_{p \times c}$ (eigenvector orthogonal matrix; $p$ is the number of features). The eigenvalue vector $\lambda$ is calculated as the diagonal of $S^2$. The principal components are linear projections of the features onto the orthogonal directions that best describe the dataset variance. However, when data presents a group structure, the information provided by a component is more detailed than a variance value. Thus, the total covariance $\Psi$ can be decomposed into a sum of within $\Psi_{within}$ and between $\Psi_{between}$ group covariance parts. The pooled covariance matrix is used as an estimation of the within group covariance matrix ($\Psi_{within}$) as in equation (3):

$$\Psi_{within} = \frac{(n_1 - 1)\Psi_1 + (n_2 - 1)\Psi_2}{n_1 + n_2 - 2} . \tag{3}$$

By using the Bessel's correction, the sample covariance matrix $\Psi_i$ is weighted by $n_i{-}1$ ($n_i$ is the number of samples belonging to the $i^{th}$ group) instead of $n_i$ in order to correct the estimator bias ($\Psi_i$ has rank $n_i$-1 at most). The variance information provided by a principal component in vector notation is deduced in equation (4):

$$\lambda_j = v_j^T \Psi v_j = v_j^T \Psi_{within} v_j + v_j^T \Psi_{between} v_j. \qquad (4)$$

where $v_j$ is the $j^{th}$ eigenvector (a column of $V_{p \times c}$ matrix) and corresponds to eigenvalue $\lambda_j$. While $v_j^T \Psi_{within} v_j$ is a function of the sample distances to their respective group mean, $v_j^T \Psi_{between} v_j$ is a function of the distances between the respective group means. In the discrimination context, only the latter comprises useful variance information. Therefore, the distance between groups given by the $i^{th}$ component, normalized by its total variance, gives a relative measure to calculate the across-group variance as in equation (5):

$$AGV_i = \frac{v_i^T \Psi_{between} v_i}{\lambda_i}. \qquad (5)$$

The between group covariance matrix ($\Psi_{between}$) is calculated from $\Psi_{total} - \Psi_{within}$.

Although the principal components are organized by decreasing order of total variance (eigenvalues $\lambda$), this order is optimized for orthogonality rather than discrimination between groups. Therefore, the components are ordered according to the across group variance ($AGV$) in order to take the data group structure into account.

The dimensionality reduction results from the truncation of the $c$ principal components ranked by decreasing order of AGV. The truncation criterion ($\delta$) is a cumulative sum percentage of the descending ordered AGV scores and was assigned threshold values (typically 60%-90%) for variance truncation [20]. The optimal threshold value was found by cross-validation. If $k$ components met the truncation criterion, the truncated component matrix $U_{n \times k}$ ($k < c$) is a lower dimensional representation of the original feature space and more suitable for group discrimination. The retained variance information was transformed back to the original feature space using a modified version of the spectral decomposition property as in equation (6). In order to determine the features which resemble the retained components with minimal information loss, an across-group covariance matrix ($\Psi_{AGV}$) was calculated:

$$\Psi_{AGV} = \sum_{i=1}^{k} AGV_i v_i v_i^T. \qquad (6)$$

Note that $AGV_i$ is used instead of $\lambda_i$ in the spectral decomposition equation (6). Each diagonal value of $\Psi_{AGV}$ represents the variance of a particular variable accounted for by the $k$ retained principal components and measures feature discriminability. A ranked list with the all $p$ features in descending order of discriminability was determined. Finally, the number ($p_{opt}$) of optimal features that determine the optimal feature matrix was optimized by cross-validation. A piece of pseudocode is in appendix for algorithm illustration.

## 4  Results

The feature down-selection on synthetic data was evaluated for 4 different datasets, simulating different proportions of relevant features (see Table 1). All 4 datasets comprise 80 samples either with 79 (higher dimensional space) or 40 (lower dimensional space) variables. In the lower dimensional datasets all relevant variables (both strongly and weakly relevant) were selected. The optimal subset was found since no irrelevant variables were selected. In the higher dimensional spaces, all the predominant variables were selected. However, when 12 relevant variables were generated, 11 relevant (83% of all relevant) and 1 irrelevant variables were selected. Despite the simulated group overlapping, the selected variables kept the classification error below 12.9% in all datasets.

**Table 1.** Results for variable down-selection on artificial datasets

| Generated datasets | | Results for selected variables | | | |
|---|---|---|---|---|---|
| Variables | Relevant variables | Error (%) | Variables | Relevant var. (%) | Predominant var. (%) |
| 79 | 6 | 3.4 | 7 | 100 | 100 |
| 79 | 12 | 8.8 | 11 | 83 | 100 |
| 40 | 6 | 12.9 | 6 | 100 | 100 |
| 40 | 12 | 3.1 | 12 | 100 | 100 |

The AGV algorithm was also tested in real data from 5 subjects which achieved between 1.4 % and 11 % classification error (each subject's best time window) for left vs. right hand movement imagery (Table 2), considering event-related potentials as variables. During tongue vs. feet movement imagery, the classification error was between 7 % and 16.4 %. On average, 23 (left vs. right) and 20 (tongue vs. feet) variables were selected out of the 152 ERP variables. As illustrated on Figure 1, the best time window for either left vs. right or tongue vs. feet discrimination was 0 to 1 s after stimulus presentation. In left vs. right performance discrimination of Subject E, central and temporal (T7, C3, C4 and T8), parietal (Pz) and occipital (O1 and O2)

**Table 2.** Results summary of the event-related potential classification error and number of features selected. Both left vs. right and tongue vs. feet discriminations are here reported for all 5 subjects.

| Subject code | Left vs. Right | | Tongue vs. Feet | |
|---|---|---|---|---|
| | Error (%) | No. Variables | Error (%) | No. Variables |
| A | 1.4 | 19 | 12.7 | 17 |
| B | 9.1 | 38 | 16.4 | 19 |
| C | 2.7 | 26 | 12.0 | 29 |
| D | 11.0 | 12 | 13.2 | 12 |
| E | 10.6 | 19 | 7.0 | 24 |
| Mean | 7.0 | 23 | 12.2 | 20 |

channels were mostly selected (Figure 2a). As depicted on Figure 2b, these channels were mostly selected during the interval 0-100 ms. The occipital channels were also selected between 200 and 400 ms.

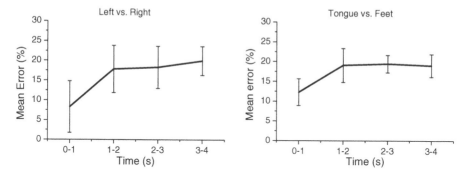

**Fig. 1.** Time variation of the across-subjects average classification error of event-related potentials in response to left vs. right hand and tongue vs. feet movement imageries.

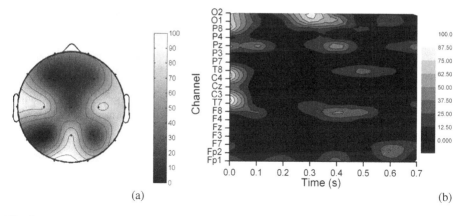

**Fig. 2.** (a) Topographic map of channel selection frequency (%) and (b) space-time map of variable (event-related potentials) selection frequency (%) for Subject E performance of left vs. right hand movement imageries during time window 0-1 s.

Considering power ratios for discrimination between left and right hand movement imageries, classification error varied between 13.2 % and 17.9 % among subjects. Tongue vs. feet discrimination retrieved classification errors between 17.5 % and 20.7 %. On average, 12 (left vs. right) and 15 (tongue vs. feet) variables were selected from 95-dimensional datasets. No particular time window was preferred among subjects. Figure 3a shows that the central channels C3 and Cz were selected frequently for left vs. right discrimination between 2 and 3 s after the stimulus presentation. Figure 3b further illustrates that the central channel were mostly selected in the frequency bands 8-12 Hz and 10-14 Hz.

**Table 3.** Summary results of the power ratio classification error and number of features selected. Both left vs. right and tongue vs. feet discriminations were reported for 4 subjects.

| Subject code | Left vs. Right | | Tongue vs. Feet | |
|---|---|---|---|---|
| | Error (%) | No. Variables | Error (%) | No. Variables |
| A | 13.2 | 16 | 20.7 | 11 |
| B | 15.9 | 10 | 20.5 | 15 |
| C | 16.6 | 8 | 17.9 | 14 |
| D | 16.4 | 10 | 17.5 | 17 |
| E | 17.9 | 16 | 18.2 | 16 |
| Mean | 16.0 | 12 | 19.0 | 15 |

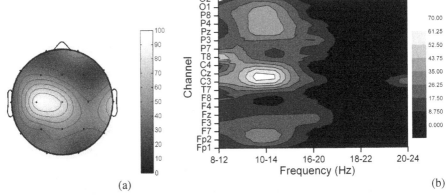

(a)                                                      (b)

**Fig. 3.** (a) Topographic map of channel selection frequency (%) and (b) space-frequency map of variable (power ratio) selection frequency (%) for Subject D performance of left vs. right hand movement imageries on the best time window (2-3 s).

## 5   Algorithm Comparison

### 5.1   Methods

Methods proposed in other studies to down-select feature sets may be categorized as *wrapper* or *filter* methods based on their dependence on a learning technique [7]. The *wrapper* methods use the predictive accuracy of a pre-selected classifier to evaluate a feature subset. Among other wrapper methods, the recursive feature elimination has been used in BCI research [11]. The *filter* methods separate feature selection from classifier training and thus produces feature subsets independent of the selected classifier. The relief algorithm is often used as a *filter* method [10]. Whether *wrapper* or *filter* typology is adopted, the genetic algorithms (search algorithms) are also very popular in BCI research [12]. Thus, the new developed algorithm, as well as 3 other representing filter, wrapper and global search methods, were compared on the data set IVa from BCI competition III [13].

### 5.2   Data

Five healthy human subjects were instructed to perform right hand vs. foot movement imagery in 4 calibration sessions with 70 trials each. Data were recorded from 118

EEG channels according to the 10-20 system at 100 samples/s [13]. The EEG signals were differently filtered for each subject in the 8-30 Hz, 8-14 Hz or 15-30 Hz frequency band depending on which band achieved the best group membership prediction. The signal epoch was defined from the cue presentation instant (i.e. 0 s) to the end of the imagery period (i.e. 3.5 s after cue presentation). The epoch data was assessed in 1 s long windows with 0.5 s overlap. In each time window, the sum of the squared filtered signals was calculated. Each feature matrix corresponding to each subject had 280 samples available with 118 features.

### 5.2.1 Relief

The RELIEF algorithm is a filter method that assigns a relevance value to each variable producing a ranking that permits the selection of the top ranked variables according to a previously chosen threshold or criterion [10]. The relevance value, or feature weight ($W$), is iteratively estimated according to how well a feature distinguishes among instances that are near each other. In each iteration, a sample $y$ is randomly selected and the weight of each feature is updated from the difference between the selected sample and two neighbouring samples: one from the same group $H(y)$ (named nearest hit) and another from a different group $M(y)$ (named nearest miss). The weight of each feature $p$ is updated as in equation (7):

$$W_p = W_p - \left| y_p - H(y)_p \right| + \left| y_p - M(y)_p \right| \tag{7}$$

The weights are calculated in $n$ (number of available training samples) sequential iterations. Iteratively, the feature with the lowest weight is removed and the classification accuracy of the resulting subset is evaluated by a linear discriminant classifier (as described in subsection 3.1.1). The selection stops when $p_{opt}$ (input parameter) features are left.

### 5.2.2 Recursive Feature Elimination (RFE)

The recursive feature elimination (RFE) algorithm based on a support vector machine classifier is a wrapper method that uses the feature weights of the SVM training process to perform backward feature elimination [11]. A linear kernel machine was used with parameters set to Matlab® Bioinformatic Toolbox defaults. RFE ranking criterion $\left| W_p^2 \right|$ for feature $p$ is calculated from equation (8) which depends on the weighted sum of support vectors that define the separation between groups as optimized by the SVM for every sample $n$.

$$W = \sum_n \alpha_n y_n x_n \tag{8}$$

where $\alpha_n$ is the sample weight, $y_n$ is the p-dimensional training sample, and $x_n$ is the group label. The samples with non-zero weights are the support vectors. The features with the lowest ranking, thus contributing less to group separation, are removed iteratively. This procedure stops when the optimum subset size ($p_{opt}$) is reached.

### 5.2.3 Genetic Algorithm (GA)

This is a wrapper method that uses a simple genetic algorithm to search the space of possible feature subsets. The genetic algorithm is a global search method based on the mechanics of natural selection and population genetics and has been successfully applied to BCI problems [12]. It starts with the generation of an initial random population, where each individual (or chromosome) encodes a candidate solution to the feature subset selection problem. The individual is constituted by various genes represented by a binary vector of dimension equal to the total number of features. A fitness measure is evaluated for each individual after which, selection and genetic operators (recombination and mutation) are applied. In this study, the classification accuracy of a linear discriminant classifier was the fitness measure. Starting with conventional values, the parameter calibration was based on empirical tests executed beforehand. The population size was 30, the number of generations was 50, the selection rate was 0.5, elite children (chromosomes that pass unchanged, without mutation, to the next generation) were 2, the mutation rate was set to 0.05, and the crossover probability was 0.5. The selection of chromosomes to be recombined was done by tournament selection (with tournament size equal to 2). Crossover and mutation were uniform. The most frequently selected features within the inner loop up to the number of features $p_{opt}$ were tested with the validation data.

### 5.2.4 Results

According to Table 4, the proposed algorithm achieved the best average performance among the tested algorithms. AGV achieved the lowest average error and standard deviation for the smallest subsets. RFE, GA and RELIEF algorithms ranked next in error increasing order. For Subject B, RFE achieved lower classification error than AGV. However, the former selected more variables. Since the AGV ranking algorithm is a filter method, it was alternatively tested with a SVM classifier (the same employed in RFE) for validation. The error average was 6.96 % thus, lower than RFE's and still maintaining a small number of variables selected. The classification error vs. number of features average curves for Subject B are presented in Figure 1. The minimum classification error achieved by backward elimination of task irrelevant features (or forward addition of task relevant features) establishes the optimum number of features to be selected. All the remaining features are deemed task relevant. Although AGV and RFE algorithms have achieved comparable minimum classification errors, the former selected subsets considerably smaller. RELIEF obtained the highest classification errors. Figure 3 does not present results for GA since it produces generation dependent feature subsets rather than nested ones.

**Table 4.** Comparison of the tested variable down-selection algorithms. The lowest classification error for each subject was printed in **bold.**

| Subject | AGV | | | RFE | | | GA | | | RELIEF | | |
|---|---|---|---|---|---|---|---|---|---|---|---|---|
| | Error (%) | SD[1] | $p_{opt}$[2] | Error (%) | SD | $p_{opt}$ | Error (%) | SD | $p_{opt}$ | Error (%) | SD | $p_{opt}$ |
| A | **25.1** | 7.2 | 21 | 27.4 | 9.5 | 103 | 31.6 | 9.7 | 58 | 34.3 | 9.9 | 55 |
| B | 9.6 | 6.0 | 37 | **7.6** | 5.3 | 76 | 13.6 | 5.8 | 58 | 16.0 | 6.6 | 53 |
| C | **28.5** | 7.4 | 9 | 33.2 | 7.8 | 78 | 37.6 | 10.0 | 58 | 36.4 | 9.2 | 62 |
| D | **19.9** | 7.1 | 20 | 29.8 | 7.4 | 93 | 28.7 | 8.6 | 59 | 30.2 | 8.3 | 70 |
| E | **9.2** | 4.8 | 8 | 14.3 | 6.4 | 101 | 17.2 | 6.9 | 58 | 21.8 | 7.8 | 89 |
| Mean | **18.5** | 6.5 | 19 | 22.5 | 7.3 | 90 | 25.7 | 8.2 | 58 | 27.7 | 8.3 | 66 |

[1] Error standard deviation.
[2] Mean number of features selected.

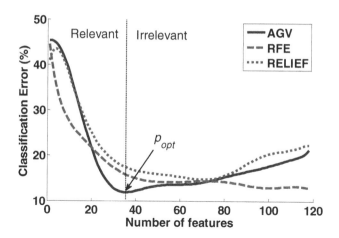

**Fig. 4.** Classification error vs. number of features average curve, for Subject B, calculated through the across-group variance (AGV) with a linear discriminant classifier, recursive feature elimination (RFE) and RELIEF algorithms. The marker $p_{opt}$ defines the optimum subset size calculated by AGV and separates relevant from irrelevant features.

## 6   Discussion and Conclusions

This study proposes a new algorithm for variable down-selection in high-dimensional variable spaces for discrimination of movement imagery tasks. This algorithm was applied to both synthetic and EEG data. Additionally, the introduced method was compared with other algorithms often employed in BCI research.

The results with synthetic datasets reveal that: (i) at least 83% of the relevant (either strong or weakly relevant) variables were selected for the optimal subsets, (ii) 100% of the predominant variables were selected for all optimal subsets; (iii) all the discriminations reached less than 12.9 % cross-validation error and were highly significant. The 3rd and 4th rows on Table 1 show that this approach is applicable for lower dimensional variable spaces ($p=40$) as well as high-dimensional ones (p=79 on 1st and 2nd rows).

AGV performed dimensionality reduction on EEG data with classification errors as low as 1.4 %. The variable selection frequencies detected on central channels, for Subject E on Figure 2, seem to manifest the lateral activation of motor-related brain regions on movement imagery preparation [3]. The relevance of parietal and occipital channels suggest simultaneous engagement of visual association areas in response to visual stimulation (i.e. cue). The best time window for ERP classification appears to be the first second after cue presentation (Figure 1) and movement-related potentials were detected as soon as 100 ms after cue presentation (Figure 2b). When power ratios were extracted from EEG data, the features in alpha frequency band (8-12 and 10-14 Hz) were mostly selected from central channels for discrimination of left vs. right hand movement imageries for Subject D. This finding is likely due to the mu rhythm contralateral reactivity to hand movement imagery [17]. Therefore, besides reducing the number of variables efficiently as well as decreasing the discrimination error, the proposed algorithm also achieves plausible results since well-known brain responses to motor imagery tasks were identified.

Among the algorithms under comparison, AGV's error curve seems most suitable for feature down-selection since it achieved the best accuracy with the fewest features. RFE ranked second best in our algorithm comparison. Although RFE's classification errors were comparable to AGV's for two subjects (A and B) the subset sizes were much larger. On the average, RFE selected 90 features out of 118, while AGV only kept 19. As suggested in [21], SVM tends to calculate similar weights ($W$) for highly correlated[3] features thus, eliminating, or keeping them simultaneously during the RFE selection. The low-slope error increase seen in the beginning of RFE backward elimination (Figure 3) suggests that the variables being discarded are reasonably irrelevant for task discrimination. The latter elimination of redundant features might promote premature elimination of more relevant ones and mislead the subset optimization. On the other hand, AGV ranks each variable based on its covariance with the truncated component space rather than its covariance with other variables. Therefore, linear correlations between variables are implicitly considered but not determinant for variable selection. Although, both RFE and AGV seem able to select strongly relevant variables, the latter seems more capable of dealing with the weakly relevant variables. Moreover, as expected for filter methods, AGV ran 8 times faster than RFE on the average. The GA results were not comparable with AGV's. Moreover, on all tested subjects, GA keeps approximately half of the original variables and the generalization error is high, thus suggesting that a premature convergence phenomenon is occurring. As expected, RELIEF achieved the poorest classification accuracy because, unlike AGV, it evaluates feature relevance independently of other features and thus is incapable of dealing with redundant features. Another drawback is that RELIEF is highly susceptible to select irrelevant channels in the presence of outliers in noisy channels.

Our findings show a novel mean to down-select variables in BCI that accomplishes both discriminative power and dimensionality reduction better than other algorithms in common use. Such a strategy is valuable in decreasing the computational complexity of neural prosthetic applications.

**Acknowledgements.** This work was supported by the portuguese Foundation for Science and Technology (FCT) fellowship SFRH/BD/21529/2005 and Center Algoritmi (NSD), the grants K25NS061001 (MK) and K02MH01493 (SJS) from the National Institute of Neurological Disorders And Stroke (NINDS) and the National Institute of Mental Health (NIMH), the Pennsylvania Department of Community and Economic Development Keystone Innovation Zone Program Fund (SJS), and the Pennsylvania Department of Health using Tobacco Settlement Fund (SJS).

# References

1. Wolpaw, J.R., McFarland, D.J., Vaughan, T.M.: Brain–Computer Interface Research at the Wadsworth Center. IEEE Transactions On Rehabilitation Engineering 8, 222–226 (2000)
2. Pfurtscheller, G., Aranibar, A.: Evaluation of event-related desynchronization (ERD) preceding and following voluntary self-paced movements. Electroencephalogr. Clin. Neurophysiol. 46, 138–146 (1979)

---

[3] Redundant variables that may be considered weakly relevant.

3. Babiloni, C., et al.: Human movement-related potentials vs desynchronization of EEG alpha rhythm: A high-resolution EEG study. NeuroImage 10, 658–665 (1999)
4. Donchin, E., Spencer, K.M., Wijesinghe, R.: The Mental Prosthesis: Assessing the Speed of a P300-Based Brain–Computer Interface. IEEE Transactions on Rehabilitation Engineering 8, 174–179 (2000)
5. Duda, R.O., Hart, P.E., Stork, D.G.: Pattern Classification. Wiley Interscience, Hoboken (2000)
6. John, G.H., Kohavi, R., Pfleger, K.: Irrelevant Features and the Subset Selection Problem. In: Cohen, W.W., Hirsh, H. (eds.) Proc. 11th Int. Conf. Machine Learning, pp. 121–129. Morgan Kaufmann, San Francisco (1994)
7. Yu, L., Liu, H.: Efficient feature Selection via Analysis of Relevance and Redundancy. Journal of Machine Learning 5, 1205–1224 (2004)
8. Dillon, W.R., Mulani, N., Frederick, D.G.: On the use of component scores in the presence of group structure. Journal of Consumer Research 16, 106–112 (1989)
9. Kamrunnahar, M., Dias, N.S., Schiff, S.J.: Model-based Responses and Features in Brain Computer Interfaces. In: Proc. 30th IEEE EMBC, pp. 4482–4485. IEEE Press, Vancouver (2008)
10. Millán, J., Franzé, M., Mouriño, J., Cincotti, F., Babiloni, F.: Relevant EEG features for the classification of spontaneous motor-related tasks. Biol. Cybern. 86, 89–95 (2002)
11. Schröder, L.T., Weston, T., Bogdan, J., Birbaumer, M., Schölkopf, N.B.: Support vector channel selection in BCI. IEEE Trans. Biomed. Eng. 51, 1003–1010 (2004)
12. Schröder, M., Bogdan, M., Rosenstiel, W., Hinterberger, T., Birbaumer, N.: Automated EEG feature selection for brain computer interfaces. In: Proc. 1st IEEE EMBS Neural Eng., pp. 626–629. IEEE Press, Capri Island (2003)
13. Blankertz, B., et al.: The BCI competition III: validating alternative approaches to actual BCI problems. IEEE Trans. Neural Syst. Rehabil. Eng. 14, 153–159 (2006)
14. Lai, C., Reinders, M.J.T., Wessels, L.: Random subspace method for multivariate feature selection. Pattern Recognition Letters 27, 1067–1076 (2006)
15. Luck, S.J.: An Introduction to the Event-Related Potential Technique. The MIT Press, Cambridge (2005)
16. Wolpaw, J.R., McFarland, D.J.: Control of a two-dimensional movement signal by a noninvasive brain-computer interface in humans. Proc. Nat. Acad. Sci. U.S.A. 101, 17849–17854 (2004)
17. Wolpaw, J.R., Birbaumer, N., McFarland, D.J., Pfurtscheller, D.J., Vaughan, T.M.: Brain-computer interfaces for communication and control. Clinical Neurophysiology 113, 767–791 (2002)
18. Schiff, S.J., Sauer, T., Kumar, R., Weinstein, S.L.: Neuronal spatiotemporal pattern discrimination: the dynamical evolution of seizures. Neuroimage 28, 1043–1055 (2005)
19. Flury, B.: A first course in multivariate statistics. Springer, New York (1997)
20. Jolliffe, I.T.: Principal Component Analysis. Springer, New York (2002)
21. Xie, Z., Hu, Q., Yu, D.: Improved feature selection algorithm based on SVM and correlation. In: Wang, J., Yi, Z., Żurada, J.M., Lu, B.-L., Yin, H. (eds.) ISNN 2006. LNCS, vol. 3971, pp. 1373–1380. Springer, Heidelberg (2006)

# Appendix

The appendix illustrates the instructions of the across-group variance (AGV) algorithm by means of pseudocode. All program variables were printed in bold and the correspondent mathematical variables were provided in parenthesis.

```
Input Data:
// FeatMatrix - Feature Matrix (Y)
// BetCovMat - Between group covariance matrix (Ψ_between)
// TrunCrit - Truncation criterion (δ)
Output Data:
// VarRank - Vector with ranked variables


[U,S,Vᵀ] = SingularValueDecomposition(FeatMatrix);
lambda = diagonal(S²); //eigenvalue calculation
// no. of principal components
nPC = min(no.columns(FeatMtrix), no.rows(FeatMtrix)-1)
// no. of variables
nV = no.columns(FeatMtrix)
for i=1 to nPC
    AGV(i) = V(i)ᵀ×BetCovMat ×V(i) / lambda(i)
end for
rAGV = sort AGV in descending order
%rAGV = 100*rAGV / sum all elements in rAGV
K = number of first %rAGV elements whose cumulative
sum > TrunCrit
AgvCovMat = empty square matrix of nV size
for i=1 to K
    AgvCovMat = AgvCovMat + AGV(i)×V(i)×V(i)ᵀ
end for
// Across-group variance of variables
AgvVar = diagonal(AgvCovMat)
VarRank = sort variable indexes by AgvVar descending
order
```

# Effect of a Simulated Analogue Telephone Channel on the Performance of a Remote Automatic System for the Detection of Pathologies in Voice: Impact of Linear Distortions on Cepstrum-Based Assessment - Band Limitation, Frequency Response and Additive Noise

Rubén Fraile[1,*], Nicolás Sáenz-Lechón[1], Juan Ignacio Godino-Llorente[1], Víctor Osma-Ruiz[1], and Corinne Fredouille[2]

[1] Department of Circuits & Systems Engineering, Universidad Politécnica de Madrid
Carretera de Valencia Km 7, 28031 Madrid
rfraile@ics.upm.es, nicolas.saenz@upm.es
igodino@ics.upm.es, vosma@ics.upm.es
[2] Laboratoire Informatique d'Avignon, Université d'Avignon
339 chemin des Meinajaris, 84911 Avignon Cedex 9, France
corinne.fredouille@univ-avignon.fr

**Abstract.** Advances in speech signal analysis during the last decade have allowed the development of automatic algorithms for a non-invasive detection of laryngeal pathologies. Performance assessment of such techniques reveals that classification success rates over 90 % are achievable. Bearing in mind the extension of these automatic methods to remote diagnosis scenarios, this paper analyses the performance of a pathology detector based on Mel Frequency Cepstral Coefficients when the speech signal has undergone the distortion of an analogue communications channel, namely the phone channel. Such channel is modeled as a concatenation of linear effects. It is shown that while the overall performance of the system is degraded, success rates in the range of 80 % can still be achieved. This study also shows that the performance degradation is mainly due to band limitation and noise addition.

## 1 Introduction

The social and economical evolution of developed countries during the last years has led to an increased number of professionals whose working activity greatly depends on the use of their voice. It has been reported that this number has reached one third of the total labor force and, in parallel, that approximately 30 % of the population suffers from some kind of voice disorder along their lives [1]. In this context, methods for

---

* This research was carried out within projects funded by the Ministry of Science and Technology of Spain (TEC2006-12887-C02) and the Universidad Politécnica de Madrid (AL06-EX-PID-033). The work has also received support from European COST action 2103.

A. Fred, J. Filipe, and H. Gamboa (Eds.): BIOSTEC 2009, CCIS 52, pp. 173–186, 2010.
© Springer-Verlag Berlin Heidelberg 2010

objective assessment of vocal function have a relevant interest [2] and, among them, speech analysis has the additional features of being non-invasive and allowing easy data colection [3].

Speech assessment for the detection of pathologies has been traditionally realised through the analysis of global distortion and noise measurements taken from records of sustained vowels [2] [3]. Classification performances over 90 % in terms of success rates have been reported for automatic pathology detection systems based on such parameters (e.g. [4]). Recently, alternative approaches based on Mel-frequency Cepstral Coefficients (MFCC) with similar performance [5] have also been proposed. These approaches have the advantage of relaying on robust parameters whose calculation does not require prior pitch estimation [6]. Moreover, analysis in cepstral domain for this application is further justified by the presence of in the cepstrum information about the level of noise [7]. Additional reasons that support the specific processing involved in MFCC calculation can be found in [6], [8] and [9].

From another point of view, remote diagnosis is one of the foreseen applications of telemedicine [10]. In this context, the use of a non-invasive diagnosis technique such as speech analysis is well suited to that application. Moreover, since the analogue wired telephone network is one of the most mature and widely extended communications infrastructures, it seems reasonable to expect that it will become one of the supporting technologies for that medical service. However, the feasibility of such application will heavily depend on the ability of voice analysis to extract significant information from speech signals even after the distortion caused by the communications channel.

Up to now, some preliminary works on this issue have been carried out and published. In the first place, pathology detection on voice transmitted over the phone has been shown to experiment a performance degradation figure around 15 % when detection is based on traditional acoustic parameters [11]. Secondly, the impact of several speech coders on voice quality has been studied, but without regarding the additional degradation introduced by communications channels [12]. Last, the problem of analysing the effect of the analogue telephone channel on a MFCC-based system for pathology detection has also been approached [13], but without differentiating among the different distortions introduced by the channel and without accounting for noise distortion.

Considering all above-mentioned aspects, that is, the adequateness of MFCC for automatic pathology detection and the interest of analyzing the impact of the analogue telephone channel on speech quality, this paper offers a detailed report on the effect of the distortions introduced by the telephone channel on the performance of automatic pathology detection based on MFCC. More specifically, a study more complete than that of [13] is provided in which the effects of band limitation, frequency response of the channel and additive noise are analysed separately. This way, the results of the study are useful, not only for remote diagnosis applications such as the one described before, but also for setting minimum conditions, in terms of bandwidth and noise levels, for speech recording in clinical applications.

The rest of the paper is organised as follows: Sect.2 contains the specific formulation of MFCC and the values for related parameters used in the study, Sect.3 describes the model of telephone channel that has been considered, in Sect.4 the database, classifier and procedure used for the experiment are detailed, results are reported in Sect.5 and, last, Sect.6 is dedicated to the conclusions.

## 2   MFCC Formulation

As argued in [6], the variability of the speech signal is specially relevant in the presence of pathologies, thus justifying the use of short-term signal processing. A framework for such short-term processing in the case of speech is provided in [14]. Within this framework, the short-time MFCC definition given in [9], which is slightly different from the original proposal in [15] but it has an easier interpretation, is used:

$$
c_p[q] = \frac{1}{M+1} \sum_{k=1}^{M} \log \left| \widetilde{S}_p(k) \right| \cdot \cos \left( \frac{\pi k}{M+1} \cdot q \right) , \tag{1}
$$

where $p$ is the frame index, $q$ is the index of the MFCC that ranges from 0 to $M$, $M$ is the number of Mel-band filters used for spectrum smoothing and $\left| \widetilde{S}_p(k) \right|$ is the estimate of the spectral energy of the speech signal in the $k^{\text{th}}$ Mel band. Specifically:

$$
\widetilde{S}_p(k) = \sum_{f_i^m \in I_k^m} \left( 1 - \frac{\left| f_i^m - F^m \cdot \frac{k}{M+1} \right|}{\Delta f^m / 2} \right) \cdot \left| S_p(i) \right| , \tag{2}
$$

where $S_p(i)$ is the $i^{\text{th}}$ element of the short-time discrete Fourier transform of the $p^{\text{th}}$ speech frame, $f_i^m$ is its associated Mel frequency,

$$
I_k^m = \left[ F^m \cdot \frac{k-1}{M+1}, F^m \cdot \frac{k+1}{M+1} \right] \tag{3}
$$

is the $k^{\text{th}}$ band in Mel-frequency scale, $\Delta f^m / 2$ is the width of these Mel bands and $F^m$ is the maximum frequency in Mel domain, which corresponds to half the sampling frequency of the speech signal. The frequency transformation that allows passing from linear to Mel scale is:

$$
f^m = 2595 \cdot \log_{10} \left( 1 + \frac{f}{700} \right) . \tag{4}
$$

For the herein reported application, speech frame duration has been chosen to be 20 ms, which allows capturing the spectral envelope of speech for fundamental frequencies above 50 Hz, thus covering the cases of both male and female voices [3]. Overlap between consecutive frames was 50 %. The number of Mel band filters $M$ has been made equal to 31, since that value has shown to exhibit good performance [9] and vectors of 21 MFCC, that is $q \in [0, 20]$, have been used as feature vectors for each speech frame.

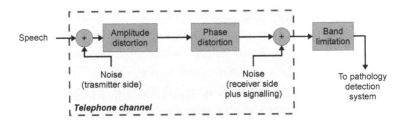

**Fig. 1.** Block diagram of the analogue telephone channel model

## 3   Telephone Channel Model

The task of assessing the impact of the analogue telephone channel on the performance of a MFCC-based pathology detector was done bearing in mind the same modeling methodology as in [13]. Such methodology comprises the main aspects of the model proposed in [16]. Namely, the linear effects of the channel have been assumed to be the dominant ones: amplitude, phase and noise distortions. Normative restrictions on amplitude and phase distortion imposed by [17] have also been taken into account. The block diagram of the overall channel model is drawn in Fig.1 and it consists of the following elements:

1. *Amplitude distortion*: Its limits are normalised in [17] for the 300–3400 Hz band and no restrictions are imposed outside that band.
2. *Phase distortion*: Its limits for the 300–3400 Hz band are also specified in [17] and they are mainly referred to the phase effects at the edges of that band.
3. *Noise distortion*: This distortion can be split in noise at the transmitter side, which undergoes the same amplitude and phase distortion as the speech signal, and noise at the receiver side that does not suffer that distortion.
4. *Bandwidth limitation*: This has to be carried out as the first stage of the detector due to the uncertainty about the distortion out of the 300–3400 Hz band. Another reason for this limitation is that the telephone network adds some signalling in the 0–300 Hz band [17].

### 3.1   Amplitude Distortion

The analogue telephone channel acts as a band-pass filter. Attenuation of high frequencies comes from the low-pass behaviour of the transmission line while attenuation of low frequencies (below 300 Hz) allows the use of out-of-band signalling. Limits recommended by [17] for the amplitude response of the channel are represented as continuous lines in Fig.2.

The simulation of the amplitude and phase distortion of the channel has been realised separately, as proposed in [16] and illustrated in Fig.1. Within such a setup, the amplitide distortion has been modeled as a band-pass linear-phase system, hence achieving null phase distortion in this stage, implemented by means of a symmetric FIR filter. Bearing in mind restrictions in [17], a 176-order filter has been designed that has the frequency response plotted in Fig. 2 (dashed line).

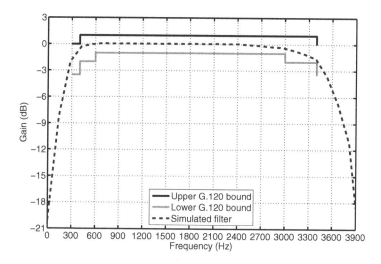

**Fig. 2.** Amplitude response of the channel: restrictions (continuous line) and model (dashed line)

## 3.2 Phase Distortion

Regarding phase distortion, [17] imposes limits to group delay variations within the pass band. Namely, different limits are specified for the low and high parts of the band, as represented by the thick lines in Fig.3. A simple procedure to obtain an all-pass filter that achieves phase distortion around certain frequencies is to design an IIR filter having zeros and poles in the frequencies at which phase distortion has to be greatest. For the filter to be all-pass, zero and pole modules must be symmetric with respect to the unit radius circle of the z-plane. Specifically, the implemented filter corresponds to the following transfer function:

$$H\left(z\right) = H_{\mathrm{ap}}\left(z; f_{\mathrm{low}}\right) \cdot H_{\mathrm{ap}}\left(z; f_{\mathrm{high}}\right) \quad, \tag{5}$$

being:

$$H_{\mathrm{ap}}\left(z; f_x\right) = \left[\frac{1 - rz^{-1}e^{j2\pi\frac{f_x}{f_{\mathrm{s}}}}}{1 - \frac{1}{r}z^{-1}e^{j2\pi\frac{f_x}{f_{\mathrm{s}}}}} \cdot \frac{1 - rz^{-1}e^{-j2\pi\frac{f_x}{f_{\mathrm{s}}}}}{1 - \frac{1}{r}z^{-1}e^{-j2\pi\frac{f_x}{f_{\mathrm{s}}}}}\right]^2 \quad, \tag{6}$$

where $r = 1.01$, $x$ may be either "low" or "high", $f_{\mathrm{low}} = 250\,\mathrm{Hz}$, $f_{\mathrm{high}} = 3450\,\mathrm{Hz}$ and $f_{\mathrm{s}}$ is the sampling frequency of the speech record. The obtained frequency-dependent group delay is depicted in Fig.3. It can be noticed that the maximum phase distortion happens at the limits of the pass band of the FIR filter, as specified by [17].

## 3.3 Band Limitation

The above-mentioned specifications for the frequency response of the telephone channel only cover the band between 300 and 3400 Hz, thus leaving uncertainty as for the distortion that the speech signal undergoes out of that band. In addition, as specified

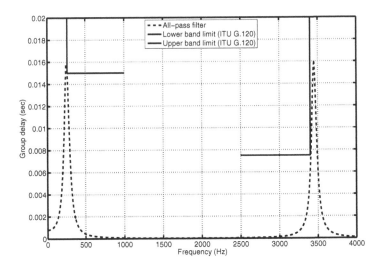

**Fig. 3.** Phase response of the channel: restrictions (continuous line) and model (dashed line)

by [17], out-of-band signalling is allowed in the 0–300 Hz band. This adds the possibility of narrow-band noise distortion to the lack of normalisation of the response of the channel within that band. These facts make it logical to perform a band limitation of the speech signal prior to its analysis, as indicated in Fig.1. In this way, only the 300–3400 Hz band of the signal is further processed. This band limitation procedure is of common use in other speech processing applications [18].

The band limitation has a direct effect on the computation of MFCC. Specifically, the $\Delta f^m$ parameter in (2) depends on both the bandwidth of the signal and the number of mel-band filters used for MFCC calculation. When limiting the frequency band of the signal, two strategies may be followed in the subsequent analysis: either maintaining the number of mel bands, hence reducing $\Delta f^m$, or keeping $\Delta f^m$ approximately equal by reducing the number of bands. The performance of these two options will be analysed in Sect.5.

### 3.4   Additive Noise

The fourth modeled distortion of the telephone channel is noise. Although more complex models exist for telephone noise modelling [16], herein a simpler approach, similar to [18], has been chosen. Namely, noise has been considered to be additive and white Gaussian (AWGN). Yet, a differentiation has been made between noise that suffers the same channel effects as the speech signal, accounting for the transmitter side, and noise that does not pass through the channel, hence the receiver side. In both cases, signal-to-noise ratio (SNR) has been controlled by tuning the power of noise to the specific power of each processed signal.

# 4    Simulation Procedure

## 4.1    Database

All the herein reported results have been obtained using a well-known database distributed by Kay Elemetrics [19]. More specifically, the utilized speech records correspond to sustained phonations of the vowel /ah/ (1 to 3 s long) from patients with normal voices and a wide variety of organic, neurological, traumatic, and psychogenic voice disorders in different stages (from early to mature). The subset taken corresponds to that reported in [20] and it corresponds to 53 records from healthy patients (normal set) and 173 to ill patients (pathological set).

The speech samples were collected in a controlled environment and sampled at sampling rates equal to either 50 or 25 kHz with 16 bits of resolution. A down-sampling with a previous half band filtering has been carried out over some registers in order to adjust every utterance to the sampling rate of 25 kHz.

## 4.2    Classifier

The chosen classifier consists of a 3 layered Multilayer Perceptron (MLP) neural-network [21] with 40 hidden nodes having logistic activation functions (as in [5]) and two outputs with linear activations. The use of two linear outputs allows obtaining two values for each speech frame, characterised by its MFCC vector $c_p$. In the training phase of the MLP, one output is trained to produce a value of "0" for pathological voice frames and "1" for normal voice frames, while the other output is trained to produce a "0" for normal data and a "1" for pathological data. In the testing phase, each output value is an estimation of the likelihood of that frame to be either normal $L_{nor}(c_p)$ (first output) or pathological $L_{pat}(c_p)$ (second output).

These likelihoods, whilst not probabilities, give an idea of how feasible is that any particular frame corresponds to each class or set. Their precise values depend on the value of the feature vector components and on the learned parameters of the MLP. Since the orders of magnitude of both likelihoods may significantly differ, it is more usual to compute log-likelihoods; the classification decision for the $p^{th}$ frame is, then, based on the difference between log-likelihoods, as described in [22]:

$$\log\left[L_{nor}\left(c_p\right)\right] - \log\left[L_{pat}\left(c_p\right)\right] > \theta \ . \tag{7}$$

If the previous condition is met, then the speech frame is classified as normal, if not, it is considered pathological. In ideal conditions, that is, if the likelihoods could be perfectly estimated by the classifier, then the value for the threshold $\theta$ should be $\theta = 0$. In practice, however, this is not the case and the choice of $\theta$ helps to make the decision system more or less conservative. Nevertheless, since decisions in this case should not be taken at the frame level, but at the record level, a mean log-likelihood difference is computed and this is the value actually compared to the threshold:

$$\frac{1}{N_{frames}} \cdot \sum_{p=1}^{N_{frames}} \log\left[L_{nor}\left(c_p\right)\right] - \log\left[L_{pat}\left(c_p\right)\right] > \theta \ , \tag{8}$$

where $N_{frames}$ is the number of frames of the speech record.

### 4.3 Testing Protocol

The testing of each detection scheme consists of an iterative process. Within each iteration 70 % of the available speech records have been randomly chosen for training the classifier, that is, to estimate the likelihood functions mentioned above. Among the remaining 30 % of records, one third (10 %) have been used for cross-validation during training in order to get an objective criterion for finishing the training phase [21]. The rest (20 %) have been used for testing. For each testing record, a decision according to the previously described framework has been taken. Last, with the decisions corresponding to all the testing records, misclassification rates for $\theta \in [-10, 10]$ and the corresponding iteration have been computed. Forty five iterations, with independently chosen training, validation and testing sets, have been repeated.

## 5   Results

There are several performance indicators for the evaluation of detection systems. A summary of the most typically used for speech applications can be found in [22]. Among these indicators, the DET plot [23] and the Equal Error Rate (EER) have been chosen for this study as graphic and quantitative indicators, respectively. For the DET plot, *false alarm* has been defined as the event of detecting a normal voice as pathological, while *miss* means the event of detecting a pathological voice as normal. In this context, the DET curve represents the relationship between miss and false alarm rates as the threshold $\theta$ in (7) and (8) changes and the EER is the point at which the DET curve crosses the diagonal of the graph, i.e. the value of miss and false alarm rates when $\theta$ is tuned so that they coincide. In all experiments, the results have been computed both at frame and record levels, corresponding to (7) and (8).

### 5.1   Effect of Band Limitation

As indicated in Fig.1, the first step in the speech analysis after transmission through the telephone channel is band limitation. This involves taking only the spectral energy between 300 Hz and 3400 Hz for spectrum smoothing using the Mel filter bank. Such bandwidth reduction can be achieved in two different ways. The first of them consists in maintaining the number of filters (M=31), thus reducing their individual widths. The second option, instead, involves maintaining the filter width by reducing the number of filters. It can be checked that if the band is split in 16 Mel bands (M=16), very similar Mel-filter widths are achieved. However, this means reducing the number of MFCC from 21 ($q \in [0, 20]$) to 16 ($q \in [0, 15]$ ), since $q < M$ due to the periodic nature of the discrete-time Fourier transform.

In Fig.4, the different performance distributions obtained for both alternatives are represented by means of box plots. The left graph corresponds to frame-level decissions and the right one to record-level decissions. In each graph, the first box ("Full bandwidth") corresponds to the original voices, the second one ("BW 31"), to band-limited voices analysed with 31 Mel filters and the third one ("BW 16"), to bandlimited voices analysed with 16 Mel filters. The results indicate, on the one hand, that a significant increase in EER is produced by the band limitation inherent to the telephonic

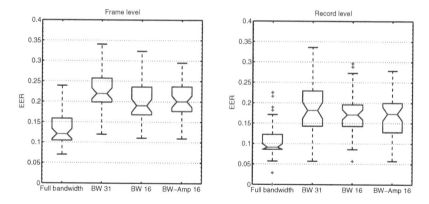

**Fig. 4.** Box plots of obtained EER distributions. Notch center in each box indicates the median value, limits of the box indicate the first and third quartiles, the dashed lines have a length equal to 1.5 times the interquartile distance and crosses indicate outlier results. The width of the notch indicates the 95% confidence interval for the median value.

channel. Such observation is complementary to results reported in [24], where it was shown that the most relevant band for dysphonia detection was between 0 and 3000 Hz. The herein reported results indicate that there is significant information within the lower part of that band, that is, below 300 Hz. On the other hand, the plots in Fig.4 also indicate that maintaining the size of the Mel-bands gives similar results to keeping the number of bands, but with the advantage of lower dimensionality. This results in the performance of the classifier exhibiting less variability and, consequently, this will be the preferred option for the next experiments.

### 5.2   Effect of Amplitude Distortion

In [13], it was shown that the amplitude distortion of the speech signal has the effect of performing a quasi-linear transformation in the MFCC values. Taking this into account and recalling (1), the transformed MFCC can be written as:

$$\tilde{c}_p[q] = A + c_p[q] + \frac{1}{M+1} \sum_{k=1}^{M} \log |\xi(k)| \cdot \cos\left(\frac{\pi k}{M+1} \cdot q\right), \qquad (9)$$

where $A$ is a constant that depends on the amplitude response of the filter and $\xi(k)$ is a variable term that depends on the relation between the spectrum of the speech signal and the response of the filter within the $k^{\text{th}}$ Mel-frequency band.

The box on the right of Fig. 4 ("BW-Amp 16") represents the obtained EER distribution when the training stage of the classifier is done with the original speech records, with band limitation and M=16, and the testing is done with the outputs of filtering those records with the filter corresponding to Fig.2. It can be noticed that the limited distortion allowed within the 300-3400 Hz band by ITU specifications [17] has the consequence of not affecting greatly the average performance of the system, it just widens the distribution. Yet, 75 % of the experiments still yield EER values below 20 %.

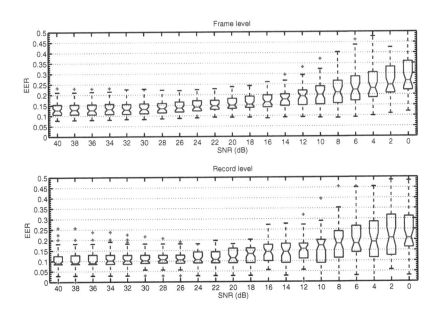

**Fig. 5.** Box plots of EER distributions obtained with noisy signals having SNR values from 40 dB down to 0 dB

### 5.3    Effect of Phase Distortion

As proven in [13], the computation of MFCC involves calculation of the modulus of the discrete Fourier transform of the signal, as indicated in (1). Consequently, MFCC are insensitive to phase distortions and there is no need to analyse this effect of the channel.

### 5.4    Effect of Noise Distortion

The last effect of the channel to be analysed is noise distortion. This has been modelled as AWGN with different power levels. The effect of noise was analysed both independently and in conjunction with the band-limiting scheme explained before. As for the independent analysis, the obtained distributions of EER for different levels of signal-to-noise ratio (SNR) are plot in Fig.5. Specifically, even SNR values from 40 dB down to 0 dB were simulated. In all cases, the training was done with the clean records and the testing with the noisy ones. The plot indicates that for SNR values around 30 dB the overall performance does not degrade significantly. However, if SNR falls below 20 dB, the median error rate at record level begins to grow significantly above 10 % and distributions tend to widen, hence indicating less robustness of the system. While the effect of noise in the case of the telephone channel is not isolated from other distortions, these results are also useful for determining the minimum required quality of speech recordings for pathology assessment. Under the AWGN assumption, SNR values below 20 dB seem not to be acceptable for this application.

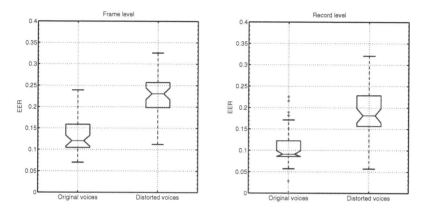

**Fig. 6.** Box plots of obtained EER distributions with original voices analysed with 31 Mel-band filters and voices undergoing all the modelled channel distortions analysed with 16 Mel-band filters

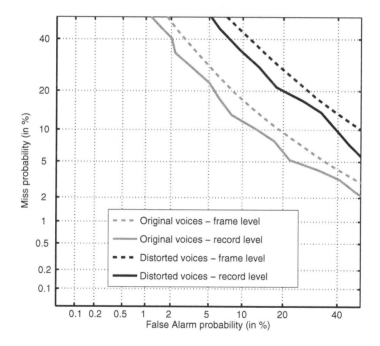

**Fig. 7.** DET plot of the pathology detection system for the original speech records (gray) and those with simulated telephone channel distortion (black)

The figure of 20 dB has been considered as a reference for the combined analysis of band limitation and amplitude and noise distortions. It has been found that, coherently with above-reported results, there is not any significant difference between adding the noise previously to the amplitude distortion (transmitter side) or after (receiver side).

For the next experiment, noise addition has been split in two parts: half of the power prior to amplitude distortion and half of the power after. Figure 6 shows the box plots of EER distributions for the original voice records and those obtained after the three distortions (band limitation, amplitude distortion and noise addition). On the whole, the median EER suffers a degradation slightly below 10 %, yielding a median success classification rate over 80 % at the record level with a 95 % confidence level. A DET plot of the sample mean error rate along the forty five experiments in each case is depicted in Fig.7. While the mean values are greater than the median values, due to the presence of outlier results, the general behaviour is the same: a degradation below 10 % at record level and an average EER that still reaches 80 %.

## 6    Conclusions

Within this paper, the performance of a speech pathology detector based on Mel Frequency Cepstral Coefficients when the speech signal has undergone the distortion of an analogue communications channel has been analysed. Namely the telephone channel has been modelled as a concatenation of linear effects: band limitation, amplitude distortion, phase distortion and noise addition. It has been shown that while the overall performance of the system is degraded, success rates over 80 % can still be achieved. This study also reveals that the performance degradation is mainly due to band limitation and noise addition. Amplitude distortion, if complying with norm [17], has little impact and phase distortion has no impact at all.

As for the most relevant sources of distortion, it has been shown that the loss of information in the 0-300 Hz band makes performance to decrease significantly. Additionally, the effect of noise degradation becomes very relevant for values of SNR below 20 dB. For SNR equal to 20 dB, and considering bandwidth limitation and amplitude distortion too, success classification rate can reach 80 %. This figure is better than the results reported in [11].

The whole set of reported results allow to conclude, in the first place, that remote pathology detection on speech transmitted through the analogue telephone channel seems feasible and, in the second place, that MFCC parameterization can provide a robust method for assessing the quality of degraded speech signals.

## References

1. Södersten, M., Lindhe, C.: Voice ergonomics - an overview of recent research. In: Berlin, C., Bligard, L.O. (eds.) Proceedings of the 39th Nordic Ergonomics Society Conference (2007)
2. Umapathy, K., Krishnan, S., Parsa, V., Jamieson, D.G.: Discrimination of pathological voices using a time-frequency approach. IEEE Transactions on Biomedical Engineering 52, 421–430 (2005)
3. Baken, R.J., Orlikoff, R.F.: Clinical Measurement of Speech and Voice. Singular Publishers, San Diego (2000)
4. Boyanov, B., Hadjitodorov, S.: Acoustic analysis of pathological voices. A voice analysis system for the screening of laryngeal diseases. IEEE Engineering in Medicine and Biology 16, 74–82 (1997)

5. Godino-Llorente, J.I., Gómez-Vilda, P.: Automatic detection of voice impairments by means of short-term cepstral parameters and neural network based detectors. IEEE Transactions on Biomedical Engineering 51, 380–384 (2004)
6. Fraile, R., Godino-Llorente, J.I., Sáenz-Lechón, N., Osma-Ruiz, V., Gómez-Vilda, P.: Use of cepstrum-based parameters for automatic pathology detection on speech. Analysis of performance and theoretical justification. In: Proceedings of Biosignals 2008, vol. 1, pp. 85–91 (2008)
7. Murphy, P.J., Akande, O.O.: Quantification of glottal and voiced speech harmonics-to-noise ratios using cepstral-based estimation. In: Proceedings of the 3rd International Conference on Non-Linear Speech Processing (NOLISP 2005), pp. 224–232 (2005)
8. Godino-Llorente, J.I., Gómez-Vilda, P., Blanco-Velasco, M.: Dimensionality reduction of a pathological voice quality assessment system based on gaussian mixture models and short-term cepstral parameters. IEEE Transactions on Biomedical Engineering 53, 1943–1953 (2006)
9. Fraile, R., Sáenz-Lechón, N., Godino-Llorente, J.I., Osma-Ruiz, V., Gómez-Vilda, P.: Use of mel-frequency cepstral coeffcients for automatic pathology detection on sustained vowel phonations: Mathematical and statistical justification. In: Proceedings of the International Symposium on Image/Video Communications over fixed and mobile networks, Bilbao (July 2008)
10. TM Alliance Team: Telemedicine 2010: Visions for a personal medical network. Technical Report BR-29, ESA Publications Division (2004)
11. Moran, R.J., Reilly, R.B., de Chazal, P., Lacy, P.D.: Telephony-based voice pathology assessment using automated speech analysis. IEEE Transactions on Biomedical Engineering 53, 468–477 (2006)
12. Jamieson, D.G., Parsa, V., Price, M.C., Till, J.: Interaction of speech coders and atypical speech ii: Effects on speech quality. Journal of Speech, Language and Hearing Research 45, 689–699 (2002)
13. Fraile, R., Godino-Llorente, J.I., Sáenz-Lechón, N., Osma-Ruiz, V., Gómez-Vilda, P.: Analysis of the impact of analogue telephone channel on MFCC parameters for voice pathology detection. In: Proceedings of the 8th INTERSPEECH Conference (INTERSPEECH 2007), pp. 1218–1221 (2007)
14. Deller, J.R., Proakis, J.G., Hansen, J.H.L.: Discrete-time processing of speech signals. Macmillan Publishing Company, New York (1993)
15. Davis, S.B., Mermelstein, P.: Comparison of parametric representations for monosyllabic word recognition in continuously spoken sentences. IEEE Transactions on Acoustics, Speech and Signal Processing ASSP-28, 357–366 (1980)
16. Dimolitsas, S., Gunn, J.E.: Modular, off line, full duplex telephone channel simulator for high speed data transceiver evaluation. IEE Proceedings 135, 155–160 (1988)
17. ITU-T: Transmission characteristics of national networks. Series G: Transmission Systems and Media, Digital Systems and Networks Rec. G.120 (12/98) (1998)
18. Reynolds, D.A., Zissman, M.A., Quatieri, T.F., O'Leary, G.C., Carlson, B.A.: The effects of telephone transmission degradations on speaker recognition performance. In: Proceedings of ICASSP 1995, Detroit, MI, USA, vol. 1, pp. 329–332 (1995)
19. Massachusetts Eye and Ear Infirmary: Voice disorders database, CD-ROM (1994)
20. Parsa, V., Jamieson, D.G.: Identification of pathological voices using glottal noise measures. Journal of Speech, Language and Hearing Research 43, 469–485 (2000)
21. Haykin, S.: Neural networks: A comprehensive foundation. Macmillan, New York (1994)
22. Bimbot, F., Bonastre, J.F., Fredouille, C., Gravier, G., Magrin-Chagnolleau, I., Meignier, S., Merlin, T., Ortega-Garcia, J., Petrovska, D., Reynolds, D.A.: A tutorial on text-independent speaker verification. EURASIP Journal on Applied Signal Processing 2004, 430–451 (2004)

23. Martin, A.F., Doddington, G.R., Kamm, T., Ordowski, M., Przybocki, M.A.: The DET curve in assessment of detection task performance. In: Proceedings of Eurospeech 1997, Rhodes, Crete, vol. IV, pp. 1895–1898 (1997)
24. Pouchoulin, G., Fredouille, C., Bonastre, J.F., Ghio, A., Giovanni, A.: Frequency study for the characterization of the dysphonic voices. In: Proceedings of the 8th INTERSPEECH Conference (INTERSPEECH 2007), pp. 1198–1201 (2007)

# A Biologically-Inspired Visual Saliency Model to Test Different Strategies of Saccade Programming

Tien Ho-Phuoc, Anne Guérin-Dugué, and Nathalie Guyader

GIPSA-Lab. 961 rue de la Houille Blanche
38402 Grenoble, France
{phuoc-tien.ho,anne.guerin,nathalie.guyader}@gipsa-lab.inpg.fr
http://www.gipsa-lab.inpg.fr

**Abstract.** Saliency models provide a saliency map that is a topographically arranged map to represent the saliency of the visual scene. Saliency map is used to sequentially select particular locations of the scene to predict a subject's eye scanpath when viewing the corresponding scene. A saliency map is most of the time computed using the same point of view or foveated point. Few models were interested in saccade programming strategies. In visual search tasks, studies shown that people can plan from one foveated point the next two saccades (and so, the next two fixations): this is called concurrent saccade programming. In this paper, we tested if such strategy occurs during natural scene free viewing. We tested different saccade programming strategies depending on the number of programmed saccades. The results showed that the strategy of programming one saccade at a time from the foveated point best matches the experimental data from free viewing of natural images. Because saccade programming models depend on the foveated point, we took into account the spatially variant retinal resolution. We showed that the predicted eye fixations were more effective when this retinal resolution was combined with the saccade programming strategies.

**Keywords:** Saccade programming, Saliency map, Spatially variant retinal resolution.

## 1 Introduction

Eye movement is a fundamental part of human vision for scene perception. People do not look at all objects at the same time in the visual field but sequentially concentrate on attractive regions. Visual information is acquired from these regions when the eyes are stabilized [4,6]. Psychophysical experiments with eye-trackers provide experimental data for both behavioral and computational models to predict the attractive regions.

Computational models are in general divided into two groups: task-independent models (bottom-up) and task-dependent models (top-down). Most models describe bottom-up influences to create a saliency map for gaze prediction. They are inspired by the concept of the Feature Integration Theory of Treisman and Gelade [17] and by the first model proposed by Koch and Ullman [9]. The most popular bottom-up saliency model was proposed by Itti [8].

Eye movement experimental data consists, in general, of fixations and saccades. Most models were usually evaluated with distribution of fixations rather than distribution of

A. Fred, J. Filipe, and H. Gamboa (Eds.): BIOSTEC 2009, CCIS 52, pp. 187–199, 2010.

saccade amplitudes (in this paper, *saccade distribution* will be used to refer to distribution of saccade amplitudes). In order to evaluate more precisely human saccades, saliency prediction must be included inside a spatio-temporal process, even if we consider only a bottom-up visual saliency map from low-level features based on still images. A question can be asked: is visual saliency evaluated again at each fixation, or not?

Some studies showed that the saccadic system can simultaneously program two saccades to two different spatial locations [11]. This means that from one foveated point, the next two saccades are programmed in parallel; this is called the *concurrent processing* of saccades. Moreover, another study [10] showed that when subjects are explicitly instructed to make a saccade only after the current information at the fovea has been analyzed, they have difficulty using this strategy. In this paper, we consider different saccade programming strategies during free viewing of natural images and we answer the questions: Do subjects make one saccade at a time, programming the next saccade from the current foveated point? Or do subjects program the next two saccades from the current foveated point? To answer these questions, we tested two different models depending on the corresponding number of programmed saccades from one foveated point (cf. section 2.4). These two models were compared with the baseline model in which all fixations were predicted from the same foveated point. The mechanism of saccade programming will be discussed through our experimental results, as it is still an open question.

Saccade programming models are greatly linked to image resolution when it is projected on the retina. The image on the fovea is viewed at higher resolution compared to the peripheral regions. This enhances the saliency of the region around the foveated point. The spatial evolution of visual resolution is one of the consequences of the non-uniform density of photoreceptors at retina level [18]. When comparing experimental and predicted scanpaths, Parkhurst [13] noticed an important bias on saccade distribution. The predicted saccade distribution is quite uniform, contrary to the experimental one where fixations are indeed located on average near the foveated point. The decrease of visual resolution on the peripheral regions contributes to an explanation of this effect. In [13], this property was integrated in the model at the final stage (multiplication of the saliency map by a gaussian function).

In our model, we implement the decreasing density of photoreceptors as the first cause of this inhomegeneous spatial representation. The output of photoreceptors encodes spatial resolution depending on eccentricity, given a parameter of resolution decay [14]. While this parameter influents low level retina processing, we intend to go further in varying its value to mesure its impact on high level visual processing, i.e. including the first steps of the visual cortex. Similarly, in [7], they noticed that the implementation of the spatially variant retinal resolution improved the ability of fixation prediction for dynamic stimuli (video games), but the question of saccade programming was not addressed. Here, we show that this positive effect of fixation prediction is even significant with static visual stimuli in combination with saccade programming models.

Consequently, the proposed models are dynamic ones as the high resolution regions, corresponding to the fovea, change temporally according to the scanpath. Three saccade

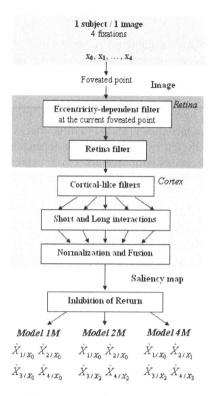

**Fig. 1.** Functional description of the three saccade programming strategies using a saliency model. For one subject and one image, the model predicts the first four fixations of the subject when viewing the image. The "1M" model uses only one foveated point to compute the four fixations, the "2M" model computes two fixations from one foveated point and the two others from another foveated point, and the "4M" model computes one fixation from one foveated point.

programming strategies are analyzed. Experimental results show the interest of this approach with spatially variant retinal resolution. The models are described in section 2. Section 3 presents the experiment of free viewing to record eye fixations and saccades, and the evaluation of the proposed models. Discussion is drawn in section 4.

## 2   Description of the Proposed Models

Our biologically-inspired models (Fig. 1) consist of one or more saliency maps in combination with saccade programming strategies. The model integrates the bottom-up pathway from low-level image properties through the retina and primary visual cortex to predict salient areas. The retina filter plays an important role: implementing non-linear response of photoreceptors and then, spatially variant resolution. The retinal image is then projected into a bank of cortical-like filters (Gabor filters) which compete and interact, to finally produce a saliency map. This map is then used with a mechanism of "Inhibition of Return" to create the saccade programming model. All these elements are described below.

## 2.1 Retina Filter

In [3] the retina is modeled quite completely according to its biological functions: spatially variant retinal resolution, luminance adaptation and contrast enhancement. These three characteristics are taken into account while usually a simple model is used such as a function of difference of Gaussians [8].

*Spatially variant retinal resolution*: Since the density of photoreceptors decreases as eccentricity from the fovea increases [18], the bluring effect increases from the fovea to the periphery and is reflected by an eccentricity-dependent filter. The cut-off frequency $f_{co}$ of this filter decreases with eccentricity $ecc$ expressed in degree (Fig. 2a). The variation of the cut-off frequency (Eq. 1) is adapted from [14,15]:

$$f_{co}(ecc) = fmax_{co} \cdot \frac{\alpha}{\alpha + |ecc|}, \tag{1}$$

where $\alpha$ is the paramater controlling the resolution decay, and $fmax_{co}$ the maximal cut-off frequency in the fovea. Biological studies showed that $\alpha$ is close to $2.3°$ [15]. Figure 2b shows an example of an image filtered spatially at the center with $\alpha = 2.3°$.

*Luminance adaptation*: Photoreceptors adapt to a varying range of luminance and increase luminance in dark regions without saturation of luminance in bright ones. They carry out a compression function (Eq. 2) :

$$y = y_{max} \cdot \frac{x}{x + x_o}, \tag{2}$$

where $x$ is the luminance of the initial image, $x_o$ represents its average local luminance, $y_{max}$ a normalization factor and $y$ the photoreceptor output [3].

*Contrast enhancement*: The output of horizontal cells, the low-pass response of photoreceptors, passes through bipolar cells and then, through different types of ganglion cells: parvocellular, magnocellular and koniocellular cells. We only consider the two principal cells: parvocellular and magnocellular. Parvocellular cells are sensitive to high spatial frequency and can be modeled by the difference between photoreceptors and horizontal cells. Therefore, they enhance the initial image contrast and whiten its energy spectrum (Fig. 5b). Magnocellular cells respond to lower spatial frequency and are modeled by a low-pass filter like horizontal cells.

In human visual perception, we know that low frequencies precede high frequencies [12]. In our model, we do not take into account this temporal aspect. However, as both low and high frequency components are necessary for a saliency map, we compute the retina output as a linear combination of the parvocellular and magnocellular outputs (Fig. 5b).

## 2.2 Cortical-Like Filters

**Gabor Filters.** Retinal image is transmitted to V1 which processes signals in different frequencies, orientations, colors and motion. We only consider the frequency and orientation decomposition carried out by complex cells. Among several works modeling responses of these cells, Gabor filters are evaluated as good candidates. A set of Gabor

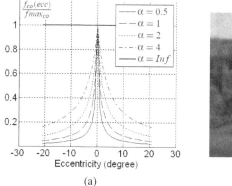

(a)                                                                    (b)

**Fig. 2.** (a) Normalized cut-off frequency as function of eccentricity for different values of $\alpha$ controlling the resolution decay. (b) An example of the output of the eccentricity-dependent filter with $\alpha = 2.3°$, the foveated point is at the image center (marked with the cross).

filters is implemented to cover all orientations and spatial frequencies in the frequential domain. A filter $G_{i,j}$ (Eq. 3) is tuned to its central radial frequency $f_j$ at the orientation $\theta_i$. There are $N_\theta$ orientations and $N_f$ radial frequencies ($N_\theta = 8$ and $N_f = 4$). The radial frequency is such as $f_{N_f} = 0.25$ and $f_{j-1} = \frac{f_j}{2}, j = N_f, .., 2$. The standard deviations of $G_{i,j}$ are $\sigma_{i,j}^f$ and $\sigma_{i,j}^\theta$ in the radial direction and its perpendicular one, respectively. $\sigma_{i,j}^f$ is chosen in such a way that filters $G_{i,j}$ and $G_{i,j-1}$ are tangent at level of 0.5. We notice that the choice of standard deviations influents the predicted saliency map. We choose $\sigma_{i,j}^f = \sigma_{i,j}^\theta$, which is justified in the next section.

$$G_{i,j}(u,v) = exp\left\{-\left(\frac{(u'-f_j)^2}{2(\sigma_{i,j}^f)^2} + \frac{v'^2}{2(\sigma_{i,j}^\theta)^2}\right)\right\} \qquad (3)$$

where $u$ (respectively $v$) is the horizontal (respectively vertical) spatial frequency and $(u', v')$ are the coordinates after rotation of $(x, y)$ around the origin by angle $\theta_i$. Then, for each channel, complex cells are implemented as the square amplitude of the Gabor filter output, providing the energy maps $e_{i,j}$.

**Interaction between Filters.** The responses of neurons in the primary visual cortex are influenced in the manner of excitation or inhibition by other neurons. As these interactions between neurons are quite complex, we only consider two types of interactions based on the range of receptive fields [5].

Short interactions are the interactions among neurons having overlapping receptive fields. They occur with the same pixel in different energy maps (Eq. 4):

$$m_{i,j}^s = 0,5.e_{i,j-1} + e_{i,j} + 0,5.e_{i,j+1} - 0,5.e_{i+1,j} - 0,5.e_{i-1,j}. \qquad (4)$$

These interactions introduce inhibition between neurons of neighboring orientations on the same scale, and excitation between neurons of the same orientation on neighboring scales. For the standard deviations of the cortical-like filters, if $\sigma_{i,j}^f > \sigma_{i,j}^\theta$ the filters

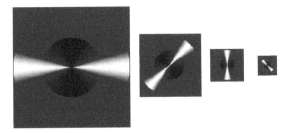

**Fig. 3.** Example of butterfly masks with different orientations and scales

are more orientation-selective. However, this choice reduces the inhibitive interaction. So, we choose $\sigma_{i,j}^f = \sigma_{i,j}^\theta$.

The second interactions are long interactions which occur among colinear neurons of non-overlapping receptive fields and are often used for contour facilitation [5]. This interaction type is directly a convolution product on each map $m_{i,j}^s$ to produce an intermediate map $m_{i,j}$. The convolution kernel is a "butterfly" mask $b_{i,j}$ (Fig. 3). The orientation of the mask $b_{i,j}$ for the map $m_{i,j}^s$ is the orientation $\theta_i$, and the mask size is inversely proportional to the central radial frequency $f_j$. The "butterfly" mask $b_{i,j}$ has two parts: an excitatory part $b_{i,j}^+$ in the preferential direction $\theta_i$ and an inhibitive one $b_{i,j}^-$ in all other directions. It is normalized in such a way that its summation is set to 1. Figure 4 (first row) shows the interaction effect for contour facilitation (Fig. 4c).

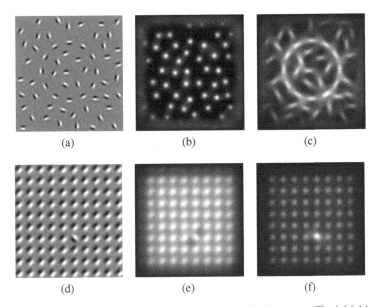

**Fig. 4.** Tests showing the influence of interaction and normalization steps. The initial images are on the left column. The corresponding saliency maps are on the center and right columns. First row, interaction effect: (a) Initial image; (b) without interaction; (c) with interaction. Second row, normalization effect: (d) Initial image; (e) without normalization; (f) with normalization.

## 2.3   Normalization and Fusion

Before the fusion, we linearly normalized each intermediate map between $[0, 1]$. We also strengthened intermediate maps, using the method proposed by Itti [8], to take into account the assumption that an object is more salient if it is different from its neighbors. This was computed as follows:

- Normalize intermediate maps in $[0, 1]$
- Let us designate $m_{i,j}^{*}$ the maximal value of map $m_{i,j}$ and $\overline{m}_{i,j}$ its average. Then, the value at each pixel is multiplied by $(m_{i,j}^{*} - \overline{m}_{i,j})^{2}$.
- Set to zero all the values which are smaller than 20% of the maximal value.

Figure 4 (second row) represents the role of the normalization step in reinforcing the filter output where a Gabor patch is different from the background (Fig. 4f).

Finally, all intermediate maps are summed up in different orientations and frequencies to obtain a saliency map. Examples of retinal image and saliency map are given in Fig. 5.

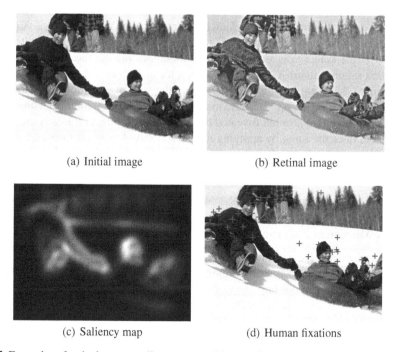

(a) Initial image                    (b) Retinal image

(c) Saliency map                    (d) Human fixations

**Fig. 5.** Examples of retinal output, saliency map and human fixations for a part of an image used in the experiment

## 2.4   Saccade Programming

Our models predict the first four fixations as an extension of the study in [11] (two saccades or two fixations). The fixations are predicted (and consequently the saccades)

from the output saliency map by implementing a simple mechanism of "Inhibition of Return" (IOR). We have defined three saccade programming models, called "1M", "2M" and "4M" (Fig. 1). The "1M" model has often been used with the hypothesis that all fixations can be predicted from the same foveated point at the image center. Here, we tested two other models where each foveated point allows the next or the next two fixations to be predicted ("4M" or "2M", respectively). In other words, to predict 4 fixations the "1M" model uses 1 foveated point; 2 and 4 foveated points for the "2M" and "4M" model. Let us explain first the IOR mechanism with the "1M" model which is considered as the baseline model. The first fixation is chosen as the pixel which has the maximal value on the saliency map. Then, this fixation becomes the foveated point to predict the second saccade. Hence, the second fixation is chosen as the pixel of maximal value after inhibiting the surface around the foveated point (radius of $1°$). This mechanism continues for the third and fourth fixations. For the "2M" and "4M" models, the IOR mechanism is applied in the same manner as for the "1M" model except that the four fixations are predicted from more than one saliency map (see below).

In the "1M" model, the eccentricity-dependent filter is applied only once to the center $X_0$ of images (because during the eye movement experiment, images appeared only if subjects were looking at the center of the screen). Then, a saliency map is computed from this foveated point to predict four fixations $\hat{X}_{1/X_0}$, $\hat{X}_{2/X_0}$, $\hat{X}_{3/X_0}$, $\hat{X}_{4/X_0}$ by using the IOR mechanism. For the "2M" model, the eccentricity-dependent filter is applied first to the center $X_0$, as for the "1M" model, to predict the first two fixations (on the first saliency map) $\hat{X}_{1/X_0}$, $\hat{X}_{2/X_0}$; then, the eccentricity-dependent filter is applied again to the second fixation for predicting the next two fixations (on the second saliency map) $\hat{X}_{3/X_2}$, $\hat{X}_{4/X_2}$. Because the second fixation is different from one subject to another we take into account foveated point $X_2$ of each subject for each image. The "4M" model does the same by applying the eccentricity-dependent filter and calculating the saliency map for each foveated point of each subject (four saliency maps are sequentially evaluated).

## 3   Experimental Evaluation

### 3.1   Eye Movement Experiment

We ran an experiment to obtain the eye scanpaths of different subjects when they were looking freely at different images. Figure 5d shows the fixations of all subjects on a part of an image. The recording of eye movements served as a method to evaluate our proposed model of visual saliency and saccade programming.

*Participants*: Eleven human observers were asked to look at images without any particular task. All participants had normal or corrected to normal vision, and were not aware of the purpose of the experiment.

*Apparatus*: Eye tracking was performed by an Eyelink II (SR Research). We used a binocular recording of the pupils tracking at 500Hz. A 9-point calibration was made before each experiment. The velocity saccadic threshold is $30°/$ s and the acceleration saccadic threshold is $8000°/$ s$^2$.

*Stimuli*: We chose 37 gray level images ($1024 \times 768$ pixels) with various contents (people, landscapes, objects or manufactural images).

*Procedure*: During the experiment, participants were seated with their chin supported in front of a 21" color monitor (75 Hz refresh rate) at a viewing distance of 57 cm (40° × 30° usable field of view). An experiment consisted in the succession of three items : a fixation cross in the center of the screen, followed by an image during 1.5 s and a mean grey level screen for 1 s. It is important to note that the image appeared only if the subject was looking at the fixation cross; we ensured the position of the eyes before the onset of images. Subjects saw the same 37 images in a random order.

We analyzed the fixations and saccades of the guiding eye for each subject and each image.

### 3.2   Criterion Choice for Evaluation

The qualities of the model, more precisely the saccade programming strategy and the eccentricity-dependent filter, are evaluated with the experimental data (fixations and saccades). This evaluation allows to test predicted salient regions and saccade distribution.

First, the three models "1M", "2M" and "4M" are used to test saccade programming strategies. The evaluation protocol as in [16] is to extract from a saliency map the most salient regions representing 20% of the map surface. Let us call each fixation "correct fixation" (respectively "incorrect fixation") if the fixation is inside (respectively outside) these predicted salient regions. The ratio $R_c(i, s)$ of "correct fixation" for an image $i$ and a subject $s$ is given below:

$$R_c(i, s) = \frac{N_{inside}}{N_{all}}.100, \qquad (5)$$

where $N_{inside}$ is the number of fixations inside the salient regions and $N_{all}$ is the total number of fixations.

The average ratio of correct fixations of all couples (subject × image) is computed. Particularly for each couple, in the "1M" model, one saliency map is used to calculate correct fixation index for the first four fixations. In the "2M" model, the first saliency map is used for the first two fixations and the second map for the next two fixations (Fig. 1). Similarly for the "4M" model, each saliency map is used for the corresponding fixation. It is noticed that in the "1M" model, the saliency map of each image is identical for all subjects (whose foveated point is always at the image center). However, it is no longer the case for the "2M" or "4M" model where a saliency map depends on the foveated points of a subject.

Second, from the most suitable saccade programming strategy chosen above, the predicted saccade distribution for the first four saccades is computed and compared with the empirical one from our experimental data. It has been shown that the parameter $\alpha$ (Eq. 1) which fits the experimental data based on contrast threshold detection when presenting eccentred gratings is around 2.3° [15]. Here, by varying the $\alpha$ values, the expected effects are: (i) this parameter must have a great influence on saccade distribution and (ii) the best value would be in the same order of magnitude as 2.3°.

## 3.3   Results

**Evaluation of the Three Saccade Programming Models.** For saccade programming, Fig. 6(a,b,c) shows the criterion of correct fixations $R_c$ as a function of fixation order with five $\alpha$ values (0.5°, 1°, 2°, 4° and infinity) for the three models. $R_c$ of the first fixation is identical for these three models because of the same starting foveated point. It is also the case for $R_c$ of the second fixation in "1M" and "2M", and $R_c$ of the third fixation in "2M" and "4M". $R_c$ for all three models has the same global trend: decrease with fixation order. The $R_c$ ratio is greater for the "4M" model in comparison to the "1M" model for all fixations. It results from the reinitialization of the foveated point. In the "2M" model, an intermediate of the two previous ones, the increase of $R_c$ from the second to third fixation is also explained by this reinitialization. Moreover, the decrease at the second and fourth fixation (in "2M") presents necessity of the reinitialization, but at each fixation. The slower decrease of $R_c$ in "4M" is also coherent with this interpretation.

The $\alpha$ parameter also influents the quality of the predicted regions. Saliency models including spatially variant retinal resolution give better results than the model with constant resolution for the first four fixations (t-test, $p < 0.005$, except the cases $\alpha = 0.5°$ and $\alpha = 1°$ for the third fixation). However, among $\alpha = 0.5°$, $\alpha = 1°$, $\alpha = 2°$ and

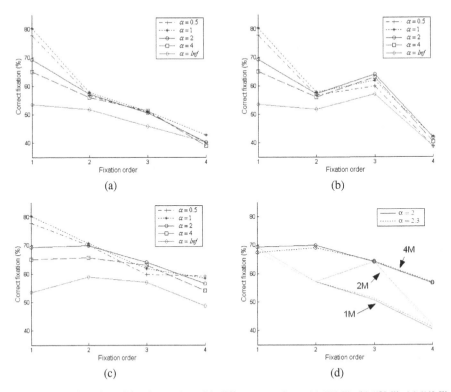

**Fig. 6.** $R_c$ as a function of fixation order with different $\alpha$ values: (a) "1M", (b) "2M", (c) "4M", (d) Comparison of the three models between $\alpha = 2°$ and $\alpha = 2.3°$

$\alpha = 4°$ there is no significant difference except for the first fixation. At this fixation, there is no difference between $R_c$ of cases $0.5°$ and $1°$ but they are significantly greater than those in cases $2°$ and $4°$ ($F(3, 1612) = 10.45, p < 0.005$).

Indeed, when varying the $\alpha$ parameter, we noticed a continuous effect on the ratio $R_c$ of "correct fixation". The results from the simulations with $\alpha = 2.3°$ are very close to those obtained with $\alpha = 2°$ (Fig. 6d).

**Configuration of the Eccentricity-dependent Filter.** Using the "4M" model, we test the role of the eccentricity-dependent filter for different values of its resolution decay $\alpha$ in comparison with the saccade amplitude distribution of experimental data . We computed the saccade amplitude distribution for human data and for the "4M" model with 5 values of the parameter $\alpha$. The saccade distribution for the model is obtained by computing the distance between two successive fixations.

First, in the context of free viewing, the empirical saccade distribution follows an exponential distribution with a mean saccade ampliture close to $7.6°$ [2,1]. Here, using the human data we found this result again with a mean value of $7°$ (from the solid line in Fig. 7a). However, this exponential distribution with only one fitting parameter is not so suitable to model the sacade distributions coming from simulations with the 4M model (solid line in Fig. 7b-f).

Second, in order to evaluate the influence of retinal resolution on saccade distribution, it is more appropriate to fit the empirical distribution (solid line in Fig. 7) with a Gamma probability density function (dashed line in Fig. 7). The Gamma distribution is more general than the exponential distribution as it depends on two parameters determining the shape ($k$) and scale ($\theta$) of a probability density function. This distribution

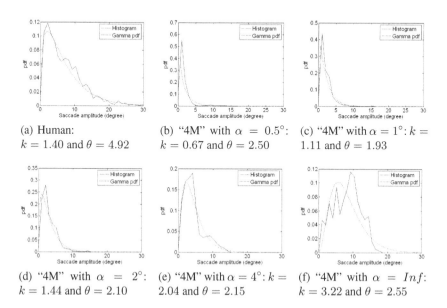

(a) Human: $k = 1.40$ and $\theta = 4.92$

(b) "4M" with $\alpha = 0.5°$: $k = 0.67$ and $\theta = 2.50$

(c) "4M" with $\alpha = 1°$: $k = 1.11$ and $\theta = 1.93$

(d) "4M" with $\alpha = 2°$: $k = 1.44$ and $\theta = 2.10$

(e) "4M" with $\alpha = 4°$: $k = 2.04$ and $\theta = 2.15$

(f) "4M" with $\alpha = Inf$: $k = 3.22$ and $\theta = 2.55$

**Fig. 7.** Saccade distributions (raw histogram and Gamma distribution) of the experimental data and the "4M" model according to different values of $\alpha$

better matches the emperical distribution. While increasing $\alpha$, we noticed the increase of the main mode and of the dispersion. These two phenomema may be quantitatively modelled by the Gamma distribution with $k$ and $\theta$ (parameters in Fig. 7). The only irregularity of $\theta$ for $\alpha = 0.5°$ may be explained by its abnormal empirical distribution. However, the values of $\theta$ in the "4M" model are smaller than that of the experimental distribution. Based on the shape of distribution, the distribution with $\alpha = 2°$ ($k = 1.44$) is chosen as the best while its shape is the most similar to the experimental one ($k = 1.40$).

## 4   Discussion

First, all three models have the ratio of correct fixation decreasing according to the fixation order. This fact can be explained by the influence of top-down mechanisms which arise late and reduce the role of bottom-up in visual attention. In reality, as time passes, fixations of different subjects are more dispersive and subject-dependent. However, the influence of bottom-up still persists and the percentage of correct fixations is much higher than by chance in all three models. This confirms the role of bottom-up in visual attention even with the increasing presence of top-down.

The result of this study seems not to support programming of several saccades in parallel. In fact, we tested three models: the "1M" model programming four saccades in parallel from the same foveated point, the "2M" model programming two saccades in parallel and the "4M" model programming only one saccade at a time. This study showed that the "1M" model is not realistic and the "4M" model seems to be the most realistic. The behaviour of the "2M" model illustrates that recomputing saliency in updating the foveated point is beneficial. The best performance comes however from the "4M" model presenting effectiveness of the reinitialization at each fixation. We can conclude that saccade programming in parallel seems not to be used by subjects when they have to look freely at natural images.

Second, this study shows the positive effect of the spatially variant retinal resolution on prediction quality. Whatever the saccade programming strategy is, the models including the spatially variant retinal resolution greatly outperform the models with constant resolution in terms of the quality of fixation prediction and saccade distribution. The parameter $\alpha$ which controls the resolution decrease has an important impact on saccade distribution (scale and shape). Moreover, we found the expected range of value for this parameter using our model to compute saliency maps. We also notice that if we have the same shape, the scale of the saccade distribution remains smaller on predicted data than on experimental data, as we only consider a bottom-up model and we have only one parameter to adjust.

In our models, a foveated point for the next saccade is selected from subjects' fixations instead of being looked for in the present saliency map. While fixations are different from one subject to another, the model is a subject-dependent model. If we want to go further in creating a more general model of predicting eye movements automatically, the model would take into account human task, for example categorization or information search, and hence passes from a region-predicting model to a scanpath-predicting model.

**Acknowledgements.** This work is partially supported by grants from the Rhône-Alpes Region with the LIMA project. T. Ho-Phuoc's PhD is funded by the French MESR.

# References

1. Andrews, T.J., Coppola, D.M.: Idiosyncratic characteristics of saccadic eye movements when viewing different visual environments. Vision Research 39, 2947–2953 (1999)
2. Bahill, A.T., Adler, D., Stark, L.: Most naturally occurring human saccades have magnitudes of 15 degrees or less. Investigative Ophthalmology 14, 468–469 (1975)
3. Beaudot, W., Palagi, P., Herault, J.: Realistic simulation tool for early visual processing including space, time and colour data. In: Mira, J., Cabestany, J., Prieto, A.G. (eds.) IWANN 1993. LNCS, vol. 686, pp. 370–375. Springer, Heidelberg (1993)
4. Egeth, H.E., Yantis, S.: Visual attention: Control, representation, and time course. Annual Review of Psychology 48, 269–297 (1997)
5. Hansen, T., Sepp, W., Neumann, H.: Recurrent long-range interactions in early vision. In: Wermter, S., Austin, J., Willshaw, D.J. (eds.) Emergent Neural Computational Architectures Based on Neuroscience. LNCS (LNAI), vol. 2036, pp. 127–153. Springer, Heidelberg (2001)
6. Henderson, J.M.: Human gaze control in real-world scene perception. Trends in Cognitive Sciences 7, 498–504 (2003)
7. Itti, L.: Quantitative modeling of perceptual salience at human eye position. Visual Cognition 14, 959–984 (2006)
8. Itti, L., Koch, C., Niebur, E.: A model of saliency-based visual attention for rapid scene analysis. IEEE Transactions on Pattern Analysis and Machine Intelligence 20, 1254–1259 (1998)
9. Koch, C., Ullman, S.: Shifts in selective visual attention: towards the underlying neural circuitry. Human Neurobiology 4, 219–227 (1985)
10. McPeek, R.M., Skavenski, A.A., Nakayama, K.: Adjustment of fixation duration in visual search. Vision Research 38, 1295–1302 (1998)
11. McPeek, R.M., Skavenski, A.A., Nakayama, K.: Concurrent processing of saccades in visual search. Vision Research 40, 2499–2516 (2000)
12. Navon, D.: Forest before trees: the precedence of global features in visual perception. Cognitive Psychology 9, 353–383 (1977)
13. Parkhurst, D., Law, K., Niebur, E.: Modeling the role of salience in the allocation of overt visual attention. Vision Research 42, 107–123 (2002)
14. Geisler, W.S., Perry, J.S.: A real-time foveated multiresolution system for low-bandwidth video communication. In: Human Vision and Electronic Imaging, Proceedings of SPIE, vol. 3299, pp. 294–305 (1998)
15. Perry, J.S.: http://fi.cvis.psy.utexas.edu/software.shtml
16. Torralba, A., Oliva, A., Castelhano, M.S., Henderson, J.M.: Contextual guidance of Eye Movements and Attention in Real-World Scenes: The Role of Global Features in Object Search. Psychological Review 113, 766–786 (2006)
17. Treisman, A., Gelade, G.: A feature integration theory of attention. Cognitive Psychology 12, 97–136 (1980)
18. Wandell, B.A.: Foundations of Vision. Stanford University (1995)

# Transition Detection for Brain Computer Interface Classification

Ricardo Aler, Inés M. Galván, and José M. Valls

Computer Science Department, Carlos III University of Madrid
Avda. Universidad, 30, 28911 Leganés, Spain
{aler,igalvan,jvalls}@inf.uc3m.es
http://www.evannai.uc3m.es

**Abstract.** This paper deals with the classification of signals for brain-computer interfaces (BCI). We take advantage of the fact that thoughts last for a period, and therefore EEG samples run in sequences belonging to the same class (thought). Thus, the classification problem can be reformulated into two subproblems: detecting class transitions and determining the class for sequences of samples between transitions. The method detects transitions when the L1 norm between the power spectra at two different times is larger than a threshold. To tackle the second problem, samples are classified by taking into account a window of previous predictions. Two types of windows have been tested: a constant-size moving window and a variable-size growing window. In both cases, results are competitive with those obtained in the BCI III competition.

## 1 Introduction

The context of this paper is the brain-computer interface (BCI), a growing research field, that would allow users to control computers and other devices by means of brain signals [1], [10] [3], [8]. One of the main problems of BCI is to accurately decode individual brain signals. Machine Learning techniques are typically applied here, by training classifiers with the brain signals of the user that is going to use the interface [2]. For instance, a BCI can be trained to recognize three different classes corresponding to three different mental states: left-hand motor imagery, right-hand motor imagery, and object rotation imagery.

Noisy weak signals and high variability between same-user sessions[1] make the classification problem difficult, resulting in many on-line classification errors, frustated users, and low transfer (speed) rates from the user to the computer. The transfer rate could be increased if three or more classes are used. However, multi-class classification problems are more difficult than two-class ones.

In this paper we aim to improve classification accuracy by taking advantage of a feature of BCIs: EEG samples (or instances) run in sequences belonging to the same class (i.e. the same thought), followed by a transition into a different class/thought. Typically, the BCI classification problem is tackled by trying to classify the EEG signal

---

[1] This means that the classifier learned during one session might not be accurate in the next session, even for the same user.

A. Fred, J. Filipe, and H. Gamboa (Eds.): BIOSTEC 2009, CCIS 52, pp. 200–210, 2010.

at every instant in time. However, given that classes run in sequences, the classification problem can be reformulated into two subproblems:

- Detecting class transitions
- Determining the class for sequences of instances between transitions.

The first problem can be approached in many different ways. In this paper we detect transitions by computing the L1 norm between the power spectra at two different times (instances) and signalling a class transition when the distance is larger than a threshold. Transition detection has also been used in [5] applying a proximity function based on three samples of the Power Spectrum Density.

By detecting transitions, the accuracy of the second problem (classification between transitions) can also be improved in two ways. First, if the class of the last sequence (before the transition) is known, that class can be discarded after the transition, hence becoming a simpler N-1 class classification problem. Also, it may be more accurate to classify an instance by taking into account several of the previous classifications. In this paper, two different types of windows are used. First, a 'moving window' with a fixed size. This kind of window only takes into account the $n$ previous instances, where $n$ has been fixed from the beginning. The second type of window is the 'growing window', with a variable size. When a transition is detected, the growing window size is initialized to zero and it keeps growing until the next transition.

The rest of the paper is organized as follows. Section 2 contains a description of the data supplied for the BCI III competition, that they are the data used in this works to validate and tested the proposed ideas. The method is described in Section 3. Section 4 shows the obtained results. Section 5 summarizes our results and draws some conclusions.

## 2   Description of EEG Data

In this paper we are going to use a high quality dataset acquired in the IDIAP Research Institute by Silvia Chiappa and José del R. Millán [9]. It was used in the BCI-III competition that took place in 2005.[2] This dataset contains data from 3 normal subjects during 4 non-feedback sessions. The subjects sat in a normal chair, relaxed arms resting on their legs. There are 3 tasks, so this is a three-class classification problem:

- Imagination of repetitive self-paced left hand movements
- Imagination of repetitive self-paced right hand movements
- Generation of words beginning with the same random letter

All 4 sessions of a given subject were acquired on the same day, each lasting 4 minutes with 5-10 minutes breaks in between them. The subject performed a given task for about 15 seconds and then switched randomly to another task at the operator's request. EEG data is not splitted in trials since the subjects are continuously performing any of the mental tasks. Data was provided in two ways: raw EEG signals, and data with precomputed features. In this paper, we use the precomputed dataset.

---

[2] $http://ida.first.fraunhofer.de/projects/bci/competition\_iii/$.

Features were precomputed as follows. The raw EEG potentials were first spatially filtered by means of a surface Laplacian. Then, every 62.5 ms (i.e., 16 times per second) the power spectral density (PSD) in the band 8-30 Hz was estimated over the last second of data with a frequency resolution of 2 Hz for the 8 centro-parietal channels C3, Cz, C4, CP1, CP2, P3, Pz, and P4. As a result, an EEG sample is a 96-dimensional vector (8 channels times 12 frequency components).

In summary, in this paper we are going to tackle a three-class classification problem with 96 input attributes, which define the PSD over 62.5 ms for 8 input channels. There are three subjects, with four datasets (sessions) for each one: three datasets for training and one dataset for testing.

## 3   Description of the Method

The method used here to improve BCI classification accuracy based on transitions and windows is based on two main ideas. First, the transition in the signal from a class to another class must be detected. This knowledge is used to discard the class assigned to the previous sequence of instances, just before the transition. Hence, the prediction problem of $N$ classes is transformed into a prediction problem of $N - 1$ classes. Usually, in classification tasks the reduction of the number of classes helps to increase the performance of the classification algorithm.

On the other hand, the proposed method is inspired in the idea that in this problem, the class to predict remains fixed for a time period that we have called 'sequence of instances between transitions'. That means that, within a sequence, the class is not continuously changing. Hence, it makes sense to try to guess the class assigned to that sequence instead of using the classifier to predict each instance of the sequence independently of each other. The simplest way to assign a class to a sequence is to compute the majority class returned by the classifier on a small set of instances at the beginning of the sequence. However, if the classifier makes many mistakes at the beginning of the sequence, the whole sequence will be missclassified. In fact, in our first approach, we tried to assign a class to whole sequences based only on a few first instances, and very frequently, complete sequences were missclassified.

Therefore, we have decided to classify the $i'th$ instance in the sequence by considering a window of the $n$ previous instances. We have tested two types of windows. In the first one, named 'moving window', the size $n$ remains constant and therefore the window moves along with time. A heuristic method for computing the window size $n$ will be explained at the end of the section. In the second case, named 'growing window', the size $n$ keeps growing with time until the next transition. This means that in order to classify the $i'th$ instance, all instances since the last transition are considered. The moving window uses fewer previous instances to classify instance $i$, hence the classification accuracy might be lower. On the other hand, the growing window has more inertia: if mistakes are made by the user, the moving window will recover eventually, once the window has gone past the mistaken instances. But the growing window considers all instances and therefore it will require more additional correct instances in order to overcome the mistaken ones.

As we described in Section 2, we assume that the original data in the time domain has been transformed to the frequency domain using the Power Spectral Density (PSD).

Thus, we assume that we have a set of samples (or instances) and every sample contains all the components of the PSD for all channels at every instant. That is:

$$sample_i = \{PSD_{r,j}(i) \; j = 1, ..., NoC, r = 1, ...R\} \tag{1}$$

where $PSD_{r,j}(i)$ is the value of the $r^{th}$ spectral component for the the $j^{th}$ channel, $NoC$ is the number of channels, and $R$ is the number of components in the PSD. For instance, if the PSD ranges from 8Hz to 30Hz and there is a component every 2Hz (a 2Hz resolution), then $R = 12$.

The classification method studied implies to detect transition and to decide the window size used to classify test instances. Hence, the method consists of different procedures which are described in detail in the next subsections and they are:

- The procedure to detect the transition
- The procedure to select the classifier once the transition is detected and one of the class is discarded
- The procedure to classify test instances with either a moving window or a growing window
- The mechanism to compute the window size of the moving window

### 3.1 Transition Detection

It is based on the observation that when a change of class occurs, a change in the frequency domain also occurs. The idea is to detect that transition by computing the L1 norm between PSDs at two consecutive samples. For every sample $i$, the distance $d_i = ||PSD(i) - PSD(i-1)||$ between the PSD of $sample_i$ and $sample_{i-1}$ is calculated as:

$$d_i = \sum_{j=1}^{Noc} \left( \sum_{r=1}^{R} |PSD_{r,j}(i) - PSD_{r,j}(i-1)| \right) \tag{2}$$

Once the distance $d_i$ is obtained, a threshold value $U$ has to be set, such that if the distance is higher than this threshold, a transition is signaled. The value of $U$ is crucial for successfully predict transitions and therefore, very important for the success of the method.

In order to determine that threshold, the following mechanism is applied. The training data set $X$ is divided into two subsets, named $X_{notransition}$ and $X_{transition}$. The first one, $X_{notransition}$, contains the samples for which there is no transition; and, the second one $X_{transition}$ contains the transition samples. This separation can be made because samples are labeled in the training set (i.e. their class is known), and therefore a transition is just a change of class. After that, the distances $d_i$ for samples in $X_{notransition}$ are calculated and the maximum of these distances, named $MaxDis_{notransition}$, is also obtained. The distances $d_i$ for samples in subset $X_{transition}$ are also obtained and ranked from low to high. Obviously, a good threshold $U$ must be larger than $MaxDis_{notransition}$. We currently define $U$ as the next $d_i$ of set $X_{transition}$ that is larger than $MaxDis_{notransition}$. This way of setting the threshold does not guarantee that all transitions will be detected. However, this method is simple and reasonable and later we will show that it performs correctly on the test data.

### 3.2  Choosing the Appropriate Classifier, Once the Transition Has Been Detected

Let us remember that when a transition is detected, the class of the previous sequence of instances is discarded (i.e. after the transition, the class must change). Therefore, after the transition, the system only needs a classifier that is able to classify samples (or instances) into $N - 1$ classes (discarding the previous class). As the system must be prepared for all classes, all $(N - 1)$-class classifiers must be trained off-line using the training data, prior to the system going on-line. For instance, if we consider tree classes $(a, b$ and,$c)$, there will be three classifiers. The first classifier named $K_{bc}$, assumes that class $a$ is discarded and will predict either class $b$ or class $c$. The second classifier, named $K_{ac}$, will discard class $b$ and classify instances in classes $a$ or $c$. The last one $K_{ab}$, will consider only classes $a$ and $b$.

Also, an $N$-class classifier must be learned, because at the beginning there is no previous sequence of samples, and therefore, no class can be discarded. In this work, we have used Support Vector Machines (SVM) because they obtain a very good performance[11], although any other machine learning algorithm could have been employed.[3]

The method is applied to predict the class in real time (on-line). The main idea is to use one of the classifiers for $N - 1$ classes, when a transition is detected, instead of using the general classifier (the classifier for $N$ classes). Usually, the performance of classifiers with fewer classes is higher because if a class is removed, uncertainty about the class is also reduced.

The procedure to select the classifier in real-time is described next: when the prediction process begins, the N-class classifier is responsible for the prediction. At the same time, the distances $d_i$ given by equation 2 are calculated. When a transition occurs (i.e. $d_i > U$), one of the $N$ classifiers for $N - 1$ classes must be chosen. To choose the most appropriate classifier, the majority class predicted in the previous sequence of instances (i.e., the set of instances between the last two transitions) must be discarded and the $N - 1$-class classifier corresponding to the remaining classes will be in charge of the predictions until the next transition is detected.

### 3.3  Procedure to Classify Test Instances with a Moving Window

As explained at the beginning of the section, we use a second idea to improve the classification accuracy: in order to classify $sample_i$, a window with the predictions of the selected classifier for the $n$ previous samples will be used, instead of just using the prediction for $sample_i$. This is similar to determining if a biased coin is biased towards heads based only on the $n$ previous few coin tosses. $Sample_i$ will be classified as the majority class of instances within the window. It is a moving window because only the last $n$ predictions just before $sample_i$ are taken into account. As the windows moves, all the samples inside it are classified by the selected classifier. Figure 1 shows how it works. The only remaining issue is to estimate a "good" window size $n$. In this paper we have applied a heuristic method that will be explained later.

---

[3] In fact, we use Weka's SMO implementation with standard parameters (a linear kernel and $C = 1$[7].

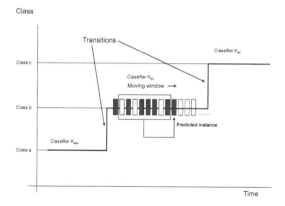

**Fig. 1.** Moving window to classify test instances

### 3.4 Procedure to Classify Test Instances with a Growing Window

We have also tried another approach named 'growing window' where the window size is not fixed, but it continues growing since the last transition. That is, all instances from the last transition to the current moment in time are included in the window. Initially the window is empty and continues growing until the next transition. Its main advantage is that it considers all possible instances. But it has an important drawback: it may take longer to recover from sequences of miss-classifications, because as the window always grows, it never forgets previous sequences of missclassified instances. This might be annoying for the user during on-line use, specially so after a long sequence of missclassifications due, for instance, to user fatigue.

### 3.5 Computing the Size of the Moving Window

Assigning the majority class of instances within the window to the next sample is reasonable, but mistakes would occur if the frequency of that class in the window is too close to 50%. This can be solved by establishing a safe confidence interval around the estimated frequency. For simplification purposes, let's suppose there are only three classes ($N = 3$, classes $a$, $b$, and $c$), and as explained before, one of them will be discarded after the transition and one of the 3 2-class classifiers will be used for the current sequence until the next transition. Let's suppose that class $a$ is discarded and therefore classifier $K_{bc}$ must be used for the current sequence ($K_{bc}$ separates class $b$ from class $c$). Although, the class of the current sequence is fixed until the next transition, the predictions of $K_{bc}$ will make mistakes. In fact, just like coin tosses, classification errors follow a Binomial distribution (with success probability $p$). If the actual class of the current sequence is $b$, the distribution of mistakes of $K_{bc}$ will follow a Binomial distribution with $p = TP^b_{K_{bc}}$, where $TP^b_{K_{bc}}$ is the True Positive rate for class $b$ and classifier $K_{bc}$ (i.e. the accuracy for class $b$ obtained by classifier $K_{bc}$). On the other hand, if the actual class is $c$, $p = TP^c_{K_{bc}}$.

If the actual class is $b$, $p$ can be estimated ($\hat{p}$) from a limited set of instances (we call it 'the window'), and from standard statistical theory (and by assuming the Binomial distribution can be approximated by a Gaussian), it is known that $\hat{p}$ belongs to a confidence interval around $p$ with confidence $\alpha$:

$$\hat{p} \in p + -z_\alpha \sqrt{\frac{p(1-p)}{n}} \tag{3}$$

where $n$ is the size of the window. From Eq 3, we can estimate the size of the window required:

$$n \geq z_\alpha^2 \frac{p(1-p)}{(p-0.5)^2} \tag{4}$$

When generating predictions, the actual class of the current sequence is not known, and therefore we have to assume the worst case, that happens when $p$ is closest to $0.5$. Therefore, $p = min(TP^b_{K_{bc}}, TP^c_{K_{bc}})$. To be in the safe side, for this paper, we have decided to make the window size independent of the classifier assigned to the current sequence. Therefore, if there are three classes $p = min(TP^b_{K_{bc}}, TP^c_{K_{bc}}, TP^a_{K_{ab}}, TP^b_{K_{ab}}, TP^a_{K_{ac}}, TP^c_{K_{ac}})$. A similar analysis could be done for more than three classes.

It is important to remark that Eq. 4 is only a heuristic approximation, since instances within a window are not independent of each other in the sense required by a Binomial distribution. The reason is that missclassifications are not uniformly distributed but happen in bursts (for example, if the user becomes tired for a period). Therefore, the probability of an instance being missclassified is not independent of previous instances. Yet, it seems heuristically appropriate that the window size is inversely proportional to $(p-0.5)^2$ (Eq. 4): the less accurate the classifier, the longer the window.

## 4   Results

The aim of this Section is to show the results of our method on the datasets described in Section 2. Let us remember that there were three subjects, and each one generated four sessions worth of data. The first three sessions are available for learning while session four is only for testing. All datasets are three-class classification problems with classes named 2, 3, and 7.

Our method computes all two-class SMO classifiers. SMO is the Weka implementation of a Support Vector Machine. Table 1 displays the results of all two-class classifiers ($K_{23}$, $K_{27}$, $K_{37}$) and the three-class classifier ($K_{237}$). The training has been made with sessions 1 and 2 and the testing with session 3. The three-class classifier accuracies can be used as a baseline to compare further results. In brackets we can observe the True Positive rate (TP) for each class. For instance, 74.7 is the True Positive rate (TP) for class 2 for the $K_{23}$ two-class classifier (i.e. $TP^2_{K_{23}}$).

Section 3 gives the details for computing the thresholds for detecting transitions. These are: $U_1 = 0.563108963$, $U_2 = 0.58576$, $U_3 = 0.587790606$, for subjects 1, 2, and 3, respectively.

Our method uses the TP rate (class accuracy), obtained with session 3, for computing the moving window size, according to Eq 4. $p$ will be set as the minimum of all TP

**Table 1.** Accuracy of two-class and three-class SMO classifiers for subjects 1, 2, and 3. Training with sessions 1 and 2, and testing with session 3.

| SMO Classifier | Subject 1 | Subject 2 | Subject 3 |
|---|---|---|---|
| $K_{23}$ | 79.0 | 71.9 | 52.3 |
| | (74.7/83.8) | (68.2/75.3) | (53.3/51.4) |
| $K_{27}$ | 82.4 | 74.3 | 57.9 |
| | (64.8/93.9) | (62.7/81.9) | (50.6/65.2) |
| $K_{37}$ | 83.0 | 76.8 | 60.4 |
| | (81.2/84.4) | (63.5/87.7) | (54.9/65.9) |
| $K_{237}$ | 73.8 | 62.0 | 40.9 |

rates. So we have $p_1 = 0.648$, $p_2 = 0.627$, and $p_3 = 0.506$ for subjects 1, 2, and 3, respectively. A confidence interval with $\alpha = 0.99$ will be used, therefore $z_\alpha = 2,5759$. Table 2 displays the windows sizes for every subject. It can be seen that the window size for subject 3 requires 46072 instances many more than available, so we apply the moving window idea only to subjects 1 and 2. This is due to the accuracy of classifiers for subject 3 are very low, in particular the accuracy of $K_{23}$ (see Table 1). We have also computed the window size for larger probabilities to check the performance of the method if smaller window sizes are used. For instance, we have also considered $p_1 = 0.80$ and $p_2 = 0.70$ (those values are approximately the accuracies of the two-class classifiers in Table 1).

**Table 2.** Window size used for subjects 1, 2, and 3

| | Probability | Window Size |
|---|---|---|
| Subject 1 | 0.648 | 78 |
| Subject 2 | 0.627 | 92 |
| Subject 3 | 0.506 | 46072 |
| Subject 1 | 0.80 | 12 |
| Subject 2 | 0.70 | 92 |

Finally, Table 3 shows the final results. The first row displays the competition results [4] on session four. [4] proposed an algorithm based on canonical variates transformation and distance based discriminant analysis combined with a mental tasks transitions detector. As required by the competition, the authors compute the class from 1 second segments and therefore no windows of samples are used. The second row, displays the best results from the competition using longer windows [6] (it reduces dimensionality of data by means of PCA and the classification algorithm is based on Linear Discrimination Analysis). No details are given for the size of the window of samples. The third row shows the results of the three-class classifier (sessions 1 and 2 were used for training and session 4, for testing).

The fourth row contains the results (on session four) of applying the transition detector only. In this case, once the transition is detected, the previous class is discarded and the prediction is made using the two-class classifier chosen. These results are better than the three-class classifier. For subject 3, the performance of the method using the transition detection is very low because some of the two-class classifiers for subject 3 have a very low accuracy (see $K_{23}$ in Table 1). Let us remember that when a transition is detected, the previous class is discarded. The previous class is computed as the majority class of the previous sequence. If the classifier used in the previous sequence is very inaccurate, the majority class might not be the actual class. Hence, the wrong class would be discarded in the current sequence, and the wrong 2-class classifier would be selected. That would generate more mistakes that would be propagated into the next sequence, and so on. Given that all samples between transitions are used to compute the previous class and that the (N-1) classifiers are better than chance, the possibility of mistaking the previous class is very low. In fact, for the data used in this article, this situation has not occurred. But it is important to remark that preventing the propagation of the missclassification of the previous class is crucial for the success of the method and we intend to improve this aspect in the future.

The fifth row in Table 3 shows the results with the method described in Section 3. In this case, both ideas, the transition detection and the moving window size, are used. Results improve significantly if the moving window is used: classification accuracy raises from 74.8 to 94.8 and from 74.6 to 86.3 (subjects 1 and 2, respectively). It can also be seen that results are also improved if a smaller window size is used (sixth row: MW with small sample). The seventh row shows the classification rate for subject 1, 2 and, 3 for the growing window.

Comparing our method with the best competition result that used a window of samples (second row of Table 3), we can see that both the moving window and the growing window are competitive with respect to the first subject and improves the performance for the second subject. Regarding to the third subject, the growing window approach does not offer an improvement in accuracy. Also, the moving window cannot be applied to the third subject due to the large number of samples required for the window.

**Table 3.** Results for subjects 1, 2, and 3

|  | Subject 1 | Subject 2 | Subject 3 |
|---|---|---|---|
| BCI competition (1 sec.) | 79.60 | 70.31 | 56.02 |
| BCI competition (long window) | 95.98 | 79.49 | 67.43 |
| 3-class classifier | 74.8 | 60.7 | 50.2 |
| Transition detector | 80.8 | 74.6 | 52.2 |
| Moving window (ws = window size) | 94.8 (ws=78) | 86.3 (ws=92) | - |
| MW small sample | 84.2 (ws=12) | 82.5 (ws=35) | - |
| Growing window | 96.9 | 88.0 | 61.7 |

It can also be seen that using windows improves accuracy significantly (second, fifth, sixth, and seventh rows versus the first one of Table 3).

## 5    Summary and Conclusions

It is known that in BCI classification problems, EEG samples run in sequences belonging to the same class (thought), and then followed by a transition into a different class. We present a method that takes this fact into consideration with the aim of improving the classification accuracy. The general classification problem is divided into two subproblems: detecting class transitions and determining the class between transitions. Class transitions are detected by computing the L1 norm between PSDs at two consecutive times; if the distance is larger than a certain threshold then a class transition is detected. Threshold values are automatically determined by the method. Once transitions can be detected, the second subproblem -determining the class between transitions- is considered. First, since the class before the transition is known, it can be discarded after the transition and therefore, a N-class problem becomes a (N-1)-class problem, which is simpler. Second, in order to determine the class between transitions, a moving window is used to predict the class of each testing instance by taking into account the $n$ last predictions. A heuristic estimation of the window size $n$ is computed based on standard statistical theory. A growing window that takes into account all instances since the last transition has also been tested.

This method has been applied to a high quality dataset with precomputed features, resulting in a three-class classification problem with 96 input attributes. This dataset corresponds to three subjects, with four sessions for each one: three for training and one for testing. Several experiments have been done in order to validate the method and the obtained results show that just by applying the transition detector, the classification rates are better than when a 3-class classifier is used. When the moving and growing windows are used, the results are further improved. Results are competitive to the best ones obtained in the BCI competition: similar for subject 1, better for subject 2, and worse for subject 3. Predictions made by the growing window are more accurate than those of the moving window. However, the growing window might take longer to recover from bursts of missclassifications and therefore be less useful for on-line use. We have also shown that even if a smaller window size is used, the classification rates are still better than those that use only the transition detector and the 3-class classifier.

## Acknowledgements

The authors wish to acknowledge the Spanish Ministry of Education and Science for financial support under the project M*::UC3M-TIN2008-06491-C04-04.

## References

1. Curran, E.A., Stokes, M.J.: Learning to control brain activity: a review of the production and control of eeg components for driving brain computer interface (bci) systems. Brain Cognition 51 (2003)
2. Dornhege, G., Krauledat, M., Muller, K.-R., Blankertz, B.: General Signal Processing and MAchine Learning Tools for BCI Analysis. In: Toward Brain-Computer Interfacing, pp. 207–234. MIT Press, Cambridge (2007)

3. Neuper, C., Pfurtscheller, G., Birbaumer, N.: In: Motor Cortex in Voluntary Movements, ch. 14, pp. 367–401. CRC Press, Boca Raton (2005)
4. Galan, F., Oliva, F., Guardia, J.: Bci competition iii. data set v: Algorithm description. In: Brain Computer Interfaces Competition III (2005)
5. Galán, F., Oliva, F., Guardia, J.: Using mental tasks transitions detection to improve spontaneous mental activity classification. Medical and Biological Engineering and Computing 45(6), 1741–0444 (2007)
6. Gan, J.Q., Tsui, L.C.S.: Bci competition iii. data set v: Algorithm description. In: Brain Computer Interfaces Competition III (2005)
7. Garner, S.: Weka: The waikato environment for knowledge analysis. In: Garner, S.R. (ed.) WEKA: The waikato environment for knowledge analysis. Proc. of the New Zealand Computer Science Research Students Conference, pp. 57–64 (1995)
8. Kubler, A., Muller, K.-R.: An Introduction to Brain-Computer Interfacing. In: Toward Brain-Computer Interfacing, pp. 1–26. MIT Press, Cambridge (2007)
9. del R. Millán, J.: On the need for on-line learning in brain-computer interfaces. In: Proceedings of the International Joint Conference on Neural Networks, Budapest, Hungary. IDIAP-RR 03-30 (July 2004)
10. Mourino, J., Millan, J., del, R., Renkens, F.: Noninvasive brain-actuated control of a mobile robot by human EEG. IEEE Trans. Biomed. Eng. 51 (2004)
11. Vapnik, V.: Statistical Learning Theory. John Wiley and Sons, Chichester (1998)

# Tuning Iris Recognition for Noisy Images

Artur Ferreira[1,2], André Lourenço[1,2], Bárbara Pinto[1], and Jorge Tendeiro[1]

[1] Instituto Superior de Engenharia de Lisboa, Lisboa, Portugal
[2] Instituto de Telecomunicações, Lisboa, Portugal
arturj@cc.isel.ipl.pt, alourenco@deetc.isel.ipl.pt,
babajp@gmail.com, jorge.tendeiro@safira.pt

**Abstract.** The use of iris recognition for human authentication has been spreading in the past years. Daugman has proposed a method for iris recognition, composed by four stages: segmentation, normalization, feature extraction, and matching. In this paper we propose some modifications and extensions to Daugman's method to cope with noisy images. These modifications are proposed after a study of images of CASIA and UBIRIS databases. The major modification is on the computationally demanding segmentation stage, for which we propose a faster and equally accurate template matching approach. The extensions on the algorithm address the important issue of pre-processing, that depends on the image database, being mandatory when we have a non infra-red camera, like a typical WebCam. For this scenario, we propose methods for reflection removal and pupil enhancement and isolation. The tests, carried out by our C# application on grayscale CASIA and UBIRIS images, show that the template matching segmentation method is more accurate and faster than the previous one, for noisy images. The proposed algorithms are found to be efficient and necessary when we deal with non infra-red images and non uniform illumination.

## 1 Introduction

Human authentication is an important problem nowadays [10]. Instead of passwords, or magnetic cards, biometric authentication is based on physical or behavioral characteristics of humans. From the set of biological characteristics, such as face, fingerprint, iris, hand geometry, ear, signature, and voice, iris is considered to be one the most accurate. It is unique to every individual (including twins) because its pattern variability is enormous among different persons, its epigenetic pattern remains static through life. This makes iris very attractive for authentication and identification.

The problem of iris recognition attracted a lot of attention in the literature: Daugman [3], Boles [2], and Wildes [16] were the first to address this field. They proposed methods composed by four stages: segmentation, normalization, feature extraction, and matching. Recently, some modifications of [3] have been proposed:

- [5] applies Hidden Markov Models to choose a set of local frequencies;
- [7] modifies segmentation stage;
- [8] modifies the normalization stage;
- [1] changes segmentation and normalization stages;
- [17] uses a different set of filters.

A. Fred, J. Filipe, and H. Gamboa (Eds.): BIOSTEC 2009, CCIS 52, pp. 211–224, 2010.
© Springer-Verlag Berlin Heidelberg 2010

In this paper we follow the approach described in [3] changing its segmentation stage and focusing the tuning of the algorithm for the noisy UBIRIS database [12][1]. These database has RGB images taken with a minimum of cooperation from the user, and thus present undesired effects. We also propose pre-processing techniques for reflection removal, and pupil enhancement and isolation. These algorithms are also evaluated on the database of infra-red images CASIA (Chinese Academy of Sciences Institute of Automation)[2]. Images of CASIA were obtained under a controlled environment.

The paper is organized as follows. Section 2 presents the stages of an iris recognition algorithm and details Daugman's approach. Section 3 describes the standard test images of CASIA and UBIRIS databases, including a statistical analysis of their features. Modifications to Daugman's method and new approaches for pre-processing and image enhancement are proposed in Section 4. Our experimental results are shown in Section 5 and concluding remarks are given in Section 6.

## 2    Iris Recognition

The process of iris recognition is usually divided into four stages [15]:

i) **Segmentation** - localization of the iris, given by its inner and outer boundaries, as shown in Fig. 1; can be preceded by a pre-processing stage to enhance image quality;
ii) **Normalization** - create a dimensionality consistent representation of the iris region (see Fig. 2);
iii) **Feature Extraction** - extraction of (texture) information to distinguish different subjects, creating a template that represents the most discriminant features of the iris;
iv) **Matching** - the computed and stored feature vectors are compared against each other using a similarity measure.

### 2.1    Daugman's Approach

Consider an intensity image $I(x, y)$, where x and y denote its rows and columns, respectively. The segmentation stage solves the problem of automatically locating the iris on the image. In [3], the inner and outer boundaries of the iris are located with the Integro-Differential (ID) operator

$$\max_{r, \, x_0, \, y_0} \left| G_\sigma(r) * \frac{\partial}{\partial r} \oint_{r, x_0, y_0} \frac{I(x, y)}{2\pi r} \, \partial s \right|, \tag{1}$$

in which $r$ represents the radius, $x_0, y_0$ the coordinates of the center of the iris and pupil and $G_\sigma(r)$ a gaussian filter used to soften the image (with $\sigma$ standard deviation). The ID operator formulates the problem as the search for the circle with center $x_0, y_0$ and radius $r$, which has a maximum change in pixel values between its adjacent circles.

---

[1] http://iris.di.ubi.pt/index.html
[2] http://www.sinobiometrics.com

Fixing different centers, first derivatives are computed varying the radius; its maximum corresponds to a boundary.

Fig. 1 shows an eye image with red and yellow circles representing the iris and pupil boundaries, respectively. On the right-side, we have the segmented iris, in which we can see the presence of the upper eyelids.

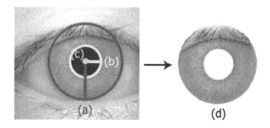

**Fig. 1.** Segmentation: (a) radius of iris (b) radius of pupil (c) center (d) iris

The normalization stage maps the iris region into a normalized image, with fixed size, allowing comparisons of different iris sizes. Iris may have different sizes due to pupil dilation caused by varying levels of illumination. The rubber sheet model [3], remaps each point $(x, y)$ of the iris image, into an image $I(r, \theta)$ where $r \in [0, 1]$ and $\theta \in ] -\pi, \pi]$, according to

$$I(x(r, \theta), y(r, \theta)) \to I(r, \theta). \tag{2}$$

Mapping from cartesian to normalized polar coordinates is done by

$$\begin{cases} x(r, \theta) = (1 - r)x_{\text{pupil}}(\theta) + rx_{\text{iris}}(\theta) \\ y(r, \theta) = (1 - r)y_{\text{pupil}}(\theta) + ry_{\text{iris}}(\theta), \end{cases} \tag{3}$$

where $(r, \theta)$ are the corresponding normalized coordinates, and $x_{\text{pupil}}$, $y_{\text{pupil}}$ and $x_{\text{iris}}$, $y_{\text{iris}}$ the coordinates of the pupil and iris boundary along $\theta$ angle. Fig. 2 shows an example of the normalized representation; it is possible to observe in the left image that the center of the pupil can be displaced with respect with the center of the iris. The right image represents the normalized image: on the x-axis we have the angles ($\theta$), and on the y-axis the radius ($r$). Observe that the upper eyelids are depicted on the lower right corner.

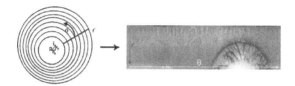

**Fig. 2.** Normalization stage: Daugman's Rubber Sheet Model

Fig. 3 shows the segmentation and normalization stages.

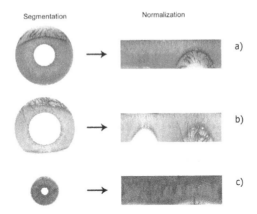

Segmentation                    Normalization

a)

b)

c)

**Fig. 3.** Segmentation and Normalization for images of different databases: a) CASIA b) CASIA c) UBIRIS

To encode the iris pattern, 2D Gabor filters are employed

$$G(r, \theta) = e^{-i\omega_0(\theta_0 - \theta)} e^{-\frac{(r_0 - r)}{\alpha^2}} \cdot e^{-\frac{(\theta_0 - \theta)^2}{\beta^2}}, \tag{4}$$

being characterized by spacial localization $x_0, y_0$, spacial frequency $w_0$ and orientation $\theta_0$, and gaussian parameters $(\alpha, \beta)$. These filters, considered very suitable to encode texture information, generate a complex image representing the relevance of the texture for a given frequency and orientation. Several filters are applied to analyze different texture information. This set of filters is defined by

$$GI = \int_\rho \int_\phi e^{-i\omega(\theta_0 - \phi)} e^{-\frac{(r_0 - \rho)}{\alpha^2} - \frac{(\theta_0 - \phi)^2}{\beta^2}} I(\rho, \phi) \, \rho \, d\rho \, d\theta. \tag{5}$$

Fig. 4 shows Gabor filters, for a given frequency $w_o$ with eight spacial orientations.

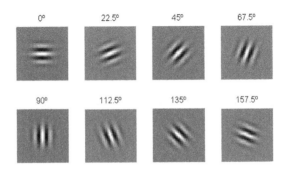

0°        22.5°        45°        67.5°

90°        112.5°        135°        157.5°

**Fig. 4.** Gabor filters for a given frequency $w_o$ with eight orientations for texture extraction

The output of the filters is quantized using the real and imaginary part, Re(GI) and Im(GI), respectively by

$$h_{Re} = \begin{cases} 1, \text{ if } \text{Re(GI)} \geq 0 \\ 0, \text{ if } \text{Re(GI)} < 0, \end{cases} \quad \text{and} \quad h_{Im} = \begin{cases} 1, \text{ if } \text{Im(GI)} \geq 0 \\ 0, \text{ if } \text{Im(GI)} < 0. \end{cases} \quad (6)$$

These four levels are quantized using two bits, obtaining a binary vector named IrisCode with 2048 bits (256 bytes), computed for each template, corresponding to the quantization of the output of the filters for different orientations and frequencies (1024 combinations). The matching step [4], computes the differences between two iriscodes (codeA and codeB) using the Hamming Distance (HD) given by

$$HD = \frac{\|(codeA \otimes codeB) \cdot maskA \cdot maskB\|}{\|maskA \cdot maskB\|}, \quad (7)$$

where $\otimes$ denotes the XOR operator, ($\|\,\|$) represents the norm of a vector, and the AND operator "." selects only the bits that have not been corrupted by eyelashes, eyelids, specular reflections, as specificed by $maskA$ and $maskB$. To take into account possible rotations between two iris images, HD is compared with HD obtained using cyclic scrolling versions of one of the images; the minimum HD gives the final matching result.

## 3  Image Databases

This section describes the main features of CASIA and UBIRIS image databases, as well as our synthetic images. The well-known CASIA-Chinese Academy of ciences Institute of Automation database from Beijing, China[3] has two versions: CASIAv1, with 756 images of 108 individuals; CASIAv3, with 22051 images of over 700 individuals. CASIAv3 is divided into the following categories:

- interval - digitally manipulated such that the pupil was replaced by a circular shape with uniform intensity, eliminating undesired illumination effects and artifacts such as reflection; CASIAv1 is a subset of this category;
- lamp - images acquired under different illumination conditions, to produce intra-class modifications (images of same eye taken in different sessions);
- twins - images of 100 pairs of twins.

These $320 \times 280$, 8 bit/pixel images were acquired with an infra-red (IR) camera; we have used their grayscale versions.

The UBIRIS database [12][4], was developed by the Soft Computing and Image Analysis Group of Universidade da Beira Interior, Covilhã, Portugal. This database was created to provide a set of test images with some typical perturbations such as blur, reflections and eyes almost shut, being a good benchmark for systems that minimize the requirement of user cooperation. Version 1 of the database (UBIRIS.v1) has 1877 images of 241 individuals, acquired in two distinct sessions:

---

[3] http://www.sinobiometrics.com
[4] http://iris.di.ubi.pt/index.html

216    A. Ferreira et al.

- session 1 - acquisition in a controlled environment, with a minimum of perturbation, noise, reflection and non-uniform illumination;
- session 2 - acquisition under natural light conditions.

These images, taken with a NIKON E5700 digital RGB camera[5], have a resolution of $200 \times 150$ pixels and a pixel-depth of 8 bit/pixel, were converted to 256-level grayscale images. Recently, a second version of the database (UBIRIS.v2) was released for the Noisy Iris Challenge Evaluation challenge - Part I 11 (NICE.I). Fig. 5 shows some test images: synthetic (with and without noise); real images from CASIA and UBIRIS databases.

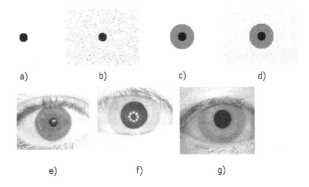

**Fig. 5.** Test images: a) simple b) simple with noise c) pupil and iris d) pupil and iris with noise e) UBIRISv1 f) CASIAv3 g) CASIAv1.

Our set of synthetic images is divided into the following subsets of 30 images each:

- simple - white background image with black circumferences, randomly placed;
- simple with noise - the same as the *simple* type, but with added noise;
- pupil and iris - white background image with two circumferences randomly placed, simulating the human eye;
- pupil and iris with noise - the same as *pupil and iris*, with added noise.

The main features of these databases are summarized in Table 1.

**Table 1.** Main features of CASIA and UBIRIS image databases

| Database | # Images | Resolution |
|---|---|---|
| CASIAv1 | 756 | $320 \times 280$ |
| CASIAv3 | 22051 | $320 \times 280$ |
| UBIRISv1 | 1877 | $200 \times 150$ |

---

[5] http://www.nikon.com/about/news/2002/e5700.htm

### 3.1   Study of UBIRIS and CASIA

The accuracy of pupil and iris detection is a crucial issue in an iris recognition system. The proposed method is based on template matching. To define the templates we carried out a statistical study over UBIRIS and CASIA databases to estimate the range of pupil diameters. For both databases, we randomly collect N=90 images computing the center and the diameter of the pupil. Table 2 shows a statistical analysis of the pupil diameter.

**Table 2.** Statistical analysis of pupil diameter, in pixels, for CASIA (v1 and v3) and UBIRISv1 image databases

| Measure | CASIA | UBIRIS |
|---|---|---|
| Mean | 86.2 | 23.7 |
| Median | 87 | 24 |
| Mode | 77 | 25 |
| Minimum | 65 | 17 |
| Maximum | 119 | 31 |
| Standard Deviation | 11.5 | 2.9 |
| Sample Variance | 132.7 | 8.3 |

For N larger than 90 there is no difference in these statistical results. Fig. 6 shows the histogram of the diameters for both databases; the UBIRIS histogram is well approximated by a normal distribution.

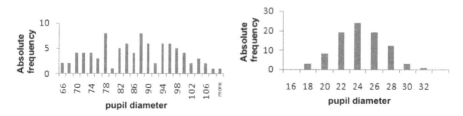

**Fig. 6.** Histogram of pupil diameter for CASIA (left) and UBIRIS (right)

## 4   Modifications and Extensions

In the case of noisy iris images, we have to use pre-processing methods before segmentation. These methods are necessary only for the UBIRIS database, since CASIAv1 images are already pre-processed (see Section 3).

### 4.1   Reflection Removal

This section describes the pre-processing techniques that we propose for grayscale images of UBIRIS. The pre-processing stage removes (or minimizes) image impairments

218    A. Ferreira et al.

such as noise and light reflections. We propose 3 methods, named A, B, and C for reflection removal and another method for pupil enhancement.

Taken a histogram analysis, from a set of 21 images, we conclude that the pupil has low intensity values, corresponding to first 7% to 10% of ocurrences in the histogram. From this range we compute a threshold between the pupil and the iris, as depicted by $T$ in Fig. 7.

The actions taken by Method A for reflection removal are as follows.

---

**Reflection Removal - Method A**

Input:    $I_{in}$ - input image, with 256 gray-levels.
Output: $I_{out}$ - image without reflections on the pupil.

---

1. From the histogram of $I_{in}$ compute a threshold $T$ (as in Fig. 7) to locate the pupil pixels.
2. Set the pupil pixels (gray level below $T$) to zero.
3. Locate and isolate the reflection area, with an edge detector and a filling morphologic filter [9].
4. $I_{out} \leftarrow$ image with the reflection area pixels set to zero.

---

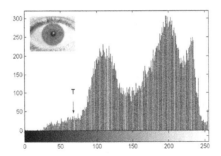

**Fig. 7.** Typical histogram of an iris image

Fig. 8 illustrates the application of Method A; on stage a) we see a white reflection on the pupil which is removed on stage d).

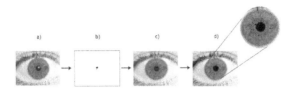

**Fig. 8.** Reflection removal by Method A: a) original image b) isolated reflection c) image without reflection d) image with an uniform pupil

Method B uses a threshold to set to a certain value the pixels on a given Region of Interest (ROI).

---

### Reflection Removal - Method B

---

Input:   $I_{in}$ - input image, with 256 gray-levels.
       $T$ - threshold for comparison.
Output: $I_{out}$ - image without reflections on the pupil.

---

1. Over $I_{in}$ locate the set of pixels below $T$.
2. Isolate this set of pixels, to form the ROI, defined by its upper-left corner $(X, Y)$ and width and height $(W, H)$.
3. Horizontally, make top-down scan of the ROI and for each line, replace each pixel in the line by the average of the pixels at both ends of that line.
4. $I_{out} \leftarrow$ image with the ROI pixels set to this average value.

---

Method C uses morphologic filters [9] to fill areas with undesired effects, such as a white circumference with a black spot. In this situation, the morphologic filter fills completely the white circumference. This filter is applied on the negative version of the image and after processing, the image is put back to its original domain.

---

### Reflection Removal - Method C

---

Input:   $I_{in}$ - input image, with 256 gray-levels.
Output: $I_{out}$ - image without reflections on the pupil.

---

1. $\tilde{I}_{in} \leftarrow$ negative version of $I_{in}$ [9].
2. $\tilde{I}_{p} \leftarrow$ output of the morphologic filling filter on $\tilde{I}_{in}$.
3. $I_{out} \leftarrow$ negative version of $\tilde{I}_{p}$.

---

Fig. 9 shows the results obtained by this method; we can see the removal of the white reflection. Among these three methods, this one is the fastest; it takes (on average) about 40 ms to run on a UBIRIS image.

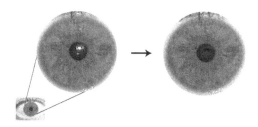

**Fig. 9.** Reflection removal by Method C

After reflection removal, we introduced a pupil enhancement algorithm to obtain better results in the segmentation phase. This way, we isolate the pupil from the rest of the image. The algorithm for pupil isolation is divided into four stages: enhancement and smooth - apply a reflection removal method and a gaussian filter to smooth the image; detection - edge and contour detection of the pupil, producing a binary image; isolation - remove the contours outside a specific area, isolating the pixels along the pupil; dilation and filling - dilate the image and fill the points across the pupil.

---

**Pupil Enhancement**

---

Input:    $I_{in}$ - input image, with 256 gray-levels.
          $\sigma$ - standard deviation for the gaussian filter.
          $T$ - minimum number of pupil pixels.
Output: $I_{out}$ - binary image with an isolated pupil.

---

1. Remove reflections (methods A, B or C) on $I_{in}$.
2. Apply a gaussian filter $G_\sigma$ to smooth the image.
3. Apply the Canny Edge [9] detector for pupil and iris detection; retain the pupil area.
4. While the number of white pixels is below $T$, dilate the detected contours.
5. $I_{out} \leftarrow$ output of the filling morphologic filter.

---

### 4.2    Segmentation

The segmentation phase is of crucial importance, because without a proper segmentation it is impossible to perform recognition. It is also the most computationally demanding stage, from the usual four stages of iris recognition algorithms. This way, we propose the following methods for segmentation: simplified versions of the ID operator; template matching (TM). For the first we propose the following options:

- version 1 - simplified version of (2) without the gaussian smoothing function;
- version 2 - finite difference approximation to the derivative and interchanging the order of convolution and integration as in [3];
- version 3 - the operator as in (2).

In order to speed up the performance of the operator we have considered a small range of angles to compute the contour integral: $\theta \in [-\pi/4, \pi/4] \cup [3\pi/4, 5\pi/4]$. For UBIRIS, we devised a new strategy for the segmentation phase, based on a template matching approach. We propose to automatically segment the image using cross-correlation between the iris images and several templates and finding the maximums of this operation. Template matching is an extensively used technique in image processing [9]. Since the iris and pupil region have a circle format (or approximate) this technique is considered very suitable, being only necessary to use circle templates with different sizes (being not necessary to take into account rotations of the templates). To cope with the range of diameters, we have used several versions of the templates. Supported by the study presented on section 3.1, we considered a range of diameters that covers 90% of the diameters displayed in Fig. 6, to narrow the number of templates. This way, we have choosen the set of diameters D = {20, 22, 24, 26, 28, 30}; four of these templates are depicted in Fig. 10. The difference between two consecutive templates is two pixels. We have found that is not necessary to consider the entire set of integers, to have an accurate estimation of the diameter (this way, we decrease the number of comparisons to half). For the CASIA database, we proceed in a similar fashion with diameters D = {70, 72, 74, . . . , 124}. The cross-correlation based template matching technique has an efficient implementation using FFT (Fast Fourier Transform) and its inverse [9].

**Fig. 10.** Templates used for the proposed template matching-based segmentation stage on UBIRIS database

## 5    Experimental Results

This section reports the experimental results obtained with our $C\#$ application. Fig. 11 shows a screen shot of this application, which has the following functionalities: enrollment - register an individual; authentication - verify the identity of an already registered user; identification - search for an individual. Regarding the pre-processing stage, we have found that the reflection removal methods A, B and C presented in section 4.1 attained good and similar results as shown in Fig. 8 and 9. The pupil enhancement algorithm also lead to improvements on the segmentation stage.

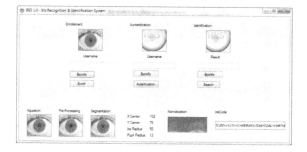

**Fig. 11.** $C\#$ prototype for iris recognition, with enrollment, authentication, and identification functionalities. On the bottom-left corner we see the results of the pre-processing algorithms and the segmentation stage. The bottom-right corner shows the normalization stage results and the IrisCode.

## 5.1  Segmentation Stage

For the segmentation stage of Daugman's algorithm, the variants described in section 4.2, regarding the ID and TM operators were evaluated. The (fast) first version of the ID operator obtained satisfactory results only for synthetic images(see Section 3 and Fig. 5). The second version performed a little better, but the results still were unsatisfactory:

- on the detection of the diameter of the pupil and iris, given the known center, attains 100 % and 90 % success for synthetic *simple* and *simple with noise* images, respectively;
- on the detection of the center of the pupil and iris, given the known diameter, we also got 100 % and 90 % success for *simple* and *simple with noise* images, respectively;
- on the detection of the center and diameter of the pupil and iris, whitout any priori knowledge, we got 90 % success for *simple* and *simple with noise* images and 64 % for UBIRIS database.

These tests showed that this simplification of the ID operator, although efficient from the computational point of view, does not obtain good results for real images. Table 3 shows the percentage of success in the detection of the diameter and center of pupil and iris for ID operator (version 3), the TM operator, and TM followed by ID. On table 3, the worst results for CASIAv3 and UBIRIS are justified by the fact that these images contain reflections. The pre-processed CASIAv1 images are easier to segment, justifying better results. In our tests, on both databases the TM approach runs about 7 to 10 times faster than the third version of the ID operator. The template matching segmentation takes 0.1 and 1.5 seconds for UBIRIS and CASIA, respectively; this is due to the larger resolution CASIA images; we have to use a larger number of templates (4 times as for UBIRIS). The use of TM followed by ID operator has a very small gain with larger computation time than the TM technique. This way, it is preferable to use solely the TM technique, which is faster and more accurate on images with noise and impairments.

**Table 3.** Percentage of success of segmentation stage, in the detection of the diameter and center of pupil and iris for: ID operator (version 3); TM; TM followed by ID. For CASIAv3, 756 images were randomly selected.

| Database | # Images | ID | TM | TM + ID |
|----------|----------|--------|--------|---------|
| UBIRISv1 | 1877 | 95.7 % | 96.3 % | 96.3 % |
| CASIAv1 | 756 | 98.4 % | 98.5 % | 98.7 % |
| CASIAv3 | 756 | 94 % | 98.8 % | 96.8 % |

## 5.2  Recognition Rate

Using Gabor filters with eight orientations (see Fig. 4) and four frequencies, our implementation got a recognition rate of 87.2% and 88%, for UBIRISv1 and CASIAv1, respectively. This recognition rate can be considerably improved; it is known that it is possible to achieve higher recognition rate with Daugman's method on CASIAv1, using a larger IrisCode [11], cyclic scrolling versions of the images and feature selection schemes [13] [3]. Our main goal in this work was to show that when we do not have infra-red already pre-processed (CASIAv1-like) images: the reflection removal pre-processing stage is necessary; sometimes pupil enhancement methods are also necessary; the segmentation stage can be performed much faster with an efficient FFT-based template matching approach.

## 6   Conclusions

We addressed the problem of iris recognition, by modifying and extending the well-known Daugman's method. We have developed a C# application and evaluated its performance on the public domain UBIRIS and CASIA databases. The study that was carried out over these databases allowed us to propose essentially two new ideas for: reflection removal; enhancement and isolation of the pupil and iris. For the reflection removal problem, we have proposed 3 different methods. The enhancement and isolation of the pupil, based on morphologic filters, obtained good results for both databases. It is important to stress that this pre-processing algorithms depend on the image database. Regarding the segmentation stage, we replaced the proposed integro-differential operator by an equally accurate and faster cross-correlation template matching criterion, which has an efficient implementation using the FFT and its inverse. This way, we have improved the segmentation stage, because the template matching algorithm is more tolerant to noisy images, than the integro-differential operator and runs faster. As future work we intend to continue tuning and optimizing the algorithm for the noisy UBIRIS database, using color images.

## References

1. Arvacheh, E.M.: A study of segmentation and normalization for iris recognition systems. Master's thesis, University of Waterloo (2006)
2. Boles, W.: A security system based on iris identification using wavelet transform. In: Jain, L.C. (ed.) First International Conference on Knowledge-Based Intelligent Electronic Systems, Adelaide, Australia, pp. 533–541 (1997)
3. Daugman, J.: High confidence visual recognition of persons by a test of statistical independence. IEEE Transactions on Pattern Analysis and Machine Intelligence 25(11), 1148–1161 (1993)
4. Daugman, J.: How iris recognition works. IEEE Transactions on Circuits and Systems for Video Technology 14(1), 21–30 (2004)
5. Greco, J., Kallenborn, D., Nechyba, M.C.: Statistical pattern recognition of the iris. In: 17th annual Florida Conference on the Recent Advances in Robotics, FCRAR (2004)
6. Jain, A.K., Ross, A., Prabhakar, S.: An introduction to biometric recognition. IEEE Transactions on Circuits and Systems for Video Technology 14(1) (2004)
7. Huang, J., Wang, Y., Tan, T., Cui, J.: A new iris segmentation method for recognition. In: 17th Int. Conf. on Pattern Recognition, ICPR 2004 (2004)
8. Joung, B.J., Kim, J.O., Chung, C.H., Lee, K.S., Yim, W.Y., Lee, S.H.: On improvement for normalizing iris region for a ubiquitous computing. In: Gervasi, O., Gavrilova, M.L., Kumar, V., Laganá, A., Lee, H.P., Mun, Y., Taniar, D., Tan, C.J.K. (eds.) ICCSA 2005. LNCS, vol. 3480, pp. 1213–1219. Springer, Heidelberg (2005)

9. Lim, J.: Two-dimensional Signal and Image Processing. Prentice-Hall, Englewood Cliffs (1990)
10. Maltoni, D., Maio, D., Jain, A., Prabhakar, S.: Handbook of Fingerprint Recognition, 1st edn. Springer, Heidelberg (2005)
11. Masek, L.: Recognition of human iris patterns for biometric identification. Master's thesis, University of Western Australia (2003)
12. Proença, H., Alexandre, L.: UBIRIS: a noisy iris image database. In: Roli, F., Vitulano, S. (eds.) ICIAP 2005. LNCS, vol. 3617, pp. 970–977. Springer, Heidelberg (2005)
13. Proença, H., Alexandre, L.: A Method for the Identification of Noisy Regions in Normalized Iris Images. In: 18th International Conference on Pattern Recognition (ICPR 2006), Hong Kong, August 20-24, vol. 4, pp. 405–408 (2006)
14. Proença, H.: Towards Non-Cooperative Biometric Iris Recognition. PhD thesis, Universidade da Beira Interior (2007)
15. Vatsa, M., Singh, R., Gupta, P.: Comparison of iris recognition algorithms. In: Proceedings of International Conference on Intelligent Sensing and Information Processing, pp. 354–358 (2004)
16. Wildes, R.: Iris recognition: an emerging biometric technology. Proceedings of the IEEE 85(9), 1348–1363 (1997)
17. Yao, P., Li, J., Ye, X., Zhuang, Z., Li, B.: Analysis and improvement of an iris identification algorithm. In: 18th International Conference on Pattern Recognition (ICPR 2006), Hong Kong, August 20-24, vol. 4, pp. 362–365 (2006)

# Three-Dimensional Reconstruction of Macroscopic Features in Biological Materials

Michal Krumnikl[1], Eduard Sojka[1], Jan Gaura[1], and Oldřich Motyka[2]

[1] Department of Computer Science, VŠB - Technical University of Ostrava
17. listopadu 15/2172, Ostrava, Czech Republic
{Michal.Krumnikl,Eduard.Sojka,Jan.Gaura}@vsb.cz
[2] Institute of environmental engineering, VŠB-Technical University of Ostrava
17. listopadu 15/2172, Ostrava, Czech Republic
Oldrich.Motyka.st@vsb.cz

**Abstract.** This paper covers the topic of three dimensional reconstruction of small textureless formations usually found in biological samples. Generally used reconstructing algorithms do not provide sufficient accuracy for surface analysis. In order to achieve better results, combined strategy was developed, linking stereo matching algorithms with monocular depth cues such as depth from focus and depth from illumination.

Proposed approach is practically tested on bryophyte canopy structure. Recent studies concerning bryophyte structure applied various modern, computer analysis methods for determining moss layer characteristics drawing on the outcomes of a previous research on surface of soil. In contrast to active methods, this method is a non-contact passive, therefore, it does not emit any kind of radiation which can lead to interference with moss photosynthetic pigments, nor does it affect the structure of its layer. This makes it much more suitable for usage in natural environment.

## 1 Introduction

Computer vision is still facing the problem of three-dimensional scene reconstruction from two-dimensional images. Not a few algorithms have been developed and published to solve this problem. These algorithms can be divided into two categories; passive and active [1]. Passive approaches such as shape from shading or shape from texture recover the depth information from a single image [2,3,4,5]. Stereo and motion analysis use multiple images for finding the object depth dependencies [6,7,8,9]. These algorithms are still developed to achieve higher accuracy and faster computation, but it is obvious that none will ever provide universal approach applicable for all possible scenes.

In this paper, we present combined strategy for reconstructing three dimensional surface of textureless formations. Such formations can be found in biological and geological samples. Standing approaches suffer mainly from the following shortcomings: high error rate of stereo correspondence in images with large disparities, feature tracking is not always feasible, samples might be damaged by active illumination system, moreover biological samples can even absorb incident light. Observing these drawbacks we have developed system linking several techniques, specially suited for our problem.

A. Fred, J. Filipe, and H. Gamboa (Eds.): BIOSTEC 2009, CCIS 52, pp. 225–234, 2010.

Presented reconstruction method was tested on bryophyte canopies surfaces. Obtaining three dimensional surface is the first stage of acquiring surface roughness, which is used as biological monitor [10,11].

## 2   Methods

In this section we will briefly describe the methods involved in our system, emphasizing improvements to increase the accuracy and the density of reconstructed points. As the base point, stereo reconstruction was chosen. Selected points from the reconstruction were used as the reference points for depth from illumination estimation. Missing points in the stereo reconstruction were calculated from the illumination and depth from focus estimation. The last technique provides only rough estimation and is largely used for the verification purposes.

### 2.1   Stereo 3D Reconstruction

The following main steps leading to a reconstruction of a sample surface are performed by particular parts of the system: (i) calibration of the optical system (i.e., the pair of cameras), (ii) 3D reconstruction of the sample surface itself. In the sequel, the mentioned steps will be described in more details.

In the calibration step, the parameters of the optical system are determined, which includes determining the intrinsic parameters of both the cameras (focal length, position of the principal point, coefficients of nonlinear distortion of the lenses) and the extrinsic parameters of the camera pair (the vector of translations and the vector of rotation angles between the cameras). For calibrating, the chessboard calibration pattern is used. The calibration is carried out in the following four steps: (1) creating and managing the set of calibration images (pairs of the images of calibration patterns captured by the cameras), (2) processing the images of calibration patterns (finding the chessboard calibration pattern and the particular calibration points in it), (3) preliminary estimation of the intrinsic and the extrinsic parameters of the cameras, (4) final iterative solution of all the calibration parameters. Typically, the calibration is done only from time to time and not necessarily at the place of measurement.

For solving the tasks that are included in Step 2, we have developed our own methods that work automatically. For the initial estimation of the parameters (Step 3), the method proposed by Zhang [12,13] was used (similar methods may now be regarded as classical; they are also mentioned, e.g., by Heikilla and Silven [14], Heikilla [15], Bouguet [16] and others). The final solution of calibration was done by the minimization approach. The sum of the squares of the distances between the theoretical and the real projections of the calibration points was minimized by the Levenberg-Marquardt method.

If the optical system has been calibrated, the surface of the observed sample may be reconstructed, which is done in the following four steps: (i) capturing a pair of images of the sample, (ii) correction of geometrical distortion in the images, (iii) rectification of the images, (iv) stereomatching, (v) reconstruction of the sample surface.

Distortion correction removes the geometrical distortion of the camera lenses. The polynomial distortion model with the polynomial of the sixth degree is used. The distortion coefficients are determined during the calibration.

Rectification of the images is a computational step in which both the images that are used for reconstruction are transformed to the images that would be obtained in the case that the optical axes of both cameras were parallel. The rectification step makes it easier to solve the subsequent step of finding the stereo correspondence, which is generally difficult. The rectification step is needed since it is impossible to guarantee that optical axes are parallel in reality. We have developed a rectification algorithm that takes the original projection matrices of the cameras determined during calibration and computes two new projection matrices of fictitious cameras whose optical axes are parallel and the projection planes are coplanar. After the rectification, the corresponding points in both images have the same $y$-coordinate.

The dense stereo matching problem consists of finding a unique mapping between the points belonging to two images of the same scene. We say that two points from different images correspond one to another if they depict a unique point of a three-dimensional scene. As a result of finding the correspondence, so called disparity map is obtained. For each image point in one image, the disparity map contains the difference of the $x$-coordinates of that point and the corresponding point in the second image. The situation for finding the correspondence automatically is quite difficult in the given context since the structure of the samples is quite irregular and, in a sense, similar to noise. We have tested several known algorithms [9,7,6] for this purpose. The results of none of them, however, are fully satisfactory for the purpose of reconstruction. Finally, we have decided to use the algorithm that was proposed by Ogale and Aloimonos [6] that gave the best results in our tests.

## 2.2  Auxiliary Depth Estimators

The depth map obtained from the 3D reconstruction is further processed in order to increase the resolution and fill the gaps of missing data. To achieve this we have implemented several procedures based on the theory of human monocular and binocular depth perception [17]. Exploited monoculars cues were depth from the focus and lighting cues.

Depth estimation based on lighting cues was presented in several papers [18,19,20]. Methods proposed in these papers were stand-alone algorithms requiring either calibrated camera or controlled light source. More simplified setup was proposed in [18] using projector mounted on a linear stage as a light source. These approaches calculate depth from multiple images taken from different angles or with varying lighting.

For our application we have developed slightly modified method based on the previous research which is capable of acquiring depths from just one image. Assuming that we have already preliminary depth map (e.g. disparity map from the stereo matching algorithm) we can find the correspondence of depth and the light intensity.

According to the inverse square law, the measured luminous flux density from a point light source is inversely proportional to the square of the distance from the source. The intensity of light $I$ at distance $r$ is $I = \frac{P}{4\pi r^2}, I \propto \frac{1}{r^2}$, here $P$ is the total radiated power from the light source. Analyzed surfaces were approximated by Lambertian reflectors.

In contrast to [18] we are assuming mostly homogeneous textureless uniform colored surfaces. Observing these assumptions we can omit the step involving the computation of the ratio of intensity from two images and alter the equation in order to exploit already known depth estimation from previous 3D reconstruction based on the stereo matching algorithm. Points that show high level of accuracy in the previous step are used as the reference points for illumination estimation. Look-up table, based on the inverse square law, mapping the computed depth to intensity is calculated.

The last method involved in the reconstruction step is the integration of focus measurement. The depth from focus approaches have been used many times for real time depth sensors in robotics [21,22,23,24]. Key to determine the depth from focus is the relationship between focused and defocused images. Blurring is represented in the frequency domain as the low-pass filter. The focus measure is estimated by frequency analysis. Discrete Laplacian is used as the focus operator. For each image segment, depth is estimated as the maximal output of the focus operator. Since the images were taken from short range, the lens focus depth was large enough to capture the scene with high details. Thus only a few focus steps were used in the experiments.

The final depth is calculated as the weighted average from values given by stereo reconstruction process, light intensity and depth from focus estimation. Appropriate weights were set according to the experimental results gained with each method. The graphical illustration (Figure 1) shows the depth of acquired data. Still the biggest disadvantage of camera based scanning technique is the occurrence of occlusions, decreasing the reconstructed point density.

**Fig. 1.** Comparison of used methods for depth estimation

## 3   Bryophyte Structure

Bryophytes are plants of a rather simple structure, they lack of roots and conductive tissues of any kind or even structural molecules that would allow establishment of their elements such as lignin [25]. The water which is generally needed for metabolic processes including photosynthesis is in the case of bryophytes also necessary for reproductive purposes, for their spermatozoids, particles of sexual reproduction, are unable to survive under dry conditions [26]. This makes them, unlike tracheophytes (vascular plants), poikilohydric – strongly dependent on the water conditions of their vicinity [27] – which led to several ecological adaptations of them. One of the most important adaptations is forming of a canopy – more or less compact layer of moss plants, frequently of a same clonal origin, which enables the plants to share and store water on higher, community level.

Recent infrequent studies concerning bryophyte canopy structure applied various modern, computer analysis methods to determine moss layer characteristics drawing on the outcomes of a research on surface of soil [28]. Surface roughness index ($L_r$) has been hereby used as a monitor of quality and condition of moss layer, other indices, i.e. the scale of roughness elements ($S_r$) and the fractal dimension ($D$) of the canopy profile have been used and found to be important as well [29]. As stated in Rice [10], contact probe, LED scanner and 3D laser scanner were used and compared in light of efficiency and serviceability in 27 canopies of different growth forms. However, none of the methods already assessed have not been found to be convenient for field research, especially due to the immobility of used equipment and therefore needed dislocation of the surveyed moss material into the laboratory. This has great disadvantage in destroying the original canopy not only due to the transfer and excision from its environment, but also due to different much dryer conditions affecting moss surface in laboratory.

## 4    Bryophyte Canopy Analysis

The former methods [29,10] are suitable and efficient for measuring structural parameters in laboratory, but generally are impracticable in the field. Despite the LED scanner is presented as a portable device, it has high demands for proper settings and conditions that has to be maintained.

The method described in this paper presents a new approach using the pair of images taken by the scanning device. Computer analysis involving 3D reconstruction and soil roughness compensation is used to calculate the canopy surface roughness. The main goal was to create a device that can be used in the field, needs a minimum time for settings and is able to operate in the variety of environments.

Our device is composed of hardware parts – optical system consisting of two cameras and software, analyzing acquired images. The images are acquired from two IDS Imaging cameras (2240-M-GL, monochromatic, 1280x1024 pixels, 1/2" CCD with lenses PENTAX, f=12 mm, F1.4) firmly mounted in a distance of 32.5 mm between them (Figure 2, left). Images has been taken in normal light, no auxiliary lamp was used.

Since the cameras mounted in a given distance produce images with relative big disparity, which might produce difficult scenery for stereomatching algorithm, we have created a laboratory device which is able to take images at variable distance. The camera mounted on a linear motion is moved over the sample, taking the images from different positions. Using Kuroda linear motion SG2605A-250P we are able to achieve position accuracy of 0.02 mm with position reproducibility better than 0.003 mm. Image capture is synchronized with movement. Linear stage is mounted on aluminum chassis with embedded computer, motor driver and additional sensor and light driving units. It was

 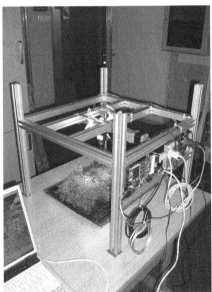

**Fig. 2.** The pair of cameras mounted on the tripod for mobile use (left) and camera mounted on linear motion with embedded computer for laboratory use (right)

meant to be used as a laboratory instrument but may be used in the field as well due to battery operation. The device is used for longitudinal studies.

### 4.1 Surface Roughness

Surface roughness is a measurement of the small-scale variations in the height of a physical surface of bryophyte canopies.

For the statistical evaluation of every selected bryophyte, we fitted all measured $z$-component values that we obtained from the 3D reconstruction (2.1) with a polynomial surface. This surface then represents the mean value of measured $z$-component. For this step, we have already a full set of $x$-, $y$-, and $z$-components (in meters) from the previous reconstruction process (Figure 3). Bottom figure shows another reconstructed sample of the same species in the original coordinates of the image.

In order to calculate the surface specific parameters, we have to minimize the impact of subsoil segmentation. We have performed a regression using the polynomial expression above to interpolate the subsoil surface. Thus we have obtained the surface that represents the average canopy level. The distances from the reconstructed $z$-coordinates and the fitting surface were evaluated statistically.

Canopy structure can be characterized by the surface characteristics. The most common measure of statistical dispersion is the standard deviation, measuring how widely $z$-values are spread in the sample.

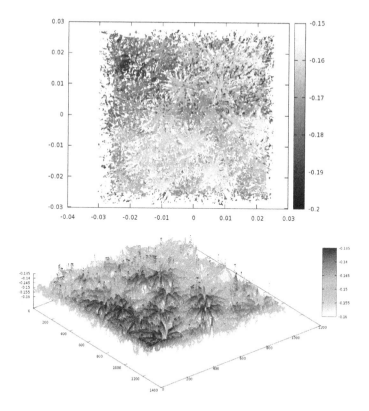

**Fig. 3.** $Z$-components from reconstructions (*Polytrichastrum formosum*)

Bryophyte canopy structure is also described by $L_r$ parameter defined by Rice as the square root of twice the maximum semivariance [29]. Semivariance is described by

$$\widehat{\gamma}(h) = \frac{1}{2n(h)} \sum_{i=1}^{n(h)} (z(x_i + h) - z(x_i))^2, \qquad (1)$$

where $z$ is a value at a particular location, $h$ is the distance between ordered data, and $n(h)$ is the number of paired data at a distance of $h$ [30].

The real surface is so complicated that only one parameter cannot provide a full description. For more accurate characteristics other parameters might be used (e.g. maximum height of the profile, average distance between the highest peak and lowest valley in each sampling length).

## 5   Results and Discussion

When applied in a study of six bryophyte species [11] surface structure (*Bazzania trilobata, Dicranum scoparium, Plagiomnium undulatum, Polytrichastrum formosum, Polytrichum commune* and *Sphagnum girgensohnii*), both in laboratory and *in situ*, mentioned approach was found to be able to obtain data suitable for surface roughness index calculation (Figure 4). Also, indices calculated in eight specimen per each species (four in laboratory and four in field measurements, total 48 specimens) were found to significantly distinguish the specimens in dependence on species kind; one-way analysis of variance showed high significance ($p = 0,000108$) when data were pooled discounting whether derived from laboratory or from field measurements. Laboratory measurements separate gave not that significant outcomes ($p = 0,0935$), for there were found distinctively different indices of *Dicranum scoparium* specimens in laboratory and in field caused probably by disturbance of their canopies when transferring and storage in laboratory. This is supported by the fact that independent two-sample $t$-test showed significant difference between laboratory and field measurements outcomes only in case of this one species ($p = 0,006976$). This approach was then found to be suitable to be utilized even under *in situ* conditions which is according to the outcomes of the mentioned study considered to be much more convenient way to study bryophyte canopy structure.

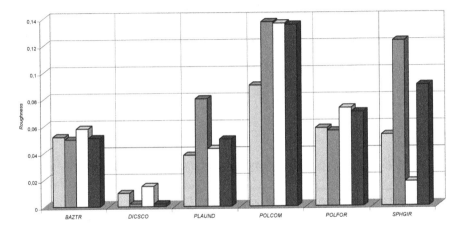

**Fig. 4.** Obtained roughness in field conditions, the variability between samples of the same species is quite high, especially in the case of *Polytrichum commune*

## 6   Conclusions

By comparing the results of pure stereo matching algorithms [9,7,6] and our combined approach we have found out our method to be more suitable for homochromatic surfaces. Biological samples we have been working with were typical by the presence of rather long narrow features (leaves or branches). Heavy density of such formations in the images is more than unusual input for the stereo matching algorithms supposing rather smooth and continuous surfaces. Segmentation used by graph cut based stereo matching algorithms usually lead in creating a great many regions without match in the second image.

Without using additional cues, described in this paper, results of reconstructed image was poor, usually similar to noise without and further value for bryophyte analysis.

Meanwhile our reconstruction process produce outputs that are sufficient for further biological investigation. Both number of analyzed specimens and number of obtained $z$-values for statistical analysis are unprecedented and this approach is so far the only one successfully used in field. Further research will be carried out in order to describe the surface more appropriately for biological purposes.

## References

1. Jarvis, R.A.: A perspective on range finding techniques for computer vision. IEEE Trans. Pattern Analysis and Machine Intelligence 5, 122–139 (1983)
2. Horn, B.: Robot Vision. MIT Press, Cambridge (1986)
3. Horn, B., Brooks, M.: The variantional approach to shape from shading. In: Computer Vision Graphics and Image Processing, vol. 22, pp. 174–208 (1986)
4. Coleman Jr., E., Jain, R.: Shape from shading for surfaces with texture and specularity. In: IJCAI 1981, pp. 652–657 (1981)
5. Prados, E., Faugeras, O.: Shape from shading: a well-posed problem? In: Computer Vision and Pattern Recognition, vol. 2, pp. 870–877 (2005)
6. Ogale, A., Aloimonos, Y.: Shape and the stereo correspondence problem. International Journal of Computer Vision 65(3), 147–162 (2005)
7. Kanade, T., Okutomi, M.: A stereo matching algorithm with an adaptive window: Theory and experiment. IEEE Transactions on Pattern Analysis and Machine Intelligence 16(9), 920–932 (1994)
8. Sun, J., Li, Y., Kang, S.B., Shum, H.Y.: Symmetric stereo matching for occlusion handling. In: CVPR 2005: Proceedings of the 2005 IEEE Computer Society Conference on Computer Vision and Pattern Recognition (CVPR 2005), Washington, DC, USA, vol. 2, pp. 399–406. IEEE Computer Society, Los Alamitos (2005)
9. Kolmogorov, V., Zabih, R.: Computing visual correspondence with occlusions using graph cuts. In: ICCV, vol. 2, p. 508 (2001)
10. Rice, S.K., Gutman, C., Krouglicof, N.: Laser scanning reveals bryophyte canopy structure. New Phytologist 166(2), 695–704 (2005)
11. Motyka, O., Krumnikl, M., Sojka, E., Gaura, J.: New approach in bryophyte canopy analysis: 3d image analysis as a suitable tool for ecological studies of moss communities? In: Environmental changes and biological assessment IV., Scripta Fac. Rerum Natur. Univ. Ostraviensis (2008)
12. Zhang, Z.: Flexible camera calibration by viewing a plane from unknown orientations. In: International Conference on Computer Vision (ICCV 1999), Corfu, Greece, pp. 666–673 (1999)
13. Zhang, Z.: A flexible new technique for camera calibration. IEEE Transactions on Pattern Analysis and Machine Intelligence 22, 1330–1334 (2000)
14. Heikilla, J., Silven, O.: A four-step camera calibration procedure with implicit image correction. In: IEEE Computer Society Conference on Computer Vision and Pattern Recognition (CVPR 1997), pp. 1106–1112 (1997)
15. Heikilla, J.: Geometric camera calibration using circular control points. IEEE Transactions on Pattern Analysis and Machine Intelligence 22(10), 1066–1077 (2000)
16. Bouguet, J.Y.: Camera calibration toolbox for matlab (2005), http://www.vision.caltech.edu/bouguetj/calib_doc/index.html
17. Howard, I.P., Rogers, B.J.: Binocular Vision and Stereopsis. Oxford Scholarship Online, 212–230 (1996)

18. Liao, M., Wang, L., Yang, R., Gong, M.: Light fall-off stereo. In: Computer Vision and Pattern Recognition, pp. 1–8 (2007)
19. Magda, S., Kriegman, D., Zickler, T., Belhumeur, P.: Beyond lambert: reconstructing surfaces with arbitrary brdfs. In: ICCV, vol. 2, pp. 391–398 (2001)
20. Ortiz, A., Oliver, G.: Shape from shading for multiple albedo images. In: ICPR, vol. 1, pp. 786–789 (2000)
21. Wedekind, J.: Fokusserien-basierte rekonstruktion von mikroobjekten. Master's thesis, Universitat Karlsruhe (2002)
22. Chaudhuri, S., Rajagopalan, A., Pentland, A.: Depth from Defocus: A Real Aperture Imaging Approach. Springer, Heidelberg (1999)
23. Nayar, S.K., Watanabe, M., Noguchi, M.: Real-time focus range sensor. IEEE Transactions on Pattern Analysis and Machine Intelligence 18(12), 1186–1198 (1996)
24. Xiong, Y., Shafer, S.: Depth from focusing and defocusing. Technical Report CMU-RI-TR-93-07, Robotics Institute, Carnegie Mellon University, Pittsburgh, PA (1993)
25. Crum, H.: Structural Diversity of Bryophytes, p. 379. University of Michigan Herbarium, Ann Arbor (2001)
26. Brodie, H.J.: The splash-cup dispersal mechanism in plants. Canadian Journal of Botany (29), 224–230 (1951)
27. Proctor, M.C.F., Tuba, Z.: Poikilohydry and homoiohydry: antithesis or spectrum of possibilites? New Phytologist (156), 327–349 (2002)
28. Darboux, F., Huang, C.: An simultaneous-profile laser scanner to measure soil surface microtopography. Soil Science Society of America Journal (67), 92–99 (2003)
29. Rice, S.K., Collins, D., Anderson, A.M.: Functional significance of variation in bryophyte canopy structure. American Journal of Botany (88), 1568–1576 (2001)
30. Bachmaier, M., Backes, M.: Variogram or semivariogram - explaining the variances in a variogram. Precision Agriculture (2008)

# Wavelet Transform Analysis of the Power Spectrum of Centre of Pressure Signals to Detect the Critical Point Interval of Postural Control

Neeraj Kumar Singh, Hichem Snoussi, David Hewson, and Jacques Duchêne

Institut Charles Delaunay, FRE CNRS 2848, Université de Technologie de Troyes
12 rue Marie Curie, BP2060, 10010 Troyes, France
{neeraj_kumar.singh,hichem.snoussi,david.hewson}@utt.fr
jacques.duchene@utt.fr

**Abstract.** The aim of this study was to develop a method to detecting the critical point interval (CPI) when sensory feedback is used as part of a closed-loop postural control strategy. Postural balance was evaluated using centre of pressure (COP) displacements from a force plate for 17 control and 10 elderly subjects under eyes open, eyes closed, and vibration conditions. A modified local-maximum-modulus wavelet transform analysis using the power spectrum of COP signals was used to calculate CPI. Lower CPI values indicate increased closed-loop postural control with a quicker response to sensory input. Such a strategy requires greater energy expenditure due to the repeated muscular interventions to remain stable. The CPI for elderly occurred significantly quicker than for controls, indicating tighter control of posture. Similar results were observed for eyes closed and vibration conditions. The CPI parameter can be used to detect differences in postural control due to ageing.

**Keywords:** Stabilogram, Centre of pressure, Postural control, Wavelet transform analysis.

## 1  Introduction

Balance is regularly studied in order to better understand postural control mechanisms. One reason for the interest in balance is its relationship with falls, which are a major problem in the elderly. Indeed, a problem with balance is one of the most commonly-cited risk factors for falls [1]. Balance is maintained using the visual, vestibular, and proprioceptive systems.

Postural degradations occur with age, and can also be artificially created by impairing one of the sensory systems, for instance by closing a subject's eyes. The proprioceptive system can also be impaired by applying vibration to the tibialis anterior tendon when subjects are in a static upright position, which creates an illusion of body inclination, thus decreasing postural stability and increasing postural sway [2].

Balance can be measured either clinically, or biomechanically, for which force-plate analysis is often used. A range of different parameters can be extracted from the

A. Fred, J. Filipe, and H. Gamboa (Eds.): BIOSTEC 2009, CCIS 52, pp. 235–244, 2010.

centre of pressure (COP) obtained from the force plate, including temporal and spectral parameters, as well as those related to the organisation of the trajectory of the COP. Pioneering work in this area was performed by Collins and De Luca [3], who hypothesised that upright stance was controlled by open-loop and closed-loop strategies, which correspond to posture control strategies without and with sensory input, respectively. In the case of an open-loop control, sensory feedback is not used, whereas closed-loop control uses feedback from the proprioceptive, vestibular, or visual systems to maintain an upright stance. Collins and De Luca identified the critical point at which open-loop and closed-loop control strategies diverged, and proposed a method by which this time could be calculated [3]. Although the proposed method is based on posture control strategies, the method used to calculate the critical time interval makes assumptions that the two points used to fit the regression lines to the data occur between one second and 2.5 seconds [3].

The present paper describes a new method that can be used to calculate the time at which a change is made between open and closed loop control strategies. The proposed method, based on wavelet analysis will be used to calculate the critical point interval of the COP signal for elderly and control subjects, as well as for eyes closed and vibration conditions.

## 2 Methods

### 2.1 Subjects

Seventeen healthy control subjects and ten healthy elderly subjects (4 males and 6 females) participated in the study. Control subjects' mean age, height and weight were 33.3 ± 7.4y, 168.0 ± 6.5cm, and 65.7 ± 17.6kg, respectively. Elderly subjects' mean age, height and weight were 80.5 ± 4.7y, 165.6 ± 7.0cm, and 71.9 ± 9.9kg, respectively. All subjects who participated gave their written informed consent. No subjects reported any previous musculoskeletal dysfunction.

### 2.2 Centre of Pressure Data

Centre of pressure data were recorded using a Bertec 4060-08 force plate (Bertec Corporation, Columbus, OH, USA), which amplifies, filters, and digitises the raw signals from the strain gauge amplifiers inside the force plate. The resulting output is a six-channel 16-bit digital signal containing the forces and moments in the x, y, and z axes. The digital signals were subsequently converted via an external analogue amplifier (AM6501, Bertec Corporation).

The coordinates of the COP signals can be calculated as follows:

$$X = AP = \frac{M_y}{F_z}; \; Y = ML = \frac{M_x}{F_z} \tag{1}$$

The initial COP signals were calculated with respect to the centre of the force-plate before normalisation by subtraction of the mean.

## 2.3  Data Acquisition and Processing

Data were recorded using the ProTags™ software package (Jean-Yves Hogrel, Institut de Myologie, Paris, France) developed in Labview® (National Instruments Corporation, Austin TX, USA). Data were sampled at 100 Hz, with an 8th-order low-pass Butterworth filter with a cut-off frequency of 10 Hz. All calculations of COP data were performed with Matlab® (Mathworks Inc, Natick, MA, USA).

## 2.4  Experimental Protocol

Subjects were tested using two experimental protocols with a Bertec 4060-80 force plate (Bertec Corporation, Colombus, OH, USA). Elderly subjects were tested with their eyes open, while control subjects were tested with their eyes open, their eyes closed, as well as with vibration (eyes closed).

For the vibration condition, vibration was applied bilaterally using the VB115 vibrator (Techno Concept, Cereste, France) to the tibialis anterior tendon for 10 s at 50, 70, and 90 Hz. Immediately after vibration, subjects were instructed to step onto the force plate, in order to ensure subjects were subjected to the post-vibratory response.

After 12 s standing on the force plate, a second verbal command was given for subjects to step down backwards. Subjects remained as still as possible with their arms placed at their sides throughout the protocol, while no constraint was given over foot position, with subjects tested barefoot. Measurements were repeated five times for each experimental condition, with 30 s between each test.

## 2.5  Identifying the Critical Point Using Wavelet Transform Analysis

### 2.5.1  Locating the Critical Point

The critical point is defined as the point at which sway is controlled by the closed-loop (feedback) system. When postural control changes from open-loop to closed-loop, a local maximum should be observable in the COP. When upright stance is under open-loop control, sway moves toward a certain direction reaching a local maximum, at which point commands from the closed-loop system pull sway away from the local maximum back to the equilibrium position (Figure 1).

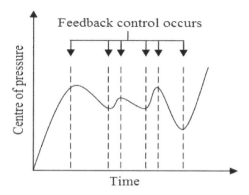

**Fig. 1.** Feedback control and the local maxima of the centre of pressure signal over time. Data are in arbitrary units.

It has been shown that local maximum modulus wavelet transform analysis is well-suited to the detection of local maxima [4]. The method used in the present study differs only in that the power spectrum is used rather than the modulus, as the power spectrum represents the energy in sway.

### 2.5.2  Wavelet Function and Frequency Bands
The wavelet transform method is particularly suitable for analyzing non-stationary signals in a multi-scale manner by varying the scale coefficient that represents frequency.

The wavelet transform formula is:

$$WT(a,b) = |a|^{-1/2} \int f(t)\varphi\left(\frac{t-b}{a}\right)dt \qquad (2)$$

where $b$ is the translation parameter and $a$ is the scale (frequency), and $WT(a,b)$ is the wavelet coefficient.

The power spectrum $PS(a,b)$ is defined as

$$PS(a,b) = |WT(a,b)|^2 \qquad (3)$$

The wavelet function $\varphi(x)$ should satisfy a number of constraints, including zero mean and orthogonality [5]. Some wavelet functions are known to distort low frequency components. In order to avoid this problem, Coiflets wavelet functions were used.

The sway energy of COP signals has been shown to be concentrated below 2 Hz [6], with the principal COP energy being distributed in the range of 0.1-0.5 Hz [7, 8]. Preliminary findings from the present study showed that most of sway energy was less than 0.5 Hz, as shown in Figure 2.

For the wavelet function $\varphi(x)$, scale $a$ determines the frequency and $b$ determines the translation. In the present study, $a$ was chosen to force the frequency of wavelet transformation to the range of 0.1-2.0 Hz.

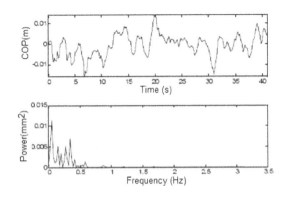

**Fig. 2.** Sample power spectrum of COP data (AP direction)

The 0.1-2.0 Hz frequency band can be divided into sub sections for which different control systems are thought to be involved. For instance, Diener and Gagey suggested that the visual system dominates frequency bands below 0.5 Hz {Diener, 1984 #1052; Gagey, 1985 # 1051}. In contrast, Thurner and colleagues reported that the visual system operates in the range of 0-0.1 Hz, the vestibular system from 0.1-0.5 Hz, somatosensory activity from 0.5-1.0 Hz, while sway over 1.0 Hz is directly controlled by the central nervous system [9]. Based on these findings, the frequencies of 0.5-1.0 Hz were chosen for the present study as the zone in which proprioceptive input predominates.

The relation between scale and frequency can be shown as:

$$F_a = \frac{F_C}{a.P} \tag{4}$$

Where $F_c$ is the centre frequency, $F_a$ the frequency for scale $a$, and $P$ is the sampling period (0.01 s).

It is evident that the scale a should be in the range 0.5-1.0 Hz for the present study, given the use of proprioceptive perturbations. To this end, a is transformed using a base-2 logarithm, and thereafter denoted as the new scale s. The lower and upper bounds of scale s were chosen as 6.5 and 8, respectively. This scale range corresponds to the frequency bands in the range of [0.44, 0.88] Hz. After determining scale $s$, translation $b$, and the wavelet function, the power spectrum can be calculated, an example of which is shown in Figure 3.

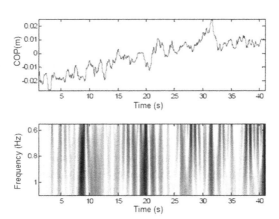

**Fig. 3.** Sample wavelet power spectrum of COP data (ML direction). Darker and lighter areas represent larger positive and negative power spectrum values, respectively.

### 2.5.3  Identifying the Critical Point Time Interval

Postural control strategies can be changed from open-loop control to closed-loop in which visual, vestibular and somatosensory feedbacks can be used. When feedback control is used, sway moves away from the local maximum and back to the equilibrium position. The critical point occurs at scale $s$ and time $t$, where the power spectrum is at the local maximum. It is simply the local maximum of the power spectrum wavelet transform method (LMPS), which can be mathematically defined as:

$$PS(s,x) < PS(s,x_0) \qquad (5)$$

Where x is either the left or right neighbourhood of $x_0$, $x \in R$, and $s$ is scale.

There are numerous local maxima within specific frequency bands, which indicates that feedback control has been started at time $t$ and scale $s$. Postural control corresponds to the frequency bands rather than a single frequency, meaning that the identification of maxima across the frequency range was required. A local maximum line $L_t$ can be defined as the line consisting of the local maxima at time $t$ across frequency bands from $a$ to $b$:

$$L_t \overset{d}{=} \{PS(s,t), s = a..b\} \qquad (6)$$

In the present study, local maxima lines were identified in the frequency bands [$a$=0.44, $b$=0.88] Hz. It is necessary to search for the local maximum around time $t$ within a $2\Delta t$ interval denoted by [$t$-$\Delta t$, $t$+$\Delta t$]. If the local maximum line $L_t$ can be identified within the time interval [$t$-$\Delta t$, $t$+$\Delta t$], it is concluded that feedback control has been used at time $t$. The time interval $\Delta t$ used to search for the local maximum line depends on the specific data. If $\Delta t$ is too small then too many local maxima will be identified. In the present study $\Delta t$ was chosen as 0.62 s. There are numerous such local maximum lines for each scale. The mean of the length of these local maxima lines was calculated and taken to be the critical point interval (CPI). The average CPI for all five trials for each subject for each experimental condition was used for subsequent statistical analysis.

## 2.6  Statistical Analysis

All statistical analyses were performed with the Statistical Package for Social Sciences (SPSS Inc., Chicago, IL, USA). Analysis of variance was used to compare results between conditions, with CPI as the dependent variable and the experimental condition as the independent variable. Data were expressed as means and 95% confidence intervals. Alpha level was set at $p<0.05$.

## 3  Results

The CPI for elderly and control subjects under the eyes open condition are presented in Figure 4. Significantly higher values for CPI can be seen for control subjects than for elderly subjects (p<0.05). The CPI for AP displacement was significantly lower than the corresponding value for ML displacement for both control and elderly subjects alike (Figure 4; p<0.05).

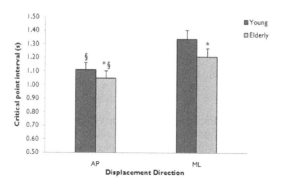

**Fig. 4.** Critical point interval for elderly subjects and control subjects in the eyes open condition. Data are means and 95% confidence intervals. *Significantly different from control subjects (p<0.05); §Significantly different from ML displacement (p<0.05).

The results for the eyes open and eyes closed conditions are presented in Figure 5. Significantly higher values for CPI were observed for the eyes open condition than for eyes closed (p<0.05). Significantly lower values of CPI were observed for AP displacement when compared to ML displacement for both eyes open and eyes closed conditions (Figure 5; p<0.05).

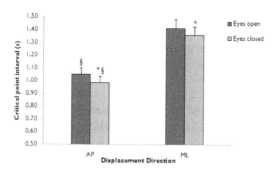

**Fig. 5.** Critical point interval for control subjects in the eyes open and eyes closed conditions. Data are means and 95% confidence intervals. *Significantly different from eyes open (p<0.05); §Significantly different from ML displacement (p<0.05).

In respect to vibration, there was no significant difference between conditions for ML displacement (Figure 6). In contrast, both 70 Hz and 90 Hz vibrations significantly decreased CPI values in comparison to the 0 Hz and 50 Hz vibration condition (Figure 6; p<0.05).

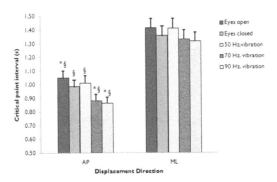

**Fig. 6.** Critical point interval for control subjects in the eyes open, eye close and vibration condition. Data are means and 95% confidence intervals. *Significantly different from eyes closed (p<0.05); §Significantly different from ML displacement (p<0.05).

## 4  Discussion

Lower values of CPI are associated with increased closed-loop postural control. In essence, the lower the CPI value, the quicker the response to sensory input, and thus the greater reliance on closed-loop control. Lower values of CPI can therefore be interpreted as a less efficient open-loop control, thus requiring an earlier intervention of the closed-loop system. This strategy of postural control will require greater energy expenditure due to the repeated muscular interventions in order to remain stable.

The CPI parameter proposed in the present paper was able to distinguish between all of the different experimental conditions tested. In respect to differences between AP and ML displacements, CPI values were lower for AP than for ML displacement for all experimental conditions indicating less postural stability. This result was expected given that ML displacement is more stable than AP displacement due to the anatomy of the ankle and knee joints, which limit movement in the ML direction. In addition, subjects' feet are placed in a series position for the ML direction, as opposed to the parallel positioning for AP.

In respect to differences with ageing, elderly subjects had lower CPI values than control subjects for both AP and ML displacement. Such results are indicative of an earlier feedback control for elderly subject, which is due to a tightly controlled posture in elderly subjects, as reported previously for other parameters such as Detrended Fluctuation Analysis (DFA). In studies of DFA, elderly subjects had less complex patterns of AP displacement, which is indicative of a tight postural control used to reduce displacement as much as possible in order to maintain a stable posture [10, 11].

The differences observed between eyes open (EO) and eyes closed (EC) conditions were in agreement with those between elderly and control subjects. The CPI values were greater for EO than for EC for both AP and ML displacements. The lack of visual information required subjects to tightly control displacement in both AP and ML directions using the closed-loop system in order to maintain equilibrium.

The effect of the vibration applied to the tibialis anterior tendon is to decrease the stability of the postural control system by invoking an illusion of tilt. The vibration required to invoke this effect must be at least 70 Hz. This effect is only seen in the AP direction in which the tendinous vibration was applied.

The results for all experimental comparisons demonstrate that a decreased ability to maintain postural stability can be identified by an increased use of the closed-loop postural control system. Such results might seem contradictory at first, in that both elderly subjects and subjects with no visual feedback can tightly control their postural equilibrium. However, similar results have been reported for COP signals for a range of parameters such as entropy and DFA [10-12], as well as for other physiological signals such as force production [13]. It has also been suggested that physiological control processes become less complex with age as well as when disease is present [14].

## 5   Conclusions

In conclusion, the CPI parameter offers a new method of detecting differences in postural control between different experimental conditions or changes due to ageing. Lower values of CPI reflect greater reliance on closed-loop postural control, which requires greater energy expenditure due to the repeated muscular interventions in order to remain stable.

**Acknowledgements.** This study was undertaken as part of the PARAChute research programme (ANR-05-RNTS-01801; RNTS-03-B-254; ESF 3/1/3/4/07/3/3/011; ERDF 2003-2-50-0014 and 2006-2-20-0011; CACR E200308251).

## References

1. Rubenstein, L.Z., Josephson, K.R.: The epidemiology of falls and syncope. Clin. Geriatr. Med. 18, 141–158 (2002)
2. Roll, J.P., Vedel, J.P.: Kinaesthetic role of muscle afferents in man, studied by tendon vibration and microneurography. Exp. Brain Res. 47, 177–190 (1982)
3. Collins, J.J., De Luca, C.J.: Random walking during quiet standing. Physical Review Letters 73, 764–767 (1994)
4. Mallat, S., Hwang, W.L.: Singularity detection and processing with wavelets. IEEE Transactions on Information Theory 38, 617–643 (1992)
5. Muzzy, J., Barcy, E., Areno, A.: Wavelets and multifractals for singular signal: application to turbulence data. Physical Review Letters 67, 3515–3519 (1991)
6. Ferdjallah, M., Harris, G.F., Wertsch, J.J.: Instantaneous postural stability characterization using time-frequency analysis. Gait & Posture 10, 129–134 (1999)
7. Schmuckler, M.: Children's postural sway in response to low- and high-frequency visual information for oscillation. Jouranl of Experimental Psychology: Human Perception and Performance 23, 528–545 (1997)
8. Loughlin, P.J., Redfern, M.S., Furman, J.M.: Time-varying characteristics of visually induced postural sway. IEEE Trans. Rehabil. Eng. 4, 416–424 (1996)
9. Thurner, S., Mittermaier, C., Hanel, R., Ehrenberger, K.: Scaling-violation phenomena and fractality in the human posture control systems. Physical Review E 62, 4018 (2000)

10. Amoud, H., Abadi, M., Hewson, D.J., Michel, V., Doussot, M., Duchêne, J.: Fractal time series analysis of postural stability in elderly and control subjects. Journal of NeuroEngineering and Rehabilitation 4, 12 (2007)
11. Norris, J.A., Marsh, A.P., Smith, I.J., Kohut, R.I., Miller, M.E.: Ability of static and statistical mechanics posturographic measures to distinguish between age and fall risk. J. Biomech. 38, 1263–1272 (2005)
12. Amoud, H., Snoussi, H., Hewson, D.J., Doussot, M., Duchêne, J.: Intrinsic mode entropy for nonlinear discriminant analysis. IEEE Signal Processing Letters 14, 297–300 (2007)
13. Challis, J.H.: Aging, regularity and variability in maximum isometric moments. Journal of Biomechanics 39, 1543–1546 (2006)
14. Goldberger, A.L., Amaral, L.A., Hausdorff, J.M., Ivanov, P., Peng, C.K., Stanley, H.E.: Fractal dynamics in physiology: alterations with disease and aging. Proceedings of the National Academy of Sciences of the United States of America 99(suppl. 1), 2466–2472 (2002)

# Early Detection of Severe Apnoea through Voice Analysis and Automatic Speaker Recognition Techniques

Ruben Fernández[1], Jose Luis Blanco[1], David Díaz[1],
Luis A. Hernández[1], Eduardo López[1], and José Alcázar[2]

[1] Signal, Systems & RadioComm, Department (GAPS), Universidad Politécnica de Madrid
Avda Complutense 30, 28040 Madrid, Spain
{ruben,jlblanco,dpardo,luis,eduardo}@gaps.ssr.upm.es
[2] Respiratory Department, Hospital Torrecardenas, 04009, Almeria, Spain
jose44@separ.es

**Abstract.** This study is part of an on-going collaborative effort between the medical and the signal processing communities to promote research on applying voice analysis and Automatic Speaker Recognition techniques (ASR) for the automatic diagnosis of patients with severe obstructive sleep apnoea (OSA). Early detection of severe apnoea cases is important so that patients can receive early treatment. Effective ASR-based diagnosis could dramatically cut medical testing time. Working with a carefully designed speech database of healthy and apnoea subjects, we present and discuss the possibilities of using generative Gaussian Mixture Models (GMMs), generally used in ASR systems, to model distinctive apnoea voice characteristics (i.e. abnormal nasalization). Finally, we present experimental findings regarding the discriminative power of speaker recognition techniques applied to severe apnoea detection. We have achieved an 81.25 % correct classification rate, which is very promising and underpins the interest in this line of inquiry.

**Keywords:** Apnoea, Automatic speaker recognition, GMM, Nasalization.

## 1 Introduction

*Obstructive sleep apnoea* (OSA) is a highly prevalent disease [1], affecting an estimated 2-4% of the male population between the ages of 30 and 60. It is characterized by recurring episodes of sleep-related collapse of the upper airway at the level of the pharynx (AHI > 15, *Apnoea Hypopnoea Index*, which represents the number of apnoeas and hypoapnoeas per hour of sleep) and it is usually associated with loud snoring and increased daytime sleepiness. OSA is a serious threat to an individual's health if not treated. The condition is a risk factor for hypertension and, possibly, cardiovascular diseases [2], it is usually related to traffic accidents caused by somnolent drivers [3], and it can lead to a poor quality of life and impaired work performance. At present, the most effective and widespread treatment for OSA is nasal CPAP (*Continuous Positive Airway Pressure*) which prevents apnoea episodes by providing a pneumatic splint to the airway. OSA can be diagnosed on the basis of a

A. Fred, J. Filipe, and H. Gamboa (Eds.): BIOSTEC 2009, CCIS 52, pp. 245–257, 2010.
© Springer-Verlag Berlin Heidelberg 2010

characteristic history (snoring, daytime sleepiness) and physical examination (increased neck circumference), but a full overnight sleep study is usually needed to confirm the disorder [1]. The procedure is known as conventional *Polysomnography*, which involves the recording of neuroelectrophisiological and cardiorespiratory variables (ECG). Excellent automatic OSA recognition performance is attainable with this method based on nocturnal ECG recordings. Nevertheless, this diagnostic procedure is expensive and time-consuming, and patients usually have to endure a waiting list of several years before the test is done, since the demand for consultations and diagnostic studies for OSA has recently increased. There is, therefore, a strong need for methods of early diagnosis of apnoea patients in order to reduce these considerable delays.

In our research we investigate the acoustical characteristics of the speech of patients with OSA for the purpose of learning whether severe OSA may be detected using *Automatic Speaker Recognition* techniques (ASR). The acoustic properties of voice from speakers suffering obstructive sleep apnoea are not well understood as not much research has been carried out in this area. However, some studies have suggested that certain abnormalities in phonation, articulation and resonance may be connected to the condition [4]. In order to have a controlled experimental framework to study apnoea voice characterization we collected a speech database [5] designed following linguistic and phonetic criteria. Our work is focused on continuous speech rather than on sustained vowels, the latter being the standard approach in pathological voice analysis [6]. The speech corpus was designed considering previous research in the field as well as an initial manual contrastive study we performed on a small group of healthy subjects and apnoea patients.

After this preliminary acoustic analysis, *GMM-based* Automatic Speaker Recognition techniques [7] were explored trying to model possible peculiarities in apnoea patients' voices. More specifically, our work is mainly focused on the nasality factor, since it has been traditionally identified as an important feature in the acoustic characteristics of apnoea speakers. Successfully detecting traits that prove to be characteristic of the voices of severe apnoea patients by applying such techniques would allow automatic (and rapid) diagnosis of the condition. To our knowledge this study constitutes pioneering research on automatic severe OSA diagnosis using speech processing algorithms on continuous speech. The proposed method is intended as complementary to existing OSA diagnosis methods (e.g. *Polysomnography*) and clinicians' judgment, as an aid for early detection of these cases. We have observed a marked inadequacy of resources that has led to unacceptable waiting periods. Early severe OSA detection can help to increase the efficiency of medical protocols by giving higher priority to more serious cases, thus optimizing both social benefits and medical resources. For instance, patients with severe apnoea have a higher risk of suffering a car accident because of somnolence caused by their condition. Early detection would, therefore, contribute to reducing the risk of suffering a car accident for these patients.

The rest of this document is organized as follows: Section 2 presents the main physiological characteristics of OSA patients and the distinctive acoustic qualities of their voices, based on the previous literature and our initial contrastive study. The speech database used in our experimental work, as well as its design criteria, is explained in Section 3. Section 4 explores the advantages that standard GMMs can bring to apnoea voice characterization. In particular, we describe how we used GMMs to study nasalization in speech, comparing the voices of severe apnoea patients with those in a 'healthy' control group. Next, in Section 5 we present a test we carried out to assess the accuracy of a GMM-based system we developed to classify speakers (apnoea/non-apnoea). Finally, conclusions and a brief outline of future research are given in Section 6.

## 2  Physiological and Acoustic Characteristics in OSA Speakers

At present neither the articulatory/physiological peculiarities nor the acoustic characteristics of speech in apnoea speakers are well understood. Most of the more valuable information in this area can be found in Fox and Monoson's work [4], a perceptual study in which skilled judges compared the voices of apnoea patients with those of a control group (referred to as "healthy" subjects). The study showed that, although differences between both groups of speakers were found, acoustic cues for these differences are somewhat contradictory and unclear. What did seem to be clear was that the apnoea group had abnormal resonances that might be due to an altered structure or function of the upper airway. Theoretically, such an anomaly should result not only in respiratory but also in speech dysfunction. Consequently, the occurrence of speech disorder in OSA population should be expected, and it could include anomalies in articulation, phonation and resonance:

- *Articulatory Anomalies.* Fox and Monoson stated that *neuromotor* dysfunction could be found in the sleep apnoea population due to a *"lack of regulated innervations to the breathing musculature or upper airway muscle hypotonus."* This dysfunction is normally related to speech disorders, especially *dysarthria.* There are several types of dysarthria, resulting in various different acoustic features. All types of dysarthria affect the articulation of consonants and vowels causing the slurring of speech.
- *Phonation Anomalies.* These may be due to the heavy snoring of sleep apnoea patients, which can cause inflammation in the upper respiratory system and affect the vocal cords.
- *Resonance Anomalies.* What seems to be clear is that the apnoea group has abnormal resonances that might be due to an altered structure or function of the upper airway causing *velopharyngeal dysfunction.* This anomaly should, in theory, result in an abnormal vocal quality related to the coupling of the vocal tract with the nasal cavity, and is revealed through two features:
  - o  Firstly, speakers with a defective velopharyngeal mechanism can produce speech with *inappropriate nasal resonance.* The term nasalization can refer to two different phenomena in the context of speech; *hyponasality* and *hypernasality.* The former is said to occur when no nasalization is produced when the sound should be nasal. Hypernasality is nasalization during the production of non-nasal (voiced oral) sounds. The interested reader can find an excellent reference in [8]. Fox and Monoson's work on the nasalization characteristics for the sleep apnoea group was not conclusive. What they could conclude was that these resonance abnormalities could have been associated with vocal tract damping features distinct from airflow in balance

between the oral and nasal cavities, affecting speech sound quality. The term applied to this speech disorder is *"cul-de-sac"* resonance, a type of hyponasality that causes the sound to be perceived as if it were resonating in a blind chamber. Perhaps more importantly, it seems that speakers with apnoea may exhibit smaller intra-speaker differences between non-nasal and nasal vowels due to this velopharyngeal dysfunction (vowels ordinarily acquire either a nasal or a non-nasal quality depending on the presence or absence of adjacent nasal consonants).

o    Secondly, due to the pharyngeal anomaly, differences in formant values can be expected. This is confirmed in Robb's work [9], in which vocal tract acoustic resonance was evaluated in a group of OSA males. Statistically significant differences were found in formant frequency and bandwidth values between apnoea and healthy groups. In particular, the results of the formant frequency analysis showed that F1 and F2 values among the OSA group were generally lower than those in the non-OSA groups. The lower formant values were attributed to greater vocal tract length.

These types of anomalies may occur either in isolation or combined. However, none of them was found to be sufficient on its own to allow accurate assessment of the OSA condition. In fact, all three descriptors were necessary to differentiate and predict whether the subject was in the normal group or in the OSA group.

## 2.1 Initial Contrastive Acoustic Study

In order to build on the relatively little knowledge available in this area, a preliminary acoustic analysis in an initial version of our apnoea speech database was made. In a related piece of research, Fiz et al. [10] applied spectral analysis on sustained vowels to detect possible apnoea-pathological cases. They used the following acoustic features: maximum frequency of harmonics, mean frequency of harmonics and number of harmonics. They found statistically significant differences between a control group (healthy subjects) and the sleep apnoea group regarding the maximum harmonic frequency for the vowels /i/ and /e/, it being lower for OSA patients. Another piece of research on the acoustic characterization of sustained vowels in apnoea patients using *Linear Predictive Coding* (LPC) can be found in [11]. However, these studies do not investigate all of the possible acoustic peculiarities that may be found in the voices of apnoea patients, since focusing solely on sustained vowels precludes the discovery of acoustic effects that occur in continuous speech only in certain linguistic contexts.

Our contrastive study consisted of a perceptual and visual comparison of frequency representations (mainly spectrographic, pitch, energy and formant analysis) of an initial apnoea and control group speakers uttering a same set of 25 phonetically balanced sentences. Following up on previous research on the acoustic characteristics of OSA speakers, we searched for articulatory and resonance anomalies in apnoea-suffering speakers.

- *Articulatory Anomalies.* An interesting conclusion from our initial perceptual contrastive study was that, when comparing the distance between the second (F2) and third formant (F3) for the vowel /i/, clear differences between the apnoea and control groups were found. For apnoea speakers the distance was greater, and this was especially clear in diphthongs with /i/ as the stressed vowel, as in the Spanish word "Suiza" (*'suj θa*). This finding may be related to the greater length of the vocal tract of OSA patients [9], but also, and perhaps more importantly, to a characteristically abnormal velopharyngeal opening which may cause a shift in the position of the third formant [12]. Indeed, a lowering of the velum (typical in apnoea speakers) is known to produce higher third formant frequencies.

- **Resonance Anomalies.** Fox and Monoson state in [4] that a common resonance feature in apnoea patients is abnormal nasality. The presence and the size of one extra low frequency formant can be considered an indicator of nasalization [13], but no perceptual differences between the groups in the overall nasality level could be found. As discussed in previous sections, this could be due to common perceptual difficulties to classify the voice of apnoea speakers as hyponasal or hypernasal. However, we did find differences in both groups (apnoea and non-apnoea) in how nasalization varied from nasal to non-nasal contexts and vice versa. Interestingly, we found variation in nasalization to be smaller for OSA speakers. One hypothesis is that the voices of apnoea speakers have a higher overall nasality level caused by velopharyngeal dysfunction, so differences between oral (no-nasal) and nasal vowels are smaller than normal because the oral vowels are also nasalized. An explanation for this could be that apnoea speakers have weaker control over the velopharyngeal mechanism, which may cause difficulty in changing nasality levels, whether absolute nasalization level is high or low. These hypotheses are intriguing and we will delve deeper into them later.

# 3   Apnoea Database

The database was recorded in the Respiratory Department at *Hospital Clínico Universitario of Málaga*, Spain. It contains the readings and facial images of 80 male subjects; half of them suffer from severe sleep apnoea (AHI > 30), and the other half are either healthy subjects or only have mild OSA (AHI < 10). Subjects in both groups have similar physical characteristics such as age and Body Mass Index (BMI). The speech material for the apnoea group was recorded and collected in two different sessions: one just before being diagnosed and the other after several months under CPAP treatment. This allows studying the evolution of apnoea voice characteristics for a particular patient before and after treatment.

## 3.1   Speech and Image Collection

Speech was recorded using a sampling frequency of 48 kHz in an acoustically isolated booth. The recording equipment consisted of a standard laptop computer with a conventional sound card equipped with a *SP500 Plantronics* headset microphone with A/D conversion and digital data exchange through a USB-port.

Additionally, for each subject in the database, two facial images (frontal and lateral views) were collected under controlled illumination conditions and over a flat white background. A conventional digital camera was used to obtain images in 24-bit RGB format, without compression and with 2272x1704 resolution. We decided to collect visual information because OSA is usually associated with a variable combination of different anatomic factors (e.g., narrowing of the upper airway, distinctive craniofacial and pharyngeal dimensions, etc) [14]. Although precise maxillary morphology analysis in OSA cases has been done using radiological analysis of lateral views, *Cephalometrics*, simple visual inspections are also considered as a first step when evaluating patients under clinical suspicion of suffering OSA [1]. Visual examination of patients includes searching for distinctive features of the facial morphology of OSA such as a short neck, characteristic mandibular distances and alterations, obesity, etc. These considerations motivated us to include photographs of the patients' faces in the database, for future research toward simple and cost-efficient automatic diagnosis of severe OSA patients using image processing techniques. To our knowledge, no research has ever been carried out to find facial features that may be extracted by image processing of both frontal and lateral views, and which may help diagnose severe apnoea cases.

## 3.2 Speech Corpus

In this sub-section, we describe the apnoea speaker database we designed with the goal of covering all the relevant linguistic/phonetic contexts in which physiological OSA-related peculiarities could have a greater impact. These peculiarities include the articulatory, phonation and resonance anomalies revealed in the previous research review (see Section 2). As we pointed out in the introduction, the central aim of our study is to apply speech processing techniques to automatically detect OSA-related traits in continuous speech. We believed that analysing continuous speech may well afford greater possibilities than working with sustained vowels because certain traits of pathological voice patterns, and in particular those of OSA patients, could then be detected in different sound categories (i.e. nasals, fricatives, etc) and also in the co-articulation between adjacent sound units [6].

The speech corpus contains readings of four sentences in Spanish repeated three times by each speaker that include instances of the following specific phonetic contexts that we derived from previous research:

- In relation to *resonance anomalies*, we designed sentences that allow intra-speaker variation measurements; that is, measuring differential voice features for each speaker, for instance to compare the degree of vowel nasalization within and without nasal contexts.
- With regard to *phonation anomalies*, we included continuous voiced sounds to measure irregular phonation patterns related to muscular fatigue in apnoea patients.
- Finally, to look at *articulatory anomalies* we collected voiced sounds affected by certain preceding phonemes that have their primary locus of articulation near the back of the oral cavity, specifically, velar phonemes such as the Spanish velar approximant 'g'. This pharyngeal region has been seen to display physical anomalies in speakers suffering from apnoea. Thus, it is reasonable to suspect that different coarticulatory effects may occur with these phonemes in speakers with and without apnoea.

All the sentences were designed to exhibit a similar melodic structure, and speakers were asked to read them with a specific rhythmic structure under the supervision of an expert. We followed this controlled rhythmic recording procedure hoping to minimise non-relevant inter-speaker linguistic variability. The sentences used were the following, with the different melodic groups underlined separately:

1. **Francia, Suiza y Hungría ya hicieron causa común.**
   *'fraN θja 'suj θa i uŋ 'gri a ya j 'θje roŋ 'kaw sa ko 'mun*
2. **Julián no vio la manga roja que ellos buscan, en ningún almacén**.
   *xu 'ljan no 'βjo la 'maŋ ga 'ɾo xa ke 'e ʎoz 'βus kan en niŋ 'gun al ma 'ken*
3. **Juan no puso la taza rota que tanto le gusta** en el aljibe.
   *xwan no 'pu so la 'ta θa 'ɾo ta ke 'taN to le 'ɣus ta en el al 'xi βe*
4. **Miguel y Manu llamarán entre ocho y nueve y media.**
   *mi 'ɣel i 'ma nu ʎa ma 'ran 'eN tre 'o t͡ʃo i 'nwe βe i 'me ðja*

The first phrase was taken from the *Albayzin* database, a standard phonetically balanced speech database for Spanish [15]. It was chosen because it contains an interesting sequence of successive /a/ and /i/ vowel sounds.

The second and third phrases, both negative, have a similar grammatical and intonation structure. They are potentially useful for contrastive studies of vowels in different linguistic contexts. Some examples of these contrastive pairs arise from comparing a nasal context, "ma**nga** roja" (*'maŋ ga 'ɾo xa*), with a neutral context, "ta**za** rota" (*'ta θa 'ɾo ta*). As we mentioned in the previous section, these contrastive analyses could be very helpful to confirm whether indeed the voices of speakers with apnoea have an altered overall nasal quality and display smaller intra-speaker differences between non-nasal and nasal vowels due to velopharyngeal dysfunction.

The fourth phrase has a single and relatively long melodic group containing mainly voiced sounds. The rationale for this fourth sentence is that apnoea speakers usually show fatigue in the upper airway muscles. Therefore, this sentence may be helpful to discover various anomalies during the sustained generation of voiced sounds. These phonation-related features of segments of harmonic voice can be characterized following any of a number of conventional approaches that use a set of individual measurements such as the Harmonic to Noise Ratio- HNR, periodicity measures and pitch dynamics (e.g. jitter). Finally, with regard to the resonance anomalies found in the literature, one of the possible traits of apnoea speakers is dysarthria. Our sentence can be used to analyse dysarthric voices that typically show differences in vowel space with respect to normal speakers.

# 4 Apnoea Voice Modelling with GMM

In this section we present an initial experimentation that sheds light on the potential of using *Gaussian Mixture Models* techniques (GMMs) over speech spectra (*cepstral* domain) to discover and model peculiarities in the acoustical signal of apnoea voices. These peculiarities might be related to the perceptually distinguishable traits described in previous research and corroborated in our preceding contrastive study. The main reason for using GMMs over the cepstral domain is related to the great potential this combination of techniques has shown for the modelling of the acoustic space of human speech. For our study we required a good modelling of the anomalies described in Section 2, which we expected to find in OSA patients. Since cepstral coefficients are related with the spectral envelope of speech signals, and therefore with the articulation of sounds, and since GMM training sets can be carefully selected in order to model specific characteristics (for instance in order to consider resonance anomalies in particular), it seems promising to combine all this information in a fused model. We should expect such a model to be useful for describing the acoustic spaces of both the OSA patient group and the healthy group, and for discriminating between them.

This approach was applied to specific linguistic contexts. In particular, as our apnoea speech database was designed to allow a detailed contrastive analysis of vowels in no-nasal (oral) and nasal phonetic contexts, we focus on reporting perceptual differences related to resonance anomalies that could be perceived as either hyponasality or hypernasality. For this purpose, sub-section 4.2 discusses how GMM techniques can be applied to study these differences in degree of nasalization in different linguistic contexts.

### 4.1 GMM Training and Testing Protocol

Gaussian Mixture Models (GMMs) and adaptation algorithms are effective and efficient pattern recognition techniques suitable for sparse speech data modelling in Automatic Speaker Recognition systems [7] that we will apply for apnoea voice modelling. In our experimental framework, the *BECARS* open source tool [16] was used. Details on parameterization, model training and classification phases for the BECARS baseline system are the following:

- Parameterization consists in extracting information from speech signal. Our speech database was processed using short-time spectral analysis with a 20 ms time frame and a 10 ms delay between frames, which gives a 50% overlap. We chose an appropriate parameterization for the information, using 39 standard components: 12 *Mel Frecuency Cepstral Coefficients* (MFCCs), plus energy, extended with their speed (delta) and acceleration (delta-delta) components.

- GMMs for apnoea and 'healthy' control groups were trained as follows. First, a universal background model (UBM) was trained from a phonetically balanced subcorpus of Albayzin Database [15]. Next, speech from apnoea and non-apnoea group speakers was used to train the apnoea and control group GMM models. Both models were trained using MAP (Maximum a Posteriori) adaptation from the UBM. This technique increases the robustness of the models especially when sparse speech material is available. Only the means were adapted, as is classically done in speaker verification. Obviously, speech data from speakers for the model training was not included in the test set.

- For testing purposes, and in order to increase the number of tests and thus to improve the statistical relevance of our results, the standard *leave-one-out* testing protocol was used. This protocol consists in discarding one sample speaker from the experimental database to train the classifier with the remaining samples. Then the excluded sample is used as the test data. This scheme is repeated until a sufficient number of tests have been performed.

## 4.2  A Study of Apnoea Speaker Resonance Anomalies Using GMMs

With the aim of testing the capabilities of using the GMM-based experimental set-up, we performed an initial study to try to model certain resonance anomalies that have already been described for apnoea speakers in preceding research [4] and revealed in our own contrastive acoustic study. In particular, as our apnoea speech database was designed to allow a detailed contrastive analysis of vowels in oral and nasal phonetic contexts, we used GMM techniques to perform a study to identify differences in degree of nasalization in different linguistic contexts.

For that purpose, we generated 40 sub-groups of apnoea speakers and 40 sub-groups of control speakers discarding one sample for each class from the whole set of pathological and healthy patients. Two GMMs for each *cluster* were trained using speech with all nasalized and non-nasalized vowels from speakers of each group. Both nasal and non-nasal GMMs were trained following the approach described in Section 4.1. MAP adaptation was carried out with a generic vowel UBM trained using Albayzin database [15]. These two nasal/non-nasal GMMs for each sub-group were used to quantify the acoustic differences between nasal and non-nasal contexts in both the apnoea and the control patients. The smaller the difference between the nasal and the non-nasal GMMs the more similar the nasalized and the non-nasalized vowels are. Unusually similar nasal and non-nasal vowels for any group of speakers reveals the presence of resonance anomalies. We took a fast approximation of the *Kullback-Leibler (KL) divergence* for Gaussian Mixture Models [17] as a measure of distance between nasal and non-nasal GMMs. This distance is commonly used in Automatic Speaker Recognition to define cohorts or groups of speakers producing similar sounds.

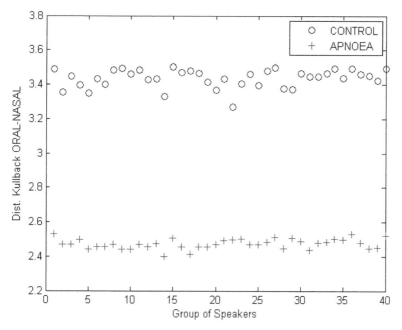

**Fig. 1.** Kullback-distance differences between nasal and non-nasal models for clusters of severe apnoea (*plus*) and 'healthy' control (*circles*) speakers

As can be seen in Figure 1, the distance between nasal and non-nasal vowel GMMs was significantly larger for the control group speakers than for the speakers with severe apnoea. This interesting result confirms that the margin of acoustic variation for vowels articulated in nasal vs. non-nasal phonetic contexts is narrower than normal in speakers with severe apnoea. It also validates the GMM approach as a powerful speech processing and classification technique for research on OSA voice characterization and the detection of OSA speakers.

## 5   Automatic Diagnosis of Severe Apnoea Using GMMs

As we have suggested in the previous section, GMM-based ASR techniques can discriminate some of the resonance anomalies of apnoea speakers that have already been described in the literature. Thus it seems reasonable to explore the possibilities of applying GMM-Based Speaker Recognition techniques for the automatic diagnosis of severe apnoea. This method could be suitable for keeping track of the progress of voice dysfunction in OSA patients, it is easy-to-use, fast, non-invasive and much cheaper than traditional alternatives. While we do not suggest it should replace current OSA diagnosis methods, we believe it can be a great aid for early detection of severe apnoea cases.

A speaker verification system is a supervised classification system capable of discriminating between two classes of speech signals (usually 'genuine' and 'impostor' speakers). For our present purposes the classes are not defined by reference to any particular speaker. Rather, we generated a general severe sleep apnoea class and a control class (speech from healthy subjects) by grouping together all the training data from speakers of each class. Thus a GMM model was trained for all the speakers belonging to the apnoea class and another GMM model for those speakers in the control class. Following a similar approach to that of other pathological voice assessment studies [17], GMMs representing the pathological and healthy classes were built as follows:

- The apnoea and control GMMs were trained from the generic UBM relying on MAP adaptation and the standard *leave-one-out* technique, similarly to how we described above (Section 4.1).
- During the apnoea/non-apnoea detection phase an input speech signal corresponding to the whole utterance of the speaker to be diagnosed is presented to the system. The parameterised speech is then processed with each apnoea and control GMM generating two likelihood scores. From these two scores an apnoea/control decision is made according to a decision threshold adjusted beforehand as a trade-off to achieve acceptable rates of both failure to detect apnoea voices (false negative) or falsely classifying healthy cases as apnoea voices (false positive).

Table 1 shows the correct classification rates we obtained when we applied the GMM control/pathological voice classification approach to our speech apnoea database [5]. We see that the overall correct classification rate was 81.25 %.

**Table 1.** Correct Classification Rate

| Correct Classification Rate in % | Control Group | Apnoea Group | Overall |
|---|---|---|---|
| | 77,50 % (31/40) | 85 % (34/40) | 81,25 % (65/80) |

We now evaluate the performance of the classifier using the following criteria:

- **Sensitivity**: ratio of correctly classified apnoea-suffering speakers (true positives) to total number of speakers actually diagnosed with severe apnoea.
- **Specificity**: ration of true negatives to total number of speakers diagnosed as not suffering from apnoea.
- **Positive Predictive Value**: ratio of true positives to total number of patients GMM-classified as having a severe apnoea voice.
- **Negative Predictive Value**: ratio of true negatives to total number of patients GMM-classified as *not* having a severe apnoea voice.

Table 2 shows the values we obtained in our test for these measures of accuracy (Fisher's exact test revealed statistically significant, p<0.0001):

**Table 2.** Sensitivity, Specificity, Positive Predictive Value and Negative Predictive Value

| Sensitivity | Specificity | Positive Predictive Value | Negative Predictive Value |
|---|---|---|---|
| 77,50 % (31/40) | 85 % (34/40) | 83,78 % (31/37) | 79,06 % (34/43) |

Some comments are in order regarding the correct classification rates obtained. The results are encouraging and they show that distinctive apnoea traits can be identified by a GMM based-approach, even when there is relatively little speech material with which to train the system. Furthermore, such promising results were obtained without paying choosing any acoustic parameters in particular on which to base the classification. Better results should be expected with a representation and parameterization of audio data that is optimized for apnoea discrimination. Obviously, our experiments need to be validated with a larger test sample. Nevertheless, our results already give us an idea of the discriminative power of this approach to automatic diagnosis of severe apnoea cases.

# 6   Conclusions and Future Research

In this paper we have presented pioneering research in the field of automatic diagnosis of severe obstructive sleep apnoea. The acoustic properties of the voices of speakers suffering from OSA were studied and an apnoea speech database was designed attempting to cover all the major linguistic contexts in which these physiological OSA features could have a greater impact. For this purpose we analyzed in depth the possibilities of applying state-of-the-art GMM techniques to the modelling of the peculiar features of the realizations of certain phonemes by apnoea patients. In relation with this issue, we focused on nasality as an important feature in the acoustic characteristics of apnoea speakers. Our state-of-the-art GMM approach has confirmed that there are indeed significant differences between apnoea and control group speakers in terms of relative levels of nasalization between different linguistic contexts. Furthermore, we tested the discriminative power of GMM-based speaker recognition techniques adapted to severe apnoea detection with promising experimental results. A correct classification rate of 81.25 % shows that GMM-based OSA diagnosis could be useful for the preliminary diagnosis of apnoea patients and, which suggests it is worthwhile to continue to explore this area.

Regarding future research, our automatic apnoea assessment needs to be validated with a larger sample from a broader spectrum of population. Furthermore, best results can be expected using a representation of the audio data that is optimized for apnoea discrimination. Regarding the significant differences between apnoea and healthy speakers on the relative nasalization degree between different linguistic contexts, an interesting study would be focused on exploiting this information by means of nasalization measures, in order to the improvement of the GMM-based apnoea/non-apnoea detector. Finally, we mention that future research will also be focused on exploiting physiological OSA features in relevant linguistic contexts in order to explore the discriminating power of each feature using linear discriminant classifiers or calibration tools. We aim to apply these findings to improve the performance of the automatic apnoea diagnosis system.

**Acknowledgements.** The activities described in this paper were funded by the Spanish Ministry of Science and Technology as part of the TEC2006-13170-C02-02 project. The authors would like to thank the volunteers at *Hospital Clínico Universitario of Málaga*, and to Guillermo Portillo who made the data collection possible.

# References

1. Puertas, F.J., Pin, G., María, J.M., Durán, J.: Documento de consenso Nacional sobre el síndrome de Apneas-hipopneas del sueño (SAHS). Grupo Español De Sueño, GES (2005)
2. Coccagna, G., Pollini, A., Provini, F.: Cardiovascular disorders and obstructive sleep apnea syndrome. Clinical and Experimental Hypertension 28, 217–224 (2006)
3. Lloberes, P., Levy, G., Descals, C., Sampol, G., Roca, A., Sagales, T., de la Calzada, M.D.: Self-reported sleepiness while driving as a risk factor for traffic accidents in patients with obstructive sleep apnoea syndrome and in non-apnoeic snorers. Respiratory Medicine 94(10), 971–976 (2000)
4. Fox, A.W., Monoson, P.K.: Speech dysfunction of obstructive sleep apnea: A discriminant analysis of its descriptors. Chest Journal 96(3), 589–595 (1993)
5. Fernández, R., Hernández, L.A., López, E., Alcázar, J., Portillo, G., Toledano, D.T.: Design of a Multimodal Database for Research on Automatic Detection of Severe Apnea Cases. In: Proceedings of 6th Language Resources and Evaluation Conference, LREC, Marrakech, Morocco (2008)
6. Parsa, V., Jamieson, D.G.: Acoustic discrimination of pathological voice: Sustained vowels versus continuous speech. Journal of Speech, Language, and Hearing Research 44(2), 327–339 (2001)

7. Reynolds, D.A., Quatieri, T.F., Dunn, R.B.: Speaker verification using adapted gaussian mixture models. Digital Signal Processing 10, 19–41 (2000)
8. Pruthi, T.: Analysis, vocal-tract modeling and automatic detection of vowel nasalization. Doctor Thesis at the University of Maryland (2007)
9. Robb, M., Yates, J., Morgan, E.: Vocal Tract Resonance Characteristics of Adults with Obstructive Sleep Apnea. Acta Otolaryngologica 117, 760–763 (1997)
10. Fiz, J.A., Morera, J., Abad, J., Belsulnces, A., Haro, M., Fiz, J.I., Jane, R., Caminal, P., Rodenstein, D.: Acoustic analysis of vowel emission in obstructive sleep apnea. Chest Journal 104, 1093–1096 (1993)
11. Obrador, A., Haro, M., Álvarez, L.l., Calderón, J.C., Sabater, G., Casamitja, M.T., Sendra, S.: El análisis de la voz y la pulsioximetria nocturna como técnicas de despistaje del síndrome de apnea-hipopnea en el sueño. In: Congreso Nacional de Neumología, Tenerife, Spain (2008)
12. Hidalgo, A., Quilis, M.: Fonética y fonología españolas. Edi. Tirant blanch (2002)
13. Glass, J., Zue, V.: Detection of nasalized vowels in American English. In: Proceedings of Acoustics, Speech, and Signal Processing, IEEE International Conference on ICASSP, Tampa, USA, vol. 10, pp. 1569–1572 (1985)
14. Cakirer, B., Hans, M.G., Graham, G., Aylor, J., Tishler, P.V., Redline, S.: The Relationship between Craniofacial Morphology and Obstructive Sleep Apnea in Whites and in African-Americans. American Journal of Respiratory and Critical Care Medicine 163(4), 947–950 (2001)
15. Moreno, A., Poch, D., Bonafonte, A., Lleida, E., Llisterri, J., Mariño, J.B., Naude, C.: ALBAYZIN Speech Database: Design of the Phonetic Corpus. In: Proceedings of Eurospeech, Berlin, Germany, vol. 1, pp. 175–178 (1993)
16. Blouet, R., Mokbel, C., Mokbel, H., Sanchez Soto, E., Chollet, G., Greige, H.: BECARS: a Free Software for Speaker Verification. In: Proceedings of The Speaker and Language Recognition Workshop, ODYSSEY, Toledo, Spain, pp. 145–148 (2004)
17. Do, M.N.: Fast approximation of Kullback-Leibler distance for dependence trees and Hidden Markov Models. IEEE Signal Processing Letter 10, 115–118 (2003)
18. Fredouille, C., Pouchoulin, G., Bonastre, J.F., Azzarello, M., Giovanni, A., Guio, A.: Application of Automatic Speaker Recognition techniques to pathological voice assessment (dysphonia). In: Proceeding of 9th European Conference on Speech Communication and Technology, Interspeech, Lisboa, Portugal, pp. 149–152 (2005)

# Automatic Detection of Atrial Fibrillation for Mobile Devices

Stefanie Kaiser, Malte Kirst, and Christophe Kunze

FZI Forschungszentrum Informatik, Karlsruhe, Germany
kirst@fzi.de
http://www.fzi.de/ess

**Abstract.** Two versions of a new detector for automatic real-time detection of atrial fibrillation in non-invasive ECG signals are introduced. The methods are based on beat to beat variability, tachogram analysis and simple signal filtering. The implementation on mobile devices is made possible due to the low demand on computing power of the employed analysis procedures. The proposed algorithms correctly identified 436 of 440 five minute episodes of atrial fibrillation or flutter and also correctly identified up to 302 of 342 episodes of no atrial fibrillation, including normal sinus rhythm as well as other cardiac arrhythmias. These numbers correspond to a sensitivity of 99.1 % and a specificity of 88.3 %.

## 1  Introduction

Atrial fibrillation (AF) is a widely spread disease and the most frequently diagnosed cardiac arrhythmia in western countries. Approximately 1–5 % of the population in such countries suffer from atrial fibrillation, with increasing percentages at higher patients' ages, reaching an incedence of almost 12 % in male patients at ages over 85. Due to the rising average age in the industrial nations and to the ascending commonness of other established risk-factors, such as hipertension or overweight, experts expect a doubling of the incidences during the next 50 years.

Whereas atrial fibrillation at itself is not a life-threatening disease, it entails dangerous secondary complications, such as embolisms and apoplectic strokes. Approximately 15 % of all strokes are caused by atrial fibrillation.

The timely diagnosis of AF proves to be complicated due to several reasons. First, atrial fibrillation implicates scant perceivable symptoms and is mostly not noticed by the patients themselves. Second, in early stages the disease occurs in irregular episodes with unpredictable times of appearance and durations. On the other hand, physicians have ever fewer time spendable on each patient, making it impossible to analyze long-term ECG manually.

Therefore, automatic detection of atrial fibrillation in electrocardiograms is and will be increasingly important and necessary during the next decades. [1,2,3]

## 2  Atrial Fibrillation

The healthy heart beats at a regular rhythm with approximately 60–80 bpm (normal sinus rhythm, NSR), where the electrical excitation for each beat starts at the sinus node and subsequently spreads over the atrium and ventricles.

A. Fred, J. Filipe, and H. Gamboa (Eds.): BIOSTEC 2009, CCIS 52, pp. 258–270, 2010.
© Springer-Verlag Berlin Heidelberg 2010

In contrast, during atrial fibrillation, the vestibules are stimulated at a frequency of 350–600 activations per minute, causing a quasi constantly circulating excitation. This condition provokes a dysfunction of the blood pumping activity in the atrium, creating the risk of blood accumulation and therefore the risk of embolisms. Also, the constant stimulation of the atrium does not allow the organized and periodic conduction of the activation toward the chambers. Rather, the points in time of the simulation propagation toward the chambers and the so induced heartbeats are random and the time intervals between two heartbeats (RR interval) become absolutely irregular and chaotic.

In the ECG, atrial fibrillation is perceptible by high disparity of the length of the RR intervals – meaning very irregular heartbeat, the absence of the P-wave and through a constant baseline fibrillation, caused by the constant activation of the atrium. Figure 1 shows the ECG of atrial fibrillation compared to the ECG of normal sinus rhythm.

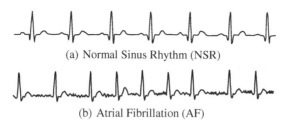

(a) Normal Sinus Rhythm (NSR)

(b) Atrial Fibrillation (AF)

**Fig. 1.** Comparison of the morphology of an ECG signal for normal sinus rhythm and atrial fibrillation

## 3   State of the Art

Automatic detection of atrial fibrillation has been base of research during many years and several methods have been developed in this field, predicated on different approaches. A distinction into methods appropriate for the detection on invasive ECG signals and those for detection on non-invasive signals has to be made, due to the discrepancy of the signal quality, the signal-to-noise ratio (SNR) and the information contained in the signal.

The two general main approaches investigated for the detection of AF are the analysis of the signal-baseline between heart-beats, including baseline variation, zero-crossing and detection of P-waves [4], and the analysis of the rhythm, including variance of RR intervals [5], density allocation of RR intervals [6], analysis with artificial neural networks [7] and analysis in frequency domain [8], between others. In the field of mobile devices, detection on non-invasive signals is required and in this range the approach of rhythm analysis is the most commonly used.

Furthermore mobile devices demand low computing power costs. Therefore complicated and computationally intensive procedures, such as artificial neural networks or transformation into the frequency domain should be avoided.

In addition, the length of the analyzed datasets has to be taken into consideration at the examination of different algorithms, since the longer the contemplated dataset, the easier will the detection of a certain signal pattern be. However, short analysis sections are preferable for prompt detection and for the detection of short arrhytmia periods.

## 4    Datasets

In the run-up to this work an adequate, statistically relevant test-database was generated, based on the following four source-databases.

- Physionet MIT-BIH Atrial Fibrillation Database [9]
- American Heart Association (AHA) Database [9]
- ECG signals recorded at the Institute for Signal Processing Technology (ITIV), Universität Karlsruhe (TH), Germany
- ECG signals recorded at the University Hospital Tübingen (UKT), Germany

These databases include recordings with normal sinus rhythm, atrial fibrillation and flutter as well as other cardiac arrhythmias, such as unifocal and multifocal premature ventricular contraction (PVC), bigeminy, trigeminy and quadrageminy, couplets, triplets and tachycardia.

The created test database consists of 782 five-minute datasets with representative ECG rhythms, that were classified into *atrial fibrillation datasets* (AF) and *no atrial fibrillation datasets* (NOAF). The created datasets contain an average of 436 heartbeats.

Finally, out of the 782 records 440 were categorized as AF and 342 as NOAF. Out of the latter 142 show normal sinus rhythm or isolated PVC (NSR) and 200 show other strong arrhythmias (OAR), see Table 1).

Table 1. Overview of the final test database

| ECG-Type | # Datasets | Total Length |
|---|---|---|
| Atrial fibrillation | 440 | 2200 min |
| Normal sinus rhythm | 142 | 710 min |
| Other strong arrhythmias | 200 | 1000 min |
| All | 782 | 3910 min |

The length of five minutes was chosen as a compromise between easier detection on longer entities and prompt detection on shorter episodes. The final decision over the contemplation period for detection was taken in collaboration with physicians of the University Hospital Tübingen.

In order to obtain a standardized database, as well as to ensure the possibility of saving other additional records, such as tachograms, results, etc. along with the original ECG signals, all 782 ECG records were converted into the Unisens format [10,11].

## 5    Algorithm

According to the requirements of a mobile device, an algorithm was developed that reliably detects atrial fibrillation on non-invasive ECG signals under adherence of low processing power costs. The proposed method rests upon the analysis of the rhythm of the heart beats and is more precisely based on the analysis of the RR interval tachogram.

Furthermore the developed algorithm divides into two separated detection methods, the PPV-Detector and the PPV-MF-Detector. Whereas the further constitutes the basic detection algorithm, the PPV-MF-Detector answers an extension, achieving an improvement of the detection quality.

## 5.1 ECG Premachining

The tachogram-based analysis requires a premachining of the ECG signal, consisting of the QRS detection and the computing of the actual tachogram.

**QRS Detection.** QRS detection creates a list containing the points in time of the heartbeats. Numerous QRS-detection-algorithms have been published. In this work the Open Source ECG Analysis algorithm has been used for the QRS detection [12].

**Tachogram Generation.** The tachogram is a heart rate variability signal (HRV), that considers not only normal heart beats but also PVC and that measures the beat-to-beat variations in the heart rate. It shows the RR interval duration between the current and the previous beat over the time of the current beat.

$$RR_i = t(R_i) - t(R_{i-1}) \qquad (1)$$

This means, for each heartbeat the time interval to the previous heartbeat is calculated. The result corresponds to the value of the tachogram at the point in time of the contemplated heartbeat.

The tachogram then provides information about the ECG rhythm and its regularity. A regular heartbeat, such as appears in NSR, will generate a flat tachogram with an almost constant value. Arrhythmias in the ECG will lead to amplitude varieties and a heart beat as irregular as it occurs during AF will lead to a tachogram with an appearance similar to white noise. Figure 2 shows the tachograms of a NSR-ECG, a NSR-ECG with PVC and an AF-ECG.

## 5.2 PPV-Detector

The PPV-Detector comprehends the basic detection method of the proposed algorithm and represents a fully functional AF detection algorithm by itself. PPV-Detector stands for *Peak-to-Peak and Variance Analysis Detector.*

Methodically, the PPV-Detector divides the 300 seconds long datasets into 30 equally long ten second segments. Each segment is further on analyzed separately and individually.

**Peak-to-Peak and Variance Reckoning.** The peak-to-peak value $PP$ defines the maximum difference between any two RR intervals in one segment, and equals the difference between the maximum and the minimum amplitude inside the examined segment. It is therefore calculated as

$$PP(s_n) = \max(RR(s_n)) - \min(RR(s_n)) \quad n = 1, .., 30, \qquad (2)$$

where $s_n$ corresponds to the segment n and $RR(s_n)$ to the set of RR interval durations of segment $s_n$.

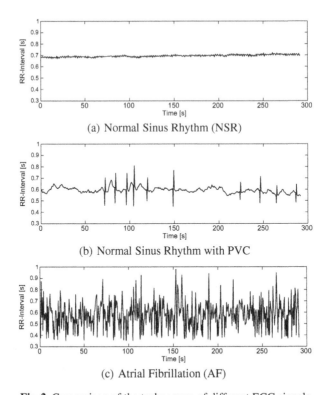

(a) Normal Sinus Rhythm (NSR)

(b) Normal Sinus Rhythm with PVC

(c) Atrial Fibrillation (AF)

**Fig. 2.** Comparison of the tachograms of different ECG signals

The variance of the set of RR interval durations of each segment $s_n$ is calculated by

$$\mathrm{var}(s_n) = \frac{\sum_{i=1}^{I}(RR_i - \mathrm{mean}(s_n))^2}{I-1}, \tag{3}$$

$$\mathrm{mean}(s_n) = \frac{\sum_{i=1}^{I} RR_i}{i}, \tag{4}$$

where $I$ corresponds to the amount of RR interval values $RR_i$ in each segment $s_n$.

**PPV-Detector Decision Tree.** The datasets are classified as AF and NOAF by using a decision tree based on threshold comparisons. This decision tree can be devided into two separate parts, where the first analyzes the individual segments, whereas the second classifies the entire dataset into either AF or NOAF.

In a first step the classification for every single segment is made. Thereby, for each segment $s_n$ the $PP(s_n)$ is calculated. Each segment with a $PP$ higher than 0.2 is classified as *AF-typical* whereas those with a $PP$ smaller than this are classified as *not-AF-typical*. A decision for not-AF-typical leads to the increment of a counter in order to keep track of the number of not-AF-typical segments in the dataset. In addition, only in this case the variance Var of the concerned segment is calculated and saved in a buffer. Figure 3(a) shows this first part of the flow chart for the PPV-Detector decisions.

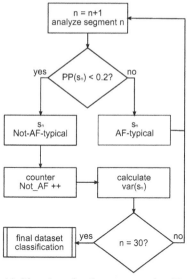

(a) Flowchart for the segment classification.

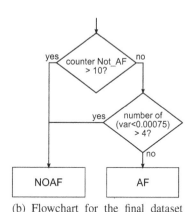

(b) Flowchart for the final dataset classification.

**Fig. 3.** Flowcharts of the algorithm

Once every segment in the dataset has been treated according to the explained method, a final diagnosis of the dataset is reached by the second part of the detection algorithm as follows.

If the number of not-AF-typical segments exceeds ten, the dataset is immediately classified as NOAF. If this is not the case, the buffered variances $Var(s_n)$ are taken in consideration and are compared to another threshold. If more than four of the buffered variances out of the dataset segments are smaller than 0.00075, the dataset will again be classified as NOAF.

Only if the number of not-AF-typical segments is smaller or equal to ten and less than four of the buffered variance are smaller than the set threshold, the dataset will be diagnosed as AF. Figure 3(b) shows the second part of the flow chart for the PPV-Detector, in which the final diagnosis decision for the dataset is made.

### 5.3   PPV-MF-Detector

Whereas the PPV-Detector algorithm consists of a very simple AF analysis focused on the reckoning of peak-to-peak values and variances, the PPV-MF-Detector understands a further analysis, that is based on the PPV-Detector, but includes a second analysis helped by morphological filters (MF).

The fundamentals of this method lie in the fact, that strong structural differences, that do not show in the analysis of peak-to-peak values and variances, can be found between tachograms of atrial-fibrillation ECG signals and those obtained from other strong arrhythmias, such as bigeminal premature ventricular contractions (bigeminy), trigeminal premature ventricular contractions (trigeminy), quadrageminal premature ventricular contractions (quadrageminy) or series of couplets and series of triplets. These structural differences consist of the existence of repeating morphologies or structures in tachograms of such other arrhythmias. Figure 4 clearly shows these differences in the tachograms of atrial fibrillation compared to bigeminy and quadrageminy.

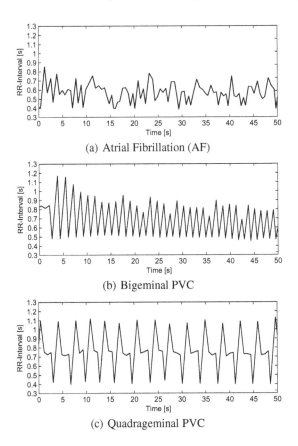

(a) Atrial Fibrillation (AF)

(b) Bigeminal PVC

(c) Quadrageminal PVC

**Fig. 4.** Comparison of the tachograms of different arhythmias

The idea of the PPV-MF-Detector is to use morphological filters in order to suppress these repeating structures in the tachogram. The result of this procedure is a second, modified tachogram, that is then again analyzed by a basic PPV-Detector.

**Morphological Filtering Process.** Morphological filters find their origin in the area of image processing, where they still today find the most frequent use. Nevertheless these filters have also found applications in signal processing, mainly in the field of noise reduction but also to suppress specific signal structures.

The basic idea of morphological filtering consists of the sum or rest of a structuring element with the signal that is to be filtered, being the two most important operations *Opening* and *Closing*. Morphological filtering has been previously applied in the field of biosignal processing as in [13].

Since the form and length of the structuring element of a morphological filter influence the result of the filtering process in an essential way, an adequate structuring element has to be found for each tachogram. It has been proved during this work, that in general a structuring element with a length of four datapoints and with the morphology of a trapezoid function is the most adequate choice for this filtering. The exact values of the four points of the structuring element are however adapted to each dataset that is to be analyzed.

In our method, the amplitude of the first and the fourth point of the structuring element correspond to the mean value of the 75 lowest points of the original tachogram, while the amplitude of the second and the third point of the structuring element correspond the mean value of the 75 highest points of the original tachogram.

$$SE(1) = SE(4) = \frac{\sum_{i=1}^{75} \min_i(RR(s_n))}{75}, \tag{5}$$

$$SE(2) = SE(3) = \frac{\sum_{i=1}^{75} \max_i(RR(s_n))}{75}, \tag{6}$$

Figure 5 shows the general appearance of form of the structuring element chosen for the morphological filtering of the tachograms.

**Fig. 5.** Form chosen for the structuring element of the morphological filter

**Creation of the Alternative Tachogram.** The alternative tachogram is created through a two step procedure. Previous to the morphological filtering, the mean value of the original tachogram signal is calculated.

Then, the original tachogram is morphologically filtered by sequential implementing a closing and an opening MF operator. For each of these two morphological operations the same, previously calculated, structuring element is used.

Figure 6 shows two tachograms, one of quadragemi025y and one of atrial fibrillation in the uppermost graphs. The output signal of this step is shown in the second graphs of the same figure.

In a following step, the resulting signal is rested from the original tachogram, resulting in the $\Delta$-signal (Graph 3 in Figure 6 for quadrageminy and AF respectively).

Finally, for each signal point at which the $\Delta$-signal reaches an amplitude higher then 0.3, the value of the original tachogram is substituted by the mean value of the tachogram. The result signal of this step constitutes the new, alternative tachogram. This result signal can be seen in the lowest graph of Figure 6. It is clearly observable that the new tachogram for the quadrigeminy signal shows important disparities with the original tachogram, appearing now as a very constant signal. On the other hand, the new tachogram of the AF signal shows relatively very little differences compared to the original tachogram.

Figure 7 shows the flowchart for the creation of the alternative tachogram signal.

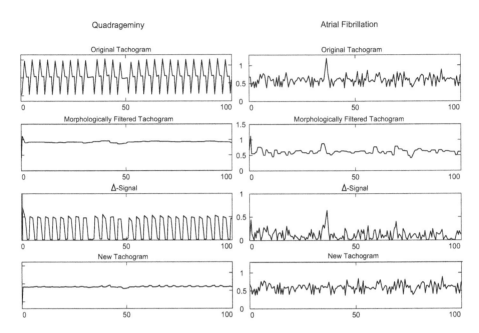

**Fig. 6.** Output signals of the different steps of the creation of the new tachogram. Quadrageminy on the right side and AF on the left side.

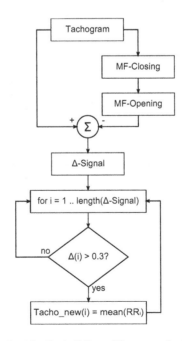

**Fig. 7.** Flowchart for the building of the new, adapted tachogram

**PPV-MF-Detector Decision Tree.** The final MF detection algorithm combines the original PPV-Detector with the morphological filtering and creation of the new, adapted tachogram.

The detector diagnoses the ECG signal in two steps. In the first step, the original PPV-Detector diagnoses the ECG signal. Only if the first diagnosis is AF, the PPV-MF-Detector continues with the creation of the alternative tachogram, which is then, once again, analyzed again by a second, slightly adapted PPV-Detector.

For the first, initial PPV-Detecor, the standard PPV-thresholds are to be used. For the second PPV-Detector, the thresholds have been slightly adapted. The threshold values for both PPV-Detectors are listed in table 2. All threshold values have been determined empirically. Figure 8 shows the decision tree equivalent to the PPV-MF-Detector.

**Table 2.** Thresholds for the PPV decisions

|  | $PP(s_n)$ | # PP | $Var(s_n)$ | # Var |
|---|---|---|---|---|
| basic PPV | 0.2 | 10 | 0.00075 | 4 |
| PPV-MF | 0.2 | 10 | 0.0006 | 4 |

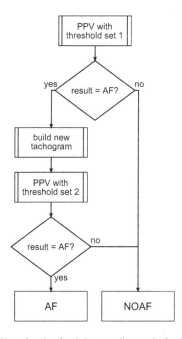

**Fig. 8.** Decision Tree for the final dataset diagnosis for PPV-MF-Detector

## 6   Results

An overview of the detection results obtained with the basic PPV-Detector and for the PPV-MF-Detector is displayed in Table 3. Here, sensitivity (Se) indicates the portion of

AF signals that have been correctly detected as AF, whereas the specificity (Sp) denotes the percentage of NOAF signals, this is to say NSR signals and other arrhythmias, that have been correctly detected as NOAF.

**Table 3.** Detection qualities for the proposed methods

| Detection Algorithm | Sensitivity | Specificity |
|---|---|---|
| PPV-Detector | 99.1 % | 80.1 % |
| PPV-MF-Detector | 99.1 % | 88.3 % |

As Table 3 shows, both, the PPV-Detector and the PPV-MF-Detector reach the same, very high level of sensitivity. The difference in the qualities of both algorithms rests in the specificity. Here, the PPV-MF-Detector achieves a notably higher percentage than the PPV-Detector, at the expense of an increased necessary computing power in relation to the first.

A more precise look at the results for the different datasets is presented in Table 4. It reveals that not only the sensitivity of the two detector versions is equal, but also the specificity for normal sinus rhythm signals is equal and very high for both algorithms. The results differ only in reference to the specificity for ECG signals with arrhythmias other than atrial fibrillation. The quality increase of the specificity between the two algorithms in this domain is of 14 %.

**Table 4.** Detection qualities for the different ECG signal conditions

| ECG | TP | FN | TN | FP | Se | Sp |
|---|---|---|---|---|---|---|
| | | | PPV-Detector | | | |
| ECG | TP | FN | TN | FP | Se | Sp |
| All | 436 | 4 | 274 | 68 | 99.1 % | 80.1 % |
| AF | 436 | 4 | — | — | 99.1 % | — |
| NSR | — | — | 137 | 5 | — | 96.5 % |
| OAR | — | — | 137 | 63 | — | 68.5 % |
| | | | PPV-MF-Detector | | | |
| ECG | TP | FN | TN | FP | Se | Sp |
| ECG | TP | FN | TN | FP | Se | Sp |
| All | 436 | 4 | 302 | 40 | 99.1 % | 88.3 % |
| AF | 436 | 4 | — | — | 99.1 % | — |
| NSR | — | — | 137 | 5 | — | 96.5 % |
| OAR | — | — | 165 | 35 | — | 82.5 % |

## 7    Conclusions and Discussion

Within this article, two alternatives of an algorithm for the detection of atrial fibrillation in five minutes long ECG signals have been proposed.

As it can be observed by means of the results exposed in Section 6, the two algorithm reach very satisfying detection qualities in terms of sensitivity for atrial fibrillation and specificity of normal sinus rhythm. On the other hand, the methods differ noticeably in the specificity regarding ECG signals with strong arrhythmias other than atrial fibrillation. At the expense of a higher demand on computing power, the PPV-MF-Detector produces better results than the PPV-Detector itself.

Both alternatives have been developed focusing on the intention of detecting episodes of AF in a long term electrocardiogram, that are to be flagged for the later revision by a physician and the ultimate diagnosis. On the other hand the algorithms have been developed under the constraints of mobile devices. This is, in the first place, low processing power. Due to the characteristics named earlier, each of the two proposed algorithms has different advantages and disadvantages, so that the ideal choice depends on the precise utilization environment.

In summary the two versions of the detector that have been proposed, provide very high sensitivity being the algorithms based on very simple basic principles, such as threshold comparisons and therefore on very low computing power demands.

In order to lower computing power demands even more, an alternative modification of the proposed method is currently studied, which consists of the analysis of ECG episodes containing a fixed amount of beats instead of a fixed time period. This allows to simplify the implementation of the methods specially on mobile devices due to non-variable memory allocations. The first results obtained for this approach reach a sensitivity of 98.6 % and a specificity of 83.6 % for a dataset length of 300 beats. The number of heartbeats in a five minute period varies with a mean value of 436.0 and a standard deviation of 95.4. The detection quality on a fixed number of heartbeats can be improved by using datasets with more than 300 samples.

# 8    Outlook

A further improvement of the specificity in the area of other strong arrhythmias, such as bigeminy, trigeminy, couplets, etc., may be reached under a slight increase of computing power demands.

One alternative approach to the detection consists in the suppression of PVC beats previous to the analysis with the PPV- and the PPV-MF-Detectors. A new determination of the thresholds would be necessary in this case.

Further on, the possibility of distinguishing and diagnosing not only between "atrial fibrillation" and "not atrial fibrillation", but also between the different other arrhythmias should be taken in consideration. Another approach in this area could be the intent of delivering the exact number of PVC beats occurred in one certain ECG segment.

**Acknowledgements.** This work has been kindly supported by the German government BMBF project MµGUARD. The authors thank the University Hospital Tübingen (UKT) for their time and support with medical questions and supply of clinical ECG records. We also want to thank the Institute for Signal Processing Technology (ITIV), Universität Karlsruhe (TH), Germany for providing further ECG records.

# References

1. Heeringa, J., van der Kuip, D., Hofman, A., Kors, J., van Herpen, G., Stricker, B., Stijnen, T., Lip, G., Witteman, J.: Prevalence, incidence and lifetime risk of atrial fibrillation: the Rotterdam study. Europace (2006)
2. Ringborg, A., Nieuwlaat, R., Lindgren, P., Jönsson, B., Fidan, D., Maggioni, A., Lopez-Sendon, J., Stepinska, J., Cokkinos, D., Crijns, H.: Costs of atrial fibrillation in five European countries: results from the Euro Heart Survey on atrial fibrillation. Europace 10, 403–411 (2008)
3. Hohnloser, P.D.S., Grönefeld, P.D.D.G., Israel, P.D.D.C.: Prophylaxe und Therapie von Vorhofflimmern, 1st edn. UNI-MED Verlag, Bremen (2005)
4. Kim, J., Bocek, J., White, H., Crone, B., Alferness, C., Adams, J.: An atrial fibrillation detection algorithm for an implantable atrial defibrillator, 169–172 (1995)
5. Logan, B., Healey, J.: Robust Detection of Arial Fibrillation for a a Long Term Telemonitoring System. IEEE Computers for Cardiology, 391–394 (2005)
6. Tateno, K., Glass, L.: A Method for Detection of Atrial Fibrillation using RR Intervals. IEEE Computers for Cardiology, 391–394 (2000)
7. Artis, S., Mark, R., Moody, G.: Detection of Atrial Fibrillation using Artificial Neural Networks. IEEE Proceedings on Computers in Cardiology, 173–176 (1991)
8. Sadek, L.E., Ropella, K.M.: Detection of Atrial Fibrillation from the Surface Electrocardiogram using Magnitude-Squared Coherence. IEEE Engineering in Medicine and Biology Society 1, 179–180 (1995)
9. Goldberger, A.L., Amaral, L.A.N., Glass, L., Hausdorff, J.M., Ivanov, P.C., Mark, R.G., Mietus, J.E., Moody, G.B., Peng, C.K., Stanley, H.E.: PhysioBank, PhysioToolkit, and PhysioNet: Components of a new research resource for complex physiologic signals. Circulation 101(23), e215–e220 (2000); Circulation Electronic Pages, http://circ.ahajournals.org/cgi/content/full/101/23/e215
10. Kirst, M., Ottenbacher, J., Nedkov, R.: UNISENS – Ein universelles Datenformat für Multisensordaten. In: Biosignalverarbeitung : Innovationen bei der Erfassung und Analyse bioelektrischer und biomagnetischer Signale, pp. 106–108 (2008)
11. Kirst, M., Ottenbacher, J.: Unisens (2008), http://www.unisens.org
12. Hamilton, P.: Open Source ECG Analysis. IEEE Computers in Cardiology, 101–104 (2002)
13. Chu, C.H.H., Delp, E.J.: Impulsive noise suppression and background normalization of electrocardiogram signals using morphological operators. IEEE Transactions on Biomedical Engineering 36(2), 262–273 (1989)

# Speaker-Adaptive Speech Recognition Based on Surface Electromyography

Michael Wand and Tanja Schultz

Universität Karlsruhe (TH), Germany
mwand@ira.uka.de, tanja@ira.uka.de
http://csl.ira.uka.de

**Abstract.** We present our recent advances in *silent speech* interfaces using electromyographic signals that capture the movements of the human articulatory muscles at the skin surface for recognizing continuously spoken speech. Previous systems were limited to speaker- and session-dependent recognition tasks on small amounts of training and test data. In this article we present speaker-independent and speaker-adaptive training methods which allow us to use a large corpus of data from many speakers to train acoustic models more reliably. We use the speaker-dependent system as baseline, carefully tuning the data preprocessing and acoustic modeling. Then on our corpus we compare the performance of speaker-dependent and speaker-independent acoustic models and carry out model adaptation experiments.

## 1 Introduction

Automatic Speech Recognition (ASR) has now matured to a point where it is successfully deployed in a wide variety of every-day life applications, including telephone-based services and speech-driven applications on all sorts of mobile personal digital devices.

Despite this success, speech-driven technologies still face two major challenges: first, recognition performance degrades significantly in the presence of noise. Second, confidential and private communication in public places is difficult due to the clearly audible speech.

In the past years, several alternative techniques were proposed to tackle these obstacles, among them the recognition of whispered speech with a throat microphone [1] or non-audible murmur with a special stethoscopic microphone [2]. Other approaches include using optical or ultrasound images of the articulatory apparatus, i.e. [3], or sub-vocal speech recognition [4].

In this article, we present our most recent investigations in electromyographic (EMG) speech recognition, where the activation potentials of the articulatory muscles are directly recorded from the subject's face via surface electrodes[1].

In contrast to many other technologies, the major advantage of EMG is that it allows to recognize *non-audible*, i.e. *silent* speech. This makes it an interesting technology not

---

[1] Strictly spoken, the technology is called *surface electromyography*, however we use the abbreviation EMG for simplicity.

A. Fred, J. Filipe, and H. Gamboa (Eds.): BIOSTEC 2009, CCIS 52, pp. 271–285, 2010.

only for mobile communication in public environments, where speech communication may be both a confidentiality hazard and an annoying disturbance, but also for people with speech pathologies.

Research in the area of EMG-based speech recognition has only a short history. In 2002, [5] showed that myoelectric signals can be used to discriminate a small number of words. In 2006, [6] showed that speaker dependent recognition of continuous speech via EMG is possible. The recognition accuracy in this task could be improved by a careful design of acoustic features and signal preprocessing [7], and advances in acoustic modeling using phonetic features in combination with phone models [8]. However, the described experiments were based on relatively small amounts of data, and consequently were limited to speaker-dependent modeling schemes. In [9], first results on EMG recognition across recording sessions were reported, however these experiments were run on a small vocabulary of only 10 isolated words.

This article reports EMG-based recognition results on continuously spoken speech comparing speaker-dependent, speaker-adaptive, and speaker-independent acoustic models. We use recognition results from the speaker-dependent system as baseline and show that the accuracy of this system improves by appropiately tuning the data preprocessing and adapting the acoustic modeling. For the speaker-independent and speaker-adaptive experiments we first develop generic speaker independent acoustic models based on a large amount of training data from many speakers and then adapt these models based on a small amount of speaker specific data. The baseline performance of the speaker-dependent EMG recognizer is 47.15% Word Error Rate on a testing vocabulary of 108 words.

The article is organized as follows: In section 2, we describe the data acquisition and the resulting data corpus *EMG-PIT*. In section 3, we explain the setup of the EMG recognizer, the feature extraction methods, as well as the different training and adaptation variants. In section 4, we present the recognition results of the different methods and section 5 concludes the article.

## 2   The EMG-PIT Data Corpus

During the years 2007 - 2008 we collected a large database of EMG signals from 78 native speakers of American English. This collection was done in a joint effort with colleagues from the Department of Communication Science and Disorders at University of Pittsburgh [10]. The resulting data corpus bears the name *EMG-PIT*; to the best of our knowledge it is the largest corpus of EMG recordings of speech so far.

The collection was done in two phases, a pilot study with 14 speakers, and the final collection of 64 speakers. The 14 pilot study subjects participated in two recording sessions, the other speakers participated in one recording session. All participants were female adults between 18 and 35 years of age with normal vocal qualities. The subjects were recruited primarily from the student population of Pittsburgh (University of Pittsburgh and Carnegie Mellon University).

To further study the similarities and differences of audible and silent speaking mode [9], the database covers both speaking modes with parallel utterances, i.e. each speaker read the same sentences in both silent and audible speaking mode. The audible utterances were simultaneously recorded with a conventional air-transmission microphone.

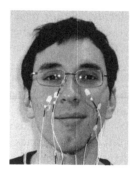

**Fig. 1.** Electrode Positioning

For EMG recording we used a computer-controlled 8-channel EMG recorder (Vario-port, Becker-Meditec, Germany), together with a self-developed recording tool. Technical specifications of the Varioport system include an amplification factor of 1170, 16 bits A/D conversion, a step size (resolution) of 0.033 microvolts per bit, and a frequency range of 0.9-295 Hz. All EMG signals were sampled at 600 Hz. To allow for backward compatibility with our former experiments in [9], we adopted the electrode positioning which yielded optimal results (see figure 1). This electrode setting uses five channels and captures signals from the *levator angulis oris*, the *zygomaticus major*, the *platysma*, the *anterior belly* of the *digastric* and the *tongue*. To also be able to experiment with new electrode positions, we applied one unused channel to collect new electrode positions. The acoustic data was recorded at 16kHz, 16bit resolution and stored in PCM encoding. All subjects were recorded with a close-up video Camcorder while producing audible and silent speech.

To get good phone coverage and to avoid transcription work, the subjects read phonetically balanced English sentences in a controlled setting rather than to record conversational, unplanned speech. These sentences were taken from the Broadcast News domain. To cover large amounts of context but at the same time allow for mode and variability comparisons, the speaker read one batch of 10 BASE utterances, which are the same for each speaker, and one batch of 40 speaker specific SPEC utterances, only read by one speaker. The vocabulary of the BASE sentences consists of 108 words. Each recording session consisted of two parts, one part audible and one part silent speech. In each part we recorded one BASE set and one SPEC set. The total of 50 BASE and SPEC utterances in each part were recorded in random order. For the pilot study, subjects recorded two sessions, where the order of the audible and silent parts was reversed after the first session to control effects from utterance repetitions between the parts. In the main study, each subject recorded first the audible part and then the silent part. The following table shows the statistics from the EMG-PIT corpus.

| Phase | Speakers | Sessions | Utterances | | Duration [min] | |
|---|---|---|---|---|---|---|
| | | | Audible | Silent | Audible | Silent |
| Pilot | 14 | 28 | 1400 | 1400 | 108 | 110 |
| Main | 64 | 64 | 3200 | 3200 | 287 | 251 |
| Total | 78 | 92 | 4600 | 4600 | 395 | 361 |

This article reports results on the *audible utterances* of the *pilot study* only, leaving the remaining data as verification set for future studies. Thus the corpus of utterances which was used for this study has the following properties:

| Speakers | 14 females speakers |
|---|---|
| Sessions | 2 sessions per speaker |
| Average Length (total) (training set) (test set) | 231 seconds per session 180 seconds 51 seconds |
| Decoding vocabulary | 83 words (108 words including pronunciation variants) |

# 3    EMG-Based Speech Recognizer

The initial EMG recognizer was taken from [7], which in turn was set up according to [6]. It used an HMM-based acoustic modeling, which was based on fully continuous Gaussian Mixture Models. For the initial context-independent phoneme recognizer there were 136 codebooks (three per phoneme, modeling the beginning, middle and end of a phoneme, and one silence codebook). It should be noted that due to the small amount of training data, most speaker dependent codebooks ended up with about one to four Gaussians after the initial automatic merge-and-split codebook generation.

The training concept worked as follows: The time-aligned training data (see section 3.1) was used either for a full training run (see section 3.3), or we applied MLLR adaptation on models which were pre-trained on a large set of speakers to adapt them to the current speaker and session (see section 3.4). The latter is especially important in practical applications since it allows setting up a recognizer with a very small amount of individual training data: in section 4.2 we describe how the recognition results change when the size of the set of speaker-specific training data is reduced.

During the decoding, we used the trained acoustic model together with a trigram language model trained on Broadcast News data. The testing process consisted of an initial testing run followed by a lattice rescoring in order to obtain optimal results. See section 3.6 for details.

In section 3.5 we present our investigations on using bundled phonetic feature models for the EMG recognizer.

## 3.1    Initial Time Alignment

In order to find a time alignment for the training sentences, the *audio data* which had been simultaneously recorded was used. The audio data was forced-aligned with an English Broadcast News (BN) speech recognizer trained with the Janus Recognition Toolkit (JRTk). This recognizer is HMM-based, and makes use of quintphones with 6000 distributions sharing 2000 codebooks. The baseline performance of this system is 10.2% WER on the official BN test set (Hub4e98 set 1), F0 condition [11].

The resulting time-alignment can not be mapped directly to the EMG data since the EMG signal precedes the audio signal by about 30ms - 60ms [8]. Accordingly, we modeled this effect by delaying the EMG signal for an amount of 0 ms to 90 ms (in steps of 10 ms). Additionally, in this article we demonstrate that considering a large frame context during acoustic modeling makes the acoustic models more robust with respect to the time delay and yields better recognition results. The effect of the EMG signal delay and of considering a large frame context is charted in section 4.1.

## 3.2  Feature Extraction

We compare two methods for feature extraction, which are both based on *time-domain (TD) features*. Their only difference is the amount of context which is considered for the final features.

We use the following definitions [6]: For any feature $f$, $\bar{f}$ is its frame-based time-domain (amplitude) mean, $P_f$ is its frame-based power, and $z_f$ is its frame-based zero-crossing rate. $S(f,n)$ is the stacking of adjacent frames of feature $f$ in the size of $2n+1$ ($-n$ to $n$) frames.

For an EMG signal with normalized mean x[n], the nine-point double-averaged signal $w[k]$ is defined as

$$w[n] = \frac{1}{9} \sum_{n=-4}^{4} v[n], \quad \text{where} \quad v[n] = \frac{1}{9} \sum_{n=-4}^{4} x[n].$$

The rectified high-frequency signal is $r[n] = |x[n] - w[n]|$. In [6], the best WER was obtained with the following feature:

$$\textbf{TD5} = S(\textbf{f2},5), \text{where } \textbf{f2} = [\bar{\textbf{w}}, \textbf{P}_\textbf{w}, \textbf{P}_\textbf{r}, \textbf{z}_\textbf{r}, \bar{\textbf{r}}].$$

We use this feature as baseline in this article and call it TD5, where the number 5 stands for the stacking width. We found that we got optimal results by increasing the context width to 15 frames, yielding a total of 31 frames to be stacked. The resulting feature is called TD15 and is defined as

$$\textbf{TD15} = S(\textbf{f2},15), \text{where } \textbf{f2} = [\bar{\textbf{w}}, \textbf{P}_\textbf{w}, \textbf{P}_\textbf{r}, \textbf{z}_\textbf{r}, \bar{\textbf{r}}].$$

In section 4.1, we compare the features TD5 and TD15.

In these computations, we used a frame size of 27 ms and a frame shift of 10 ms since we found earlier that these values give optimal results [12]. In both cases, the features from the five EMG channels are stacked to create a final "joint" feature consisting of the synchronized data from all channels, On the resulting joint feature vector, Linear Discriminant Analysis (LDA) is applied to reduce the dimensionality of the final feature vectors to 32 according to [6].

We compare the performance of features TD5 and TD15 in section 4.1 and demonstrate that TD15 performs better than the original TD5 feature. Therefore in the later sections we only use the TD15 feature.

### 3.3   Training Process

A full training run consisted of the following steps: First, an LDA transformation matrix for feature dimensionality reduction was calculated based on the time-aligned data. Initial codebooks were created by a merge-and-split algorithm in order to adapt to the small amount of training data and to compensate for differences in the available number of samples per phoneme. After this, four iterations of Viterbi EM training were performed to improve the initial models.

### 3.4   Across-Speaker Experiments and Adaptation

Speaker independent acoustic models were obtained by initially training acoustic models based on the training data of all speakers *but* the two sessions of the test speaker. On the trained models, we tested with the test set of the respective test speaker ("cross-speaker training"). In the adaptation experiments, we performed MLLR-based speaker adaptation of the models prior to the test ("speaker-adaptive training"). The results of these experiments are charted in section 4.2.

### 3.5   Bundling of Phonetic Features

In the first batch of experiments, we consider (speaker-dependent and speaker-independent) *phoneme models* of the EMG signal, i.e. we regard each frame of the EMG signal as the representation of the beginning, middle, or end state of a phoneme. However, it has been shown in acoustic speech recognition that the recognizer may benefit from additionally modeling *phonetic features (PFs)*, which represent properties of a given phoneme, such as the place of articulation or the manner of articulation [13].

Note that in some previous works, i.e. [13,14] these models are titled "Articulatory Features". Since this modeling approach does *not* reflect the movements of the articulators, but rather represents phonetic properties of phonemes, we use the term "Phonetic Features" (PFs) in our work.

We derive the PFs from phonemes as described in [15], i.e. we use the IPA phonological features for PF derivation. In this work, we use PFs that have binary values. For example, each of the articulation places Glottal, Palatal and Labiodental is a PF that has a value either present or absent. These PFs do intentionally not form an orthogonal set because we want the PFs to benefit from redundant information. In the experiments reported in this article, we use nine different PFs, namely the set { Consonant, Vowel, Alveolar, Voiced, Fricative, Affricate, Glottal, Labiodental, Palatal }, since on the relatively small vocabulary of the speaker-dependent systems these PFs are found to receive sufficient training data to allow for good classification (compare [16], figure 2).

The architecture we employ for the PF-based EMG decoding system is a *multistream* architecture [15,17], see figure 2. This essentially means that the models draw their *acoustic probabilities* not from one single source (or stream) but from a weighted sum of various sources. These additional sources correspond to acoustic models representing substates of PFs, like "middle of a vowel" or "end of a non-fricative". The conventional EMG phoneme-based recognizer contributes as well.

It was suggested by [18] that one major shortcoming of previous PF recognition systems was that features were modeled as statistically independent. The independence

«HELLO WORLD»    Pronunciation Dictionary Lookup

| Phonemes | h | e | l | ou | w | er | l | d |
|---|---|---|---|---|---|---|---|---|
| | | | Phonetic Features | | | | | |
| Alveolar | | | x | | | | x | x |
| Glottal | x | | | | | | | |
| Plosive | | | | | | | | x |
| Fricative | x | | | | | | | |
| | | | | ... | | | | |
| Vowel | | x | | x | | x | | |
| Front (Vowel) | | x | | | | | | |

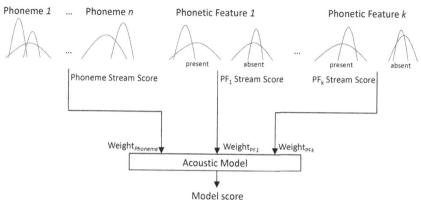

Fig. 2. The Multi-Stream Phonetic Features Decoding Architecture. The upper part shows how the PFs are obtained from the phonetic information, the lower part shows the weighting of the various information sources.

assumption is not correct since physiologically every phonetic feature describes the interplay of various articulators, i.e. the interdependent activity of several facial muscles.

We described a data-driven algorithm for finding dependencies between phonetic features in [19]. We call the process of pooling dependent features together "feature bundling", since eventually we will end up with a set of PF acoustic models which represent *bundles* of PFs, like "voiced fricative" or "rounded front vowel". We additionally allow these bundled phonetic features (BDPFs) to depend not only on properties of the current phoneme, but also on the right and left context phonemes. In [19] we reported that an EMG recognizer based on bundled phonetic features outperforms a recognizer based on context-independent phonemes only by more than 30%.

The algorithm which performs this pooling is a standard decision-tree based clustering approach [20], as it is successfully used in large vocabulary acoustic speech recognition to determine phoneme context clusters. This clustering works by creating a *context decision tree*, which classifies phonemes by asking linguistic questions about the current phoneme and its left and right context. The set of all possible questions is predefined, examples of these categorical questions are: *Is the current phone voiced?* or *Is the right-context phone a fricative?*.

The context tree is created separately for each PF stream, from top to bottom. This means that the initial set of acoustic models e.g. for the stream "FRICATIVE" consists of six models: namely the beginning, middle and end of a "FRICATIVE" or "NON-FRICATIVE". Each context question splits one acoustic model into two new models. The splitting criterion is maximizing the loss of entropy caused by the respective split. Note that both the models representing the *presence* and *absence* of a phonetic feature take part in the splitting process. The process ends when a pre-determined termination condition is met. This condition must be chosen based on the properties of the available data to create a good balance between the accuracy and the trainability of the context-dependent models.

Our termination criterion is that a fixed number of 70 tree leaves for each phonetic feature, corresponding to 70 independent acoustic models, is generated for each PF stream, since this number was experimentally found to yield optimal results. Then the general training process is as follows:

– First, an ordinary context-independent EMG recognizer is trained on the given training data. This recognizer uses both phoneme and PF models, but *no* PF bundling yet.
– In a second step, the context decision tree is grown as described above, and a set of bundled phonetic features (BDPFs) is generated.
– Finally, the BDPF EMG recognizer is trained using the acoustic models defined in the previous step.

With the BDPF recognizer, we perform the same set of cross-speaker and adaptation experiments as with the phoneme-based recognizer, see section 3.4. The results are charted in section 4.2.

### 3.6   Testing

For decoding, we use the trained acoustic model together with a trigram BN language model. We restricted the decoding vocabulary to the words appearing in the test set. This resulted in a test set of 10 sentences per speaker with a vocabulary of 108 words. On the test sentences, the trigram-perplexity of the language model is 24.24.

The testing process used lattice rescoring in order to determine the optimal weighting of the language model compared to the acoustic model.

## 4   Experimental Results

### 4.1   Preprocessing for the Speaker-Dependent System

Figure 3 compares the word error rates (WER) of the speaker-dependent recognition systems of the feature preprocessings TD5 and TD15 and phoneme-based and BDPF modeling. The results were obtained on speaker-dependent systems, i.e. by training on

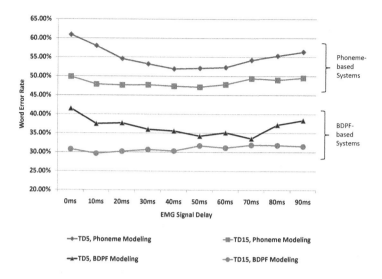

**Fig. 3.** Average Speaker-dependent Word Error Rate for Different Time Delays

the training data of *one* session and tested on test data from *the same* session. Note that we give the averages over all 28 sessions.

The average WER of the baseline recognizer, equipped with the TD5 feature, is 54.89%, where the WER ranges from 51.92% to 60.90%, with a noticeable minimum between 30 ms and 60 ms. Using the TD15 feature, we get an average WER of 48.34%, where for EMG signal delays between 0 ms and 90 ms, the WER ranges from 47.15% to 49.85%. Thus one can see that the wider context yields not only a performance improvement of about 12% relative, but also a much higher robustness when the delay between audio and EMG signal is varied. This can be important in settings where the exact synchronization between EMG and audio signal may not be exactly determined during the training run.

Similarly, the average word error rate of the BDPF recognizer with the TD5 feature is 36.67%, which by using the TD15 feature is reduced by about 11.5% relative to 30.95%. Again, one can see that the graph becomes "flatter" by using a higher context width, i.e. the performance becomes less dependent on the EMG signal delay.

Experiments showed that increasing the context width consistently increases the recognition performance until a critical width of about 15 frames to each side, which corresponds to a total context window of about 300 ms. Beyond that value no more significant improvement occurs. One can conclude from this that speech-relevant observable patterns in the EMG signal may have a length of up to 300 ms and that a purely frame-based preprocessing, which only considers a window of about 27 ms, does not fully capture the discriminating properties of the EMG signal. Also note that the optimal delay for the TD5 preprocessing still lies around 50ms, which is consistent with the results in [6].

In accordance with the results given above, for all further experiments we used the following recognizer setup:

– Feature Preprocessing: Time-domain feature TD15
– Modeling: Bundled Phonetic Features (BDPFs)
– EMG signal delay: 50ms.

### 4.2  Cross-Speaker and Adaptive Experiments

In the following experiments we compare three training scenarios:

– Speaker-Dependent Training: As above, the system is trained and tested with data from one speaker and one session only.
– Cross-Speaker Training: The system is trained on all sessions from all speakers *except* the two sessions from the test speaker. The system is tested on the test data of one session.
– Speaker-Adaptive Training: We use the trained models from the cross-speaker training step, but the resulting system is then adapted toward the test speaker using MLLR adaptation [21] on the training data from one session. As above, testing is done on the test data from the same session.

Figure 4 shows a breakdown of the results of these experiments for each speaker and indicates that the speaker-dependent and adaptive systems clearly outperform the cross-speaker system. This is not very surprising as the speaker independent models have to capture speaker variabilities but at the same time suffer from slight variations in the electrode positioning across speakers. Furthermore, we see that speaker dependent

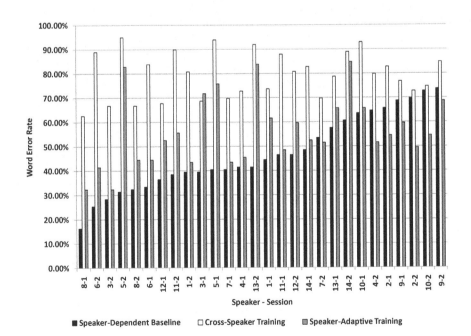

**Fig. 4.** Comparison of Word Error Rates for the Phoneme-based Recognizer

model training achieves better results than MLLR adaptation for most of the speakers and sessions. However for sessions where speaker-dependent training performs badly, particularly for speakers 2 and 9 and to some extent 4 and 10, the performance of the adapted system does not degrade similarly and may outperform the speaker-dependent system.

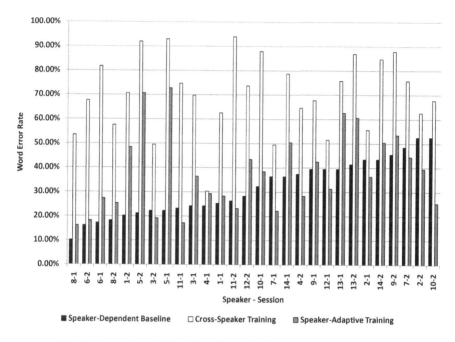

**Fig. 5.** Comparison of Word Error Rates for the BDPF-based Recognizer

We repeated the above experiment, comparing the training scenarios *Speaker-Dependent Training*, *Cross-Speaker Training* and *Speaker-Adaptive Training*, now using bundled phonetic features (BDPFs) as acoustic models. The results are charted in figure 5 and indicate that BDPF modeling brings a clear performance improvement in all the training scenarios. Beyond this, the same pattern as for the phoneme-based recognizer holds: Speaker-dependent recognition generally still achieves better results than both cross-speaker and speaker-adaptive recognition, however for most speakers, with the notable exception of speaker 5, the adaptive system has a performance close to the one of the speaker-dependent system and even outperforms it for 10 out of the 28 sessions.

As a final experiment, we investigated whether MLLR adaptation is applicable for very small sets of speaker-specific training data. For this purpose we took subsets of 10, 20 and 30 sentences out of the full training sets of 40 sentences for each speaker and used each of these reduced sets to train a speaker-dependent system and to create a speaker-adaptive system by performing MLLR adaptation on the *original* cross-speaker system, trained on the full set of training data from all speakers except the one to be tested. Note that the test set remained unchanged in these experiments.

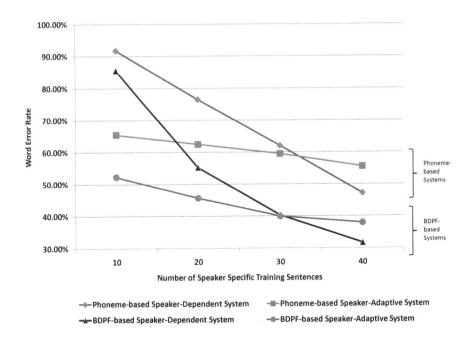

**Fig. 6.** Comparison of Word Error Rates for speaker-dependent and speaker-adaptive systems on different amounts of training data

Figure 6 displays the average word error rate of these recognizers and clearly shows that while for the full set of training sentences the average WER is better for the speaker-dependent systems, the situation is very different for smaller sets of training data: With 10 adaptation sentences, the best speaker-adaptive system yields an average WER of 52.26% (with no adaptation at all, i.e. for the corresponding cross-speaker system, the average WER is 70.27%), while a speaker-dependent system yields a high average WER of 85.49%. When the training set grows, all systems quickly improve, but for up to 30 training sentences, the speaker-adaptive systems on average have a better performance than the speaker-dependent systems.

## 4.3   Summary

The following table summarizes the average Word Error Rates of the different recognizers we presented in the above sections. Note that all values are based on the TD15 preprocessing, and that we always give results for the full set of 40 speaker-specific training/adaptation sentences.

| System | Word Error Rate | | Rel. Gain by BDPF |
|---|---|---|---|
| | Phonemes | BDPF | |
| Speaker-Dependent | 47.15% | 31.68% | 32.8% |
| Cross-Speaker | 79.50% | 70.27% | 11.6% |
| Speaker-Adaptive | 56.34% | 37.92% | 32.7% |

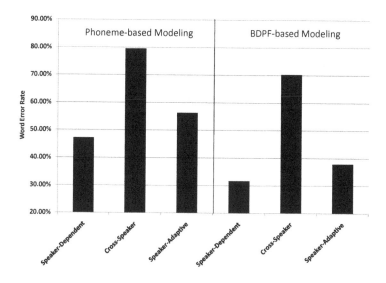

**Fig. 7.** Comparison of Average Word Error Rates for All Systems

The results are also charted in figure 7 and clearly show that MLLR model adaptation is applicable to the task of EMG speech recognition. Furthermore it can be seen that BDPF modeling yields a clear improvement in all training scenarios. In the speaker-dependent and speaker-adaptive cases, the improvements are both about 33%, whereas in the cross-speaker case, the gain is at about 11.5%. In particular, BDPF modeling improves the performance of the adaptation step, making this modeling approach a very interesting perspective for further investigations in speaker-adaptive EMG signal processing.

## 5    Conclusions

We have compared speaker-dependent, speaker-independent and speaker-adaptive systems for EMG speech recognition, reporting results on the performance of EMG speech recognition across multiple speakers and sessions of the EMG-PIT data corpus. We found that while for the full training sets of the EMG-PIT corpus the speaker- and session-dependent EMG system still performs best, for small speaker-specific training sets of up to approximately 30 utterances, on average a speaker-adaptive system outperforms a speaker- and session-dependent EMG recognizer; the speaker-adaptive system still yields acceptable recognition results with a set of only 10 adaptation sentences. This shows that the MLLR adaptation method is feasible for EMG speech recognition and that adaptation methods may be a lever for increasing the usability of EMG-based speech recognition.

We also showed that phonetic feature bundling consistently outperforms phoneme-based systems and in particular significantly increases the performance of MLLR adaptation.

**Acknowledgements.** We would like to thank Maria Dietrich and Katherine Verdolini Abbott for the cooperation in recording the EMG data as part of a larger study which was supported in part through funding received from the SHRS Research Development Fund, School of Health and Rehabilitation Sciences, University of Pittsburgh to Maria Dietrich and Katherine Verdolini Abbott.

# References

1. Jou, S.-C., Schultz, T., Waibel, A.: Whispery Speech Recognition Using Adapted Articulatory Features. In: Proc. ICASSP (2005)
2. Nakajima, Y., Kashioka, H., Shikano, K., Campbell, N.: Non-Audible Murmur Recognition. In: Proc. Eurospeech (2003)
3. Hueber, T., Chollet, G., Denby, B., Dreyfus, G., Stone, M.: Continuous-Speech Phone Recognition from Ultrasound and Optical Images of the Tongue and Lips. In: Proc. Interspeech, pp. 658–661 (2007)
4. Jorgensen, C., Binsted, K.: Web Browser Control Using EMG Based Sub Vocal Speech Recognition. In: Proceedings of the 38th Hawaii International Conference on System Sciences (2005)
5. Chan, A., Englehart, K., Hudgins, B., Lovely, D.: Hidden Markov Model Classification of Myolectric Signals in Speech. IEEE Engineering in Medicine and Biology Magazine 21(9), 143–146 (2002)
6. Jou, S.-C., Schultz, T., Walliczek, M., Kraft, F., Waibel, A.: Towards Continuous Speech Recognition using Surface Electromyography. In: Proc. Interspeech, Pittsburgh, PA (September 2006)
7. Wand, M., Stan Jou, S.-C., Schultz, T.: Wavelet-based Front-End for Electromyographic Speech Recognition. In: Proc. Interspeech (2007)
8. Jou, S.-C., Maier-Hein, L., Schultz, T., Waibel, A.: Articulatory Feature Classification Using Surface Electromyography. In: Proceedings of the IEEE International Conference on Acoustics, Speech, and Signal Processing (ICASSP 2006), Toulouse, France, May 15-19 (2006)
9. Maier-Hein, L., Metze, F., Schultz, T., Waibel, A.: Session Independent Non-Audible Speech Recognition Using Surface Electromyography. In: Proc. ASRU (2005)
10. Dietrich, M.: The Effects of Stress Reactivity on Extralaryngeal Muscle Tension in Vocally Normal Participants as a Function of Personality. PhD thesis, University of Pittsburgh (2008)
11. Yu, H., Waibel, A.: Streamlining the Front End of a Speech Recognizer. In: Proc. ICSLP (2000)
12. Walliczek, M., Kraft, F., Jou, S.-C., Schultz, T., Waibel, A.: Sub-Word Unit Based Non-Audible Speech Recognition Using Surface Electromyography. In: Proc. Interspeech, Pittsburgh, PA (September 2006)
13. Kirchhoff, K.: Robust Speech Recognition Using Articulatory Information. PhD thesis, University of Bielefeld (1999)
14. Metze, F.: Articulatory Features for Conversational Speech Recognition. PhD thesis, University of Karlsruhe (2005)
15. Metze, F., Waibel, A.: A Flexible Stream Architecture for ASR Using Articulatory Features. In: Proc. ICSLP (September 2002)
16. Stan Jou, S.-C., Schultz, T.: Automatic Speech Recognition based on Electromyographic Biosignals, page accepted for publication. In: Communications in Computer and Information Science (CCIS), BIOSTEC - BIOSIGNALS 2008 best papers, pp. 305–320. Springer, Heidelberg (2009)

17. Jou, S.-C.S., Schultz, T., Waibel, A.: Continuous Electromyographic Speech Recognition with a Multi-Stream Decoding Architecture. In: Proceedings of the IEEE International Conference on Acoustics, Speech, and Signal Processing (ICASSP 2007), Honolulu, Hawaii, US, April 15-20 (2007)
18. Frankel, J., Wester, M., King, S.: Articulatory Feature Recognition Using Dynamic Bayesian Networks. In: Proc. ICSLP (2004)
19. Schultz, T., Wand, M.: Modeling Coarticulation in Large Vocabulary EMG-based Speech Recognition. Speech Communication Journal (to appear, 2009)
20. Bahl, L.R., de Souza, P.V., Gopalakrishnan, P.S., Nahmoo, D., Picheny, M.A.: Decision Trees for Phonological Rules in Continuous Speech. In: Proc. ICASSP (1991)
21. Leggetter, C.J., Woodland, P.C.: Maximum Likelihood Linear Regression for Speaker Adaptation of Continuous Density Hidden Markov Models. Computer Speech and Language 9, 171–185 (1995)

# Towards the Development of a Thyroid Ultrasound Biometric Scheme Based on Tissue Echo-morphological Features

José C.R. Seabra[1,3] and Ana L.N. Fred[2,3]

[1] Institute for Systems and Robotics (ISR)
jseabra@isr.ist.utl.pt
http://users.isr.ist.utl.pt/~jseabra
[2] Institute of Telecommunications (IT)
[3] Instituto Superior Técnico (IST), Technical University of Lisbon, Lisbon, Portugal

**Abstract.** This paper proposes a biometric system based on features extracted from the thyroid tissue accessed through 2D ultrasound. Tissue echo-morphology, which accounts for the intensity (echogenicity), texture and structure has started to be used as a relevant parameter in a clinical setting. In this paper, features related to texture, morphology and tissue reflectivity are extracted from the ultrasound images and the most discriminant ones are selected as an input for a prototype biometric identification system. Several classifiers were tested, with the best results being achieved by a combination of classifiers (k-Nearest Neighbors, MAP and entropy distance). Using leave-one-out cross-validation method the identification rate was up to 94%. Features related to texture and echogenicity were tested individually with high identification rates up to 78% and 70%, respectively. This suggests that the acoustic impedance (reflectivity or echogenicity) of the tissue as well as texture are feasible parameters to discriminate between distinct subjects. This paper shows the effectiveness of the proposed classification, which can be used not only as a new biometric modality but also as a diagnostic tool.

## 1 Introduction

The thyroid is one of the largest endocrine glands in the body (see Fig.1). It controls how quickly the body burns energy, makes proteins and how sensitive the body should be to other hormones [15]. Thyroid ultrasonography is a non-invasive diagnostic exam, which provides immediate information on the structure and the characteristics of thyroid glands. This imaging modality is widely used in clinical practice because it combines low cost, short acquisition time, absence of ionizing radiations and sensitivity in ascertaining the morphology of the thyroid gland, as well as the size and number of thyroid nodules.

Ultrasound images usually present a low *signal to noise ratio* (SNR) and are characterized by a type of multiplicative noise called *speckle* that accompanies all coherent imaging modalities. It appears when images are obtained by using coherent radiation and is the result of the constructive and destructive interference of the echoes scattered from heterogeneous tissues and organs [1].

A. Fred, J. Filipe, and H. Gamboa (Eds.): BIOSTEC 2009, CCIS 52, pp. 286–298, 2010.

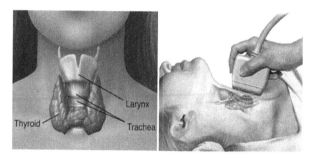

**Fig. 1.** Illustration of the thyroid gland's anatomy and location. An ultrasound examination is performed by placing the probe on the patient's neck (Courtesy of Mayo Foundation for Medical Education and Research).

The characteristic granular speckle pattern present in the ultrasound images makes the diagnostic task harder, whereas the subjectivity involved in their interpretation can be regarded as their major drawback. A framework which could provide explicit features extracted from the images would lead to a more reliable medical diagnosis, providing the experts with a second opinion and reducing the misdiagnosis rates.

Some studies have been developed which aim at characterizing the thyroid tissue using ultrasound image processing and analysis. Image intensity information has been used for the identification of thyroid Hashimoto disease [9], for the detection of nodular thyroid lesions, and for thyroid tumor classification. Textural image information encoded by means of co-occurrence matrix features [6] have been used for identification of chronic inflammations of the thyroid gland [14,13] and for the discrimination between normal and pathologic tissues [4].

Tissue echo-morphology, which accounts for the intensity (echogenicity), texture and structure, has started to be used as a relevant parameter in a clinical setting (see Fig.2). Basically, features extracted from a given region, tissue or organ can be used to identify (classify) a patient as normal or as suffering from a pathological condition. In a classification context, this is considered to be a two-class problem.

This paper proposes a biometric system based on features extracted from the thyroid tissue accessed through 2D ultrasound. Biometrics deals with identification of individuals based on their physiological or behavioral characteristics. Identification (Who am I?) refers to the problem of establishing a subject's identity - either from a set of already known identities (closed identification problem) or otherwise (open identification problem) [7].

Thyroid tissue echo-morphology qualify to be a biometric because it is a universal feature, which means that every person has the characteristic, is distinct from one individual to another, is permanent and can be easily collected through a common ultrasound scanner.

The paper is organized as follows. Section 2 formulates the problem and section 3 describes the feature module used in the biometric system. Section 4 presents the classifiers used in the identification problem. Section 5 presents the results obtained by the biometric system and section 6 concludes the paper.

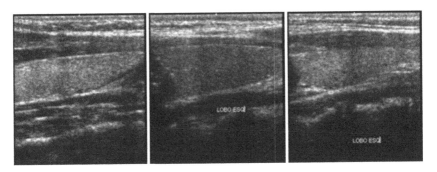

**Fig. 2.** Examples of thyroid ultrasound images, presenting different echo-morphologies. a) Hyperechogenic, b) Hypoechogenic and c) Heterogeneous thyroids.

## 2   Problem Formulation

In this paper, an analogy between two problems is made. In the context of medical diagnosis, a subject is assigned to one of two classes $N$ (normal) or $P$ (pathological). The risk of classifying pathological patients as normal (false negatives) should be penalized. Regarding a biometric identification problem, there is a class assigned to each individual. The maximum likelihood probabilities (or other types of scores) are computed in order to label the individual with its corresponding class.

The problem addressed in this paper can be stated as follows: given $C_i$ classes, each corresponding to a different individual (registered in the database), and $O_i$ observations, corresponding to 2D ultrasound sample images of the thyroid tissue recorded from each individual, establish the identity of new observations (label to the corresponding classes), which is a typical human identification problem.

The diagram block of the biometric system used in this paper is illustrated in Fig.3. It is mainly composed of three modules: (i) the sensor module, (ii) the feature extraction module, and (iii) the classification module.

The sensor module accounts for image acquisition. Ultrasound images of the thyroid gland were acquired longitudinally and transversally to the neck of 10 individuals, using

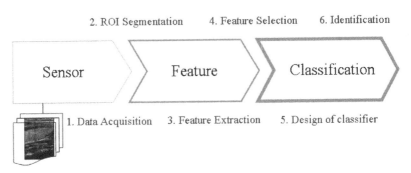

**Fig. 3.** Diagram block of the biometric identification system

an ultrasound scanner (Siemens Sonoline G50) operating in brightness (B-) mode. For each individual, the two lobes of the thyroid were scanned and one image per lobe was acquired. All thyroids were scanned under the same operating conditions in order to make the echo-morphological features extracted from the images independent of the scanner properties.

## 3    Feature Extraction Module

The feature extradion module is an important part of the biometric system because it determines which features are used for identification. In this section it is also important to consider how the thyroid glands are segmented from the ultrasound images, which features qualify for individual characterization, and from those features which of them are more relevant for discriminating between classes (subjects).

### 3.1    Segmentation

Before extracting the relevant features that describe the echo-morphology of the thyroid glands it is important to segment its anatomy from the ultrasound images. This is an important step in the development of an automatic and robust biometric tool.

The thyroid glands are the regions of interest from where the features are to be extracted. This can be done by manually outline the contours of the thyroid, which is incredibly tedious and time-consuming.

One way to circumvent this problem is to use automatic or semi-automatic methods (Active Contours [17], Level Sets [16], Graph Cuts [2,8]). In this paper, a semi-automatic method based on Gradient Vector Flow (GVF) active contours (snakes) is used.

Active contours [17], or snakes, are computer-generated curves that move within images to find the boundaries of the region of interest. The GVF snake begins with the calculation of a field of forces, called the GVF forces, over the image domain. The GVF forces are used to drive the snake, modeled as a physical object having a resistance to both stretching and bending towards the boundaries of the object. The GVF forces are calculated by applying generalized diffusion equations to both components of the gradient of an image edge map (see Fig.4). The semi-automatic nature of the segmentation process is due to user-dependent initialization: in fact, to make the method more robust, the user should provide a rough initialization of the contour by giving some initial clicks on the image.

### 3.2    Feature Extraction

After obtaining the segmented thyroid glands, 6 rectangular windows (32 by 32 pixels) were extracted from each lobe and were used for training. Similarly, 3 other regions were also extracted (see Fig.5) and used for testing.

Three different types of features are then computed for each rectangular window: (i) 2 features associated with the Rayleigh distribution parameter, (ii) 4 wavelet energy coefficients, (iii) 4 radon transform parameters. These features are also combined with the longitudinal mid-distance measure for each thyroid gland. This distance corresponds to

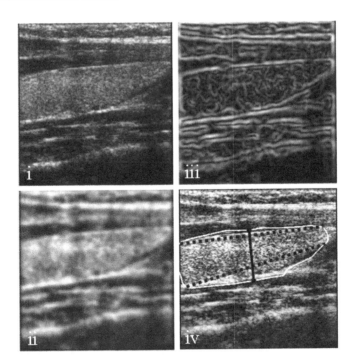

**Fig. 4.** Semi-automatic segmentation using GVF active contours: (i) original image, (ii) image convolved with gaussian mask, (iii) image edge map, (iv) segmented thyroid (the extraction of the mid-distance measure is also shown)

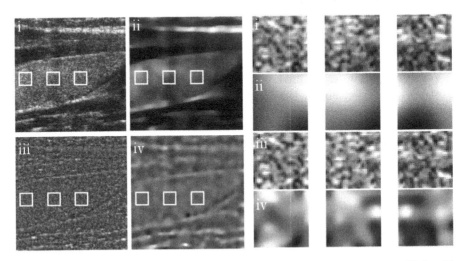

**Fig. 5.** (Left) (i) Original image, (ii) reconstructed image (local rayleigh parameters), (iii) Speckle field, and (iv) Rayleigh estimation of the speckle field. (Right) Illustration of samples used for testing (i) and their corresponding Rayleigh estimates, speckle field and Rayleigh estimation of the speckle field (ii, iii and iv).

the vertical distance measured between the borders of the thyroid at its middle section. In summary, 11 features are used to characterize each sample taken from the segmented thyroids.

### 3.3   Rayleigh Distribution Parameters from Original Sample and Speckle Field

The speckle pattern present in the ultrasound images is a result of the interference of echoes at the surface of the transducer, which emanate from the acoustic impedance of the tissues.

Several statistical models are proposed in the literature to describe this kind of pattern [10]. One of the most used in ultrasound (US), LASER and *Synthetic Aperture Radar* (SAR) is the Rayleigh distribution [3]. Commonly the speckle pattern is called speckle noise, and is often studied in de-noising problems. Another view of the problem, which is considered in this paper, is to accurately reconstruct the ultrasound images to provide a measure of the local acoustic impedance of the tissues.

In this context, a bayesian reconstruction method with a log-Euclidean prior is used [12]. In this approach, the ill-poseness nature of the reconstruction (de-noising) problem is circumvented by using *a priori* information about the unknown image to be estimated. The estimation is formulated as an optimization task where a two-term energy function is minimized. The first term pushes the solution toward the observations and the second regularizes the solution.

Let $X = \{x_{i,j}\}$ and $Y = \{y_{i,j}\}$ be a $N \times M$ image presenting the acoustic impedance of the tissue and a speckle image, respectively. The *speckle* pattern of the image $Y = \{y_{i,j}\}$ is described by a Rayleigh distribution,

$$p(y_{i,j}|x_{i,j}) = \frac{y_{i,j}}{x_{i,j}} e^{-\frac{y_{i,j}^2}{2x_{i,j}}}. \tag{1}$$

The estimation of $X$ from $Y$ is formulated as the following optimization task

$$\hat{X} = \arg\min_{X} E(X, Y), \tag{2}$$

where $E(X, Y)$ is an energy function.

The optimization problem, described by equation (2), is usually *ill-posed* in the Hadamard sense. This difficulty may be overcome by using the *maximum a posteriori* (MAP) criterion,

$$E(X, Y) = \underbrace{E_Y(X, Y)}_{\text{data fidelity term}} + \underbrace{E_X(X)}_{\text{prior term}}, \tag{3}$$

where $E_Y(X, Y)$, called *data fidelity term*, is the symmetric of the log-likelihood function

$$E_Y(X, Y) = -\log\left[\prod_{i,j=1}^{N,M} p(y_{i,j}|x_{i,j})\right], \tag{4}$$

where it is assumed statistical independence of the observations [5].

The solution to this problem (in fact, an energy minimization problem) is an image (see Fig.5 (ii)) in which the value of each pixel is the Rayleigh parameter that characterizes accurately the local reflectivity of the tissue being scanned. Thus, for each sample, one Rayleigh distribution parameter is extracted by averaging the local Rayleigh parameters inside the sampled window.

Moreover, it is known that the speckle corrupting the ultrasound images is multiplicative in the sense that its variance depends on the underlying estimated signal $X$. The image formation model may be formulated as follows: $Y = \eta\sqrt{X}$ where $X$ is the noiseless image and $\eta$ is the corresponding noise intensity (see Fig.5 (iii)). A similar reconstruction procedure was performed to extract one Rayleigh distribution parameter associated with the speckle field of each sample (see Fig.5 (iv)). The speckle field is used in the sequel of the paper to extract different textural features.

### 3.4    Wavelet Energy Coefficients

Texture information is hypothesized as being a relevant parameter to discriminate between thyroids and therefore individuals. One way to assess the texture of a given sample is to decompose its corresponding speckle field using 2D wavelets (see Fig.6(iii)). This kind of decomposition consists in using low and high pass filters onto the approximation coefficients at level j (the original image) in order to obtain the approximation at level j+1, and the details in three orientations (horizontal, vertical, and diagonal). This method is performed along 3 levels. Every subimage contains information of a specific scale and orientation, which is conveniently separated. The amount of detail for each resolution level, which accounts for the level of heterogeneity in each sample being studied, is computed as the sum of horizontal, vertical and diagonal detail energies for each level. Therefore, in this paper each sample will be also described by one feature accounting for the approximation energy as well as three different detail energies.

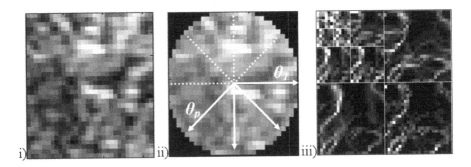

**Fig. 6.** Wavelet decomposition. Multi-resolution texture is assessed through the detail energy levels.

### 3.5    Radon Transform Features

In this paper, it is also hypothesized that the thyroid tissue may be characterized by different directionality patterns observed in the ultrasound images. The encoding of the directional patterns is realized by means of Radon Transform features [11]. The idea is to project the image intensity along a radial line oriented at different angles (0, 45, 90 and 135 degrees).

Let $(x, y)$ be the cartesian coordinates of a point in a 2D image, and $u(x, y)$ the image intensity. Then, the 2D radon transform denoted as $R_u(\rho, \theta)$ is given by

$$R_u(\rho, \theta) = \int_{-\infty}^{+\infty} \int_{-\infty}^{+\infty} u(x, y)\delta(\rho - x\cos\theta - y\sin\theta)dxdy \tag{5}$$

where $\rho$ is the perpendicular distance of a line from the origin and $\theta$ is the angle formed by the distance vector. The feature vector can be defined as

$$F = [\sigma([R_{u_1}(\rho, \theta_1)..., R_{u_n}(\rho, \theta_1)]), ..., \sigma([R_{u_1}(\rho, \theta_p), ..., R_{u_n}(\rho, \theta_p)])], \tag{6}$$

where $\sigma = \frac{R_u(\rho,\theta)}{u(x,y)}$ accounts for the contribution of the radon transform along four distinct angles $\theta_i = \{0, 45, 90, 135°\}$.

### 3.6 Dimensionality Reduction

At this point, 11 features per sample (each sample corresponding to a rectangular window) were extracted: 2 features associated with the Rayleigh distribution, 1 mid-distance measure, 4 wavelet energies, and 4 radon transform parameters. The amount of features extracted (11 features per sample, 6 samples per thyroid lobe, 2 lobes per individual, 10 individuals) makes the identification problem a complex task.

One way to deal with this problem and to eliminate the redundancy among features is to use principal component analysis (PCA). This approach is used to better handle and visualize the data by selecting the 3 most discriminating axis in the feature space and computing the 3 most relevant features (projection of the observations onto these axis). In summary, 3 features (components of the PCA) per observation sample are used in the identification problem. Fig.7(i) shows the representation of the observations (each individual sample) in the new feature space, where the 3 components of the PCA represent the 3 dimensions of the plot.

This new PCA-derived feature space can be projected onto a 2 dimensional feature space. Fig.7(ii) shows that the 2D features are able to clearly discriminate between two groups of individuals: one class addressed to men and the other to women. Even though no prior information is known about the clinical status of the individuals subject to this test it is clearly suggested that echo-morphology information might be correlated with the different types and quantities of hormones produced by men and women. This fact can lead to thyroids presenting different acoustic impedances and textures. This also explains the good discrimination between male and female populations.

At this point, we can suggest that this system might be useful as a soft biometric system for gender identification.

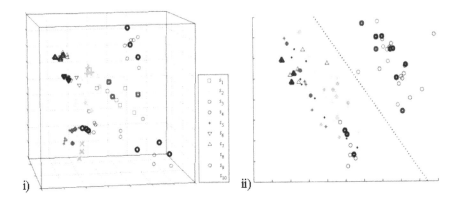

**Fig. 7.** Feature selection using PCA. (i) Representation of the observations (6 samples per individual) in the new PCA-derived feature space. Training samples (belonging to different classes) are represented with light symbols while testing samples are shown in bold. (ii) 2D feature space, showing a clear discrimination of the observed samples into two classes (gender of subjects).

## 4   Classification Module

In this paper, a closed set identification problem is addressed, which means that N possible outputs are generated for N possible models. The decision on whether to classify an observation (individual features) as being part of any of the available classes (individual database) is based on a computed score (MAP probability, distance measure, entropy). Three types of classifiers were studied:

### 4.1   K-Nearest Neighbors Classifier

The K-Nearest neighbors (KNN) classifier is based on the idea that an object is classified by a majority vote of its neighbors, with the object being assigned to the class most common amongst its k nearest neighbors. This is a common nonlinear classifier which results, when 1NN is used, in a Voronoi tesselation of the feature space.

### 4.2   MAP Classifier

The Maximum a Posteriori classifier is based on the MAP probability of a class $\omega$ given an observation $X$

$$\hat{\omega} = \arg\max p(\omega|X). \tag{7}$$

In our work we assume that the observations can be modeled by a multivariate gaussian distribution given by

$$p(X, \mu, \Sigma) = \frac{1}{2\pi^{3/2}|\Sigma|^{1/2}} \, e^{-1/2(X-\mu)'\Sigma^{-1}(X-\mu)}. \tag{8}$$

In this framework the discriminant function to be maximized is given by

$$g_i(X) = \log p(X|\omega_i) + \log p(\omega_i) \tag{9}$$

where $g_i(X) = -\frac{1}{2}\log|\Sigma_i| - \frac{1}{2}(X - \mu_i)'\Sigma_i(X - \mu_i) + \log p(\omega_i)$, and $\mu_i$ and $\sigma_i$ are maximum likelihood estimates of the mean and covariance matrices of the pdf of class i, based on the training data; $p(\omega_i) = 1/N$, being $N$ the number of individuals in the database.

### 4.3 Minimum Entropy Distance Classifier

As it was described before, the underlying observation model for each sample is described by a Rayleigh parameter (reflectivity) (see Fig.8(i)). The approximated probability density function (PDFs) generated using this Rayleigh parameter can be compared with the other PDFs in the database (Fig.8(ii)).

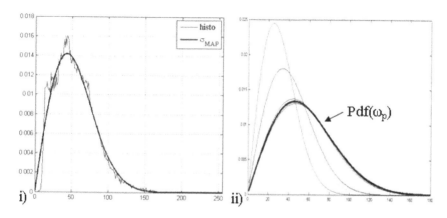

**Fig. 8.** The histogram of an observed sample can be approximated by a Rayleigh distribution with an estimated parameter which accounts for the acoustic impedance of the sample tissue. This distribution can be compared with the others in the database and entropy distance measures can be computed.

Conformity tests using the PDF for a given individual (testing distribution) and the remaining PDFs from the database (training distributions) were performed in order to assess which distribution better represents the observed one.

Considering the Kolmogorov-Smirnov conformity statistical test, $P_e = 1 - P_{H_0}$ is the probability of rejecting the null hypothesis, $H_0$, which is the hypothesis of the data have been generated by any of the distributions from the database. Here, $P_{H_0} = Q_{KS}(\lambda)$, $Q_{KS}(\lambda) = 2\sum_{j=1}^{\infty}(-1)^{j-1}e^{-2j^2\lambda^2}$, $\lambda = (\sqrt{(N)} + 0.12 + \frac{0.11}{\sqrt{N}})D$, $N$ is the number of data points and $D = max|c(n) - ch(n)|$, where $c(n)$ and $ch(n)$ are the cumulative probability functions of the testing and training distributions.

The Kullback-Leibler entropy distance is given by, $d = \sum_n p(n)\,log(\frac{p(n)}{h(n)})$. Here, $p(n)$ is the training distribution and $h(n)$ is the histogram of the observed (testing) sample.

**Table 1.** Performance of the classifiers (k-nNeighbors, MAP, SmirKolm, KullLeibler, and combination of the mentioned classifiers) for two different data samples, using all the features available, only textural features and features computed by PCA. These results refer to average values of identification rates as well as standard deviations, obtained through 10 consecutive runs of the classification methodology. Results achieved with the Leave-One-Out method are also shown.

| Classifier | Features | Identification Rate | |
| | | Exp.1 | Exp.2 |
|---|---|---|---|
| k-nNeigh | All | 0.858 ± 0.117 | 0.222 ± 0.051 |
| | PCA | 0.878 ± 0.100 | 0.267 ± 0.050 |
| | Texture | 0.789 ± 0.150 | 0.233 ± 0.088 |
| MAP | All | 0.400 ± 0.100 | 0.300 ± 0.111 |
| | PCA | 0.822 ± 0.154 | 0.656 ± 0.256 |
| | Texture | 0.722 ± 0.050 | 0.456 ± 0.077 |
| SmirKol | Rayleigh | 0.700 ± 0.067 | 0.211 ± 0.039 |
| kullLeib | Rayleigh | 0.700 ± 0.067 | 0.211 ± 0.039 |
| Combined | All | 0.878 ± 0.069 | 0.674 ± 0.084 |
| | PCA | 0.922 ± 0.069 | 0.756 ± 0.184 |
| Combined clssf. w/ LeaveOneOut | PCA | 0.937 ± 0.099 | 0.948 ± 0.175 |

The aforementioned classifiers were used individually but also in a combined scheme. The idea is to put together the classification given by each classifier and use a voting strategy to yield a final classification for each sample.

## 5   Results and Discussion

The performance of the classifiers was tested through 2 experiments and the result is summarized in Table 1. In the first experiment, 60 samples from one thyroid lobe were used as training data and 30 samples from the same lobe were used as testing data. The second experiment uses training data from one thyroid lobe (60 samples) and testing data from the opposite lobe (30 samples). Regarding the k-nearest neighbor and the MAP classifiers, tests were performed considering (i) all the features available, (ii) only the ones corresponding to the Radon transform and wavelets, which account for texture information, and (iii) the PCA derived features. The conformity tests (Kolmogorov-Smirnov, Kullback-Leibler) consider only the Rayleigh parameter (acoustic impedance or reflectivity) as describing each sample. The combined classifier was tested with all the features available as well as with the ones obtained after PCA.

In order to make the classification scenario as reliable as possible 10 runs of classification were performed, where in each run different thyroid samples were selected from the images. The identification rates shown in Table 1 are averaged values of the identification rates obtained for each run of classification.

The best performance is achieved by using a combination of all the mentioned classifiers, considering the PCA derived features, with high correct identification (ID) rates for both experiments (ID rate for Exp.1 = 0.922±0.069 and for Exp.2 = 0.756±0.184). When all the features are considered by the combined classifier, ID results are also reasonably good.

The ID rates obtained both with MAP and the K-Nearest neighbors classifier using only textural features were also high, which allows to conclude that texture information is in fact relevant for tissue characterization and differentiation. Textural features have already been shown to be relevant in a similar context [13]. The poor performance of these classifiers for Exp.2 when using these kind of features suggests that textural contents change significantly from one thyroid lobe to the other.

A good performance is also achieved with the entropy distance classifiers (KullLeib and SmiKol) for the first data set (Exp.1). This suggests that the acoustic impedance of the thyroid tissue (which is the only parameter used by these two classifiers) is indeed a good parameter for discriminating between thyroids and thus individuals. The poor performance of these classifiers when using the second data set suggests once more that the echo-morphology varies significantly from one thyroid lobe to the other.

Thus, it appears that textural and echogenicity features strongly vary from one lobe to another and therefore in order to improve the effectiveness of using echo-morphological features in both a clinical and a biometric setting, only one lobe of the thyroid should be considered for study.

Another estimate of the accuracy of the classifier uses the leave-one-out method. In this case, all but one sample from each lobe (Exp.1) or from both lobes (Exp.2) were used, thus using a larger training data set. Again, the combined classifier was used because it was the one which achieved better results in the previous experiment. Again, considering Table 1 it is observed a good performance of the classifier, in which the classifier even outperforms for Exp.2 (ID rates for Exp.1: $0.937 \pm 0.099$ and Exp.2: $0.948 \pm 0.175$). This suggests that the number of samples in the database significantly affects the performance of the classifier.

## 6   Conclusions

Computer derived features from 2D ultrasound images of the thyroid glands were used as part of a prototype biometric system. These features are related to the acoustic impedance, texture and morphology of the thyroid tissue.

Good results were achieved with all the classifiers used individually but the best performance was obtain with a combination of all the classifiers when using the three most discriminant features, computed by PCA. Moreover, reasonably high identification rates were also achieved with the entropy distance classifiers considering the Rayleigh distribution parameter, suggesting that the acoustic impedance, or reflectivity, of the tissues is a relevant feature to discriminate between individuals. Similarly, good performance was achieved when textural features computed from the speckle field were considered, which allows to conclude that the speckle field has important textural content. Analysis of thyroid echo-morphology should be further exploited because it appears to be very useful not only as a (soft) biometric system but also as a diagnostic tool.

Preliminary results, using 11 parameters extracted from ultrasound images, are encouraging. Further studies, involving larger data sets (more individuals and more samples), as well as observations taken from multiple sessions along distinct time instants, are required to better establish the accuracy of this new biometric modality.

**Acknowledgements.** The authors would like to thank Dr. Ricardo Ribeiro, from Escola Superior de Tecnologia de Saúde de Lisboa, for his help in performing the ultrasound scans and providing the thyroid images.

# References

1. Abbot, J., Thurstone, F.: Acoustic speckle: Theory and experimental analysis. Ultrasound Imaging 1, 303–324 (1979)
2. Boykov, Y., Veksler, O., Zabih, R.: Fast approximate energy minimization via graph cuts. IEEE Trans. Pattern Anal. Mach. Intell. 23(11), 1222–1239 (2001)
3. Burckhardt, C.: Speckle in ultrasound b-mode scans. IEEE Transactions on Sonics and Ultrasonics SU-25(1), 1–6 (1978)
4. Catherine, S., Maria, L., Aristides, A., Lambros, V.: Quantitative image analysis in sonograms of the thyroid gland. Nuclear Instruments and Methods in Physics Research A 569, 606–609 (2006)
5. Dias, J., Silva, T., Leitão, J.: Adaptive restoration of speckled SAR images using a compound random markov field. In: Procedings IEEE International Conference on Image Processing, Chicago, USA, vol. II, pp. 79–83. IEEE, Los Alamitos (1998)
6. Haralick, R.M., Dinstein, Shanmugam, K.: Textural features for image classification. IEEE Transactions on Systems, Man, and Cybernetics SMC-3, 610–621 (1973)
7. Flynn, P., Jain, A.K., Ross, A.A.: Handbook of biometrics. Springer, Heidelberg (2008)
8. Kolmogorov, V., Zabih, R.: What energy functions can be minimizedvia graph cuts? IEEE Trans. Pattern Anal. Mach. Intell. 26(2), 147–159 (2004)
9. Mailloux, G., Bertrand, M., Stampfler, R., Ethier, S.: Computer analysis of echographic textures in hashimoto disease of the thyroid. Journal of Clinical Ultrasound 14(7), 521–527 (1986)
10. Michailovich, O.V., Tannenbaum, A.: Despeckling of medical ultrasound images. IEEE Transactions on Ultrasonics, Ferroelectrics and Frequency Control 53(1), 64–78 (2006)
11. Savelonas, M.A., Iakovidis, D.K., Dimitropoulos, N., Maroulis, D.: Computational characterization of thyroid tissue in the radon domain. In: Twentieth IEEE International Symposium on Computer-Based Medical Systems, CBMS 2007, June 2007, pp. 189–192 (2007)
12. Seabra, J., Xavier, J., Sanches, J.: Convex ultrasound image reconstruction with log-euclidean priors. In: Proc. of the Engineering in Medicine and Biology Conference, Vancouver, Canada (2008)
13. Smutek, D., Sara, R., Sucharda, P., Tesar, L.: Different types of image texture features in ultrasound of patients with lymphocytic thyroiditis. In: ISICT 2003: Proceedings of the 1st international symposium on Information and communication technologies, pp. 100–102. Trinity College Dublin (2003)
14. Smutek, D., Sara, R., Sucharda, P., Tesar, L.: Image texture analysis of sonograms in chronic inflammations of thyroid gland. Ultrasound in Medicine and Biology 29, 1531–1543 (2003)
15. Gerard, J.T., Gerard, J.: Principles of anatomy and physiology (2000)
16. van Bemmel, C.M., Spreeuwers, L., Viergever, M.A., Niessen, W.J.: Level-set based carotid artery segmentation for stenosis grading. In: Dohi, T., Kikinis, R. (eds.) MICCAI 2002. LNCS, vol. 2489, pp. 36–43. Springer, Heidelberg (2002)
17. Xu, C., Prince, J.L.: Snakes, shapes, and gradient vector flow. IEEE Transactions on Image Processing 7(3) (March 1998)

# Part III
# HEALTHINF

# Collecting, Analyzing, and Publishing Massive Data about the Hypertrophic Cardiomyopathy*

Lorenzo Montserrat[1], Jose Antonio Cotelo-Lema[2], Miguel R. Luaces[2], and Diego Seco[2]

[1] Health in Code, Ed. El Fortín, Hospital Marítimo de Oza
As Xubias, s/n, A Coruña, Spain
lorenzo.monserrat@healthincode.com
http://www.healthincode.com
[2] Databases Laboratory, University of A Coruña
Campus de Elviña, 15071 A Coruña, Spain
joseantonio@enxenio.es, {luaces,dseco}@udc.es

**Abstract.** We present in this paper the architecture and some implementation details of a Document Management System and Workflow to help in the diagnosis of the hypertrophic cardiomyopathy, one of the most frequent genetic cardiovascular diseases. The system allows a gradual and collaborative creation of a knowledge base about the mutations associated with this disease. The system manages both the original documents of the scientific papers and the data extracted from these papers by the experts. Furthermore, a semiautomatic report generation module exploits this knowledge base to create high quality reports about the studied mutations.

**Keywords:** e-Health, Document Management System, Hypertrophic Cardiomyopathy.

## 1 Introduction

In the last decades, Document Management Systems (DMS) have become indispensable for many organizations. The majority of organizations need to access and consult stored information frequently. Thus, efficiency requirements must be considered by the Document Management Systems in order to provide a fast access to the information. Furthermore, documents need to pass from one person to another in many of these organizations. Therefore, DMS must define a set of rules for this process (a workflow process).

We present in this paper a Document Management System and Workflow for a medical organization (*Health In Code*). This system allows the creation, management, and

---

* This work has been partially supported by "Centro para el desarrollo tecnológico industrial (CDTI) del Ministerio de Industria" ref. Neotec RD 1406/1986 (IDI "20070178", title "Plataforma de diagnóstico genético de Cardiopatías") and by "Xunta de Galicia" ref. 08SIN008E. Other institutions collaborating in the support of the researchers are "Ministerio de Educación y Ciencia" (PGE y FEDER) ref. TIN2006-15071-C03-03 and "Xunta de Galicia" ref. 2006/4.

A. Fred, J. Filipe, and H. Gamboa (Eds.): BIOSTEC 2009, CCIS 52, pp. 301–313, 2010.
© Springer-Verlag Berlin Heidelberg 2010

exploitation of a knowledge base about genetic mutations associated with the hypertrophic cardiomyopathy.

*Health In Code* is a medical company dedicated to the identification of health problems that can benefit from a genetic diagnosis. Its main field of work are familial cardiovascular diseases, including cardiomyopathies and channelopaties. Cardiomyopathy means *disease of the cardiac muscle or myocardium*. Cardiomyopathies are therefore diseases characterised by the cardiac muscle having an abnormal structure and function. Although all diseases affecting the heart can damage the myocardium, cardiomyopathies do not include alterations due to ischaemic heart disease (myocardial infarction and related problems), to diseases of the cardiac valves, to disturbances produced by hypertension, or to congenital heart diseases. The term cardiomyopathy is reserved for a group of diseases in which the disturbance to the myocardium is the primary or fundamental alteration. The main cardiomyopathies are hypertrophic cardiomyopathy (affecting 1 out of every 500 adults), dilated cardiomyopathy, restrictive cardiomyopathy, arrhythmogenic right ventricle cardiomyopathy, and left ventricular non-compaction or spongiform cardiomyopathy.

Cardiomyopathies, above all hypertrophic cardiomyopathy and arrhythmogenic right ventricle dysplasia, are among the main causes of sudden death in young people and sportsmen and women, and are also an important cause in more elderly patients. Sudden death can be the first manifestation of these diseases. It is therefore fundamental to make an early diagnosis. Family check-ups and the systematic check-up of sportsmen and women play a fundamental role in this regard. The early identification of individuals affected by these diseases permits a suitable evaluation to be made of the risk of sudden death and the taking of effective prevention measures.

Cardiomyopathies are family diseases with a genetic cause. When a cardiomyopathy is diagnosed it must always be remembered that these diseases have a family background. In the case of hypertrophic cardiomyopathy and arrhythmogenic right ventricle dysplasia, the cause of the disease is practically always genetic and inheritable. In dilated cardiomyopathy, up to 50% of cases have a family background and are genetically caused, though it can also be due to infections, toxins, metabolic disturbances, or other causes.

Many different genetic alterations have been described associated with the development of one of the types of cardiomyopathy. For example, hypertrophic cardiomyopathy is caused by mutations in at least 10 genes of proteins forming part of the contraction machinery of the muscle cells. On the other hand, arrhythmogenic right ventricle dysplasia is produced by mutations in different genes related to the machinery of union and transmission of force between different cells.

Hundreds of different mutations have been described associated with these diseases and in fact each one of the mutations could be considered as a different disease, hence the interest in finding a genetic diagnosis that permits a better individualising not just of the diagnosis but also the prognosis and treatment of these diseases.

The Document Management System that we present in this paper makes possible a comprehensive genetic diagnosis of all the mutations and genetic variants associated with the hypertrophic cardiomyopathy (from now on HCM), where the differential diagnosis can be confuse. Furthermore, this system makes possible to identify many genetic interactions. Thousands of papers related to the HCM are published each year, and many of these papers are contradictory. Therefore, most of this information is completely ignored by people in charge of the diagnosis of the disease. The system that we present can help clinicians to take into account this information, and thus, it can help to improve the clinical diagnosis. There are a lot of people that can be benefit from this improvement:

- Patients with a clinical diagnosis about HCM (80000 people in Spain, more than 500000 people in the EU, etc.).
- Relatives of these patients (a mean of 4 per patient).
- Patients with left ventricular hypertrophy with doubts about the possibility of HCM. We consider in this group hypertensive patients with moderate to severe hypertrophy (5% of the hypertensives), differential diagnosis with athlete's heart, obese patients with hypertrophy, etc.
- Patients with an abnormal electrocardiogram (ECG) without apparent cause.
- Sudden death patients.

The HCM is probably the most studied cardiomyopathy and thus it was selected as the main goal of the organization. However, the system, and the overall operative process in general, was designed considering other cardiomyopathies, and even channelopaties. Hence, the resulting system has a robust design that can be easily extended to collect data about other possible diseases.

The rest of the paper is organized as follows. First, the operative process of the organization is described in Section 2. Second, in Section 3 the architecture of the system and some technical details are described. Then, in Section 4, we present the user interface of the collaborative web-based application to create the knowledge base. After that, the interest of its use is emphasized in Section 5. Finally, Section 6 presents the conclusions and some ideas for future lines of work.

## 2    Operative Process

Figure 1 shows the operative process supported by the developed system. First, interesting scientific papers are collected by a group of documentalists from different databases, conferences, journals, etc. After that, the scientific committee evaluates these papers. Selected papers that pass the quality controls are assigned to the experts who are in charge of reviewing them. A critical reading of the paper is the first task that must be done by these experts. After that, they have to analyze the data about patients described in the paper and they have to type these data in the application. An evaluation of the quality of the information is done by the scientific committee at this stage of the process. When the process passes these controls, the information is marked as valid and it becomes part of the knowledge base.

This operative process ends with an e-commerce stage where clients provide samples about patients and the company analyzes these samples and provides all the relevant clinical information available about the identified mutations. A semi-automatic subsystem to generate reports about mutations is the base of the business model of the

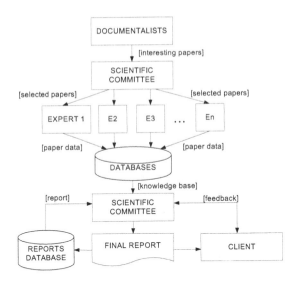

**Fig. 1.** Operative Process

company. In this context, a report is a set of pre-processed information from the knowledge base or a meta-analysis of the information about each mutation. These reports must be reviewed and edited to elaborate the final version of them. These final reports are elaborated for all the mutations identified in the samples provided by the clients. The company does not replace clinicians, but it supplies them with the best tools to take the best decisions based on the available knowledge of each mutation. Furthermore, the operative process contemplates a *feedback offer* where clients can provide clinical information about new mutation carriers and their relatives. This information is evaluated and introduced in the knowledge base if it passes the quality controls. The company offers the possibility of reinterpreting the mutation implications when clients provide clinical details about these new mutation carriers.

Therefore, the operative process requires the ordered execution of a set of activities, with several people participating in each of them. In such a complex process, the lack of control on the workflow can result in dead times, errors in the obtained results, and loss of data. In general, an unsatisfactory coordination of the people increases the overall cost and decreases the quality of the results. Because this process requires significant effort, the more automated tools that can be built and used, the better the use of human resources will be. The control of the workflow inside this work team is a key factor in the success of the process. This control can be achieved by the use of a workflow management tool specially designed for this process. That is, a system which allows to coordinate and control all the involved people, monitor and manage factors as the current state of each scientific paper, store intermediate results, control the average time to process each document, record all the people who have worked in each scientific paper, etc.

# 3  System Architecture and Technology

## 3.1  System Architecture

According to [1], workflow is concerned with the automation of procedures where documents, information, or tasks are passed between participants following a defined set of rules to achieve or contribute to an overall business goal; the computerized facilitation or automation of a business process, in whole or part. Workflow management systems can be classified in several types depending on the nature and characteristics of the process [2][3]. Collaborative workflow systems automate business processes where a group of people participate to achieve a common goal. This type of business processes involves a chain of activities where the documents, which hold the information, are processed and transformed until that goal is achieved. As the problematic of building a repository about mutations associated with the HCM fits perfectly in this model we based the architecture of the system in this model.

In general, we can differentiate three user profiles involved in the repository building:

- Administrator. Administrators are the people responsible of the process as a whole. They are the responsible of managing (add/delete/update) users and controlling the state of the application.
- Supervisor Users. The supervisor users are the people in charge of carrying out critical activities such as the metadata storage, assigning tasks to different workers (*inspectors*), or supervise their work. They are members of the scientific committee and they are in charge of performing the quality controls over the scientific papers and the inspection process.
- Inspector Users. The inspector users are the workers who carry out tasks such as inspecting scientific papers. This task involves a critical reading of the paper and typing relevant data in the system. This role is played by users with some knowledge in the mutations associated with the HCM but without any responsibility on the management of the system.

Supervisor and inspector users are the two user profiles involved in the repository building process. Therefore, a communication protocol between these user profiles has been implemented. *Supervisor* users can publish news, send messages to the *inspector* users, and answer their questions.

Figure 2 shows the overall system architecture. When we defined it, we followed the recommendations of the Workflow Reference Model [2], a commonly accepted framework for the design and development of workflow management systems, intended to accommodate the variety of implementation techniques and operational environments which characterize this technology. Thus, although we used this architecture for the implementation of a specific system, it can be used in other environments and situations. Furthermore, design patterns [4][5][6] were used in order to obtain a modular, robust, and easy to extend architecture.

As we can see in Figure 2, the authentication and authorizing module is in charge of the authentication of the users who want to access to the system. Each user has a system

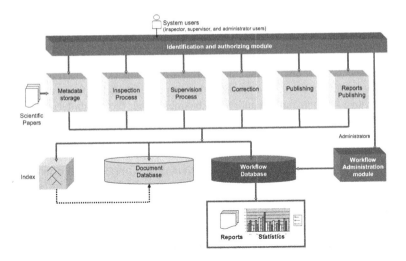

**Fig. 2.** System architecture

role depending on the tasks he/she is going to work on. According to this system role, the authorizing module only provides the user with access to the needed features. The system architecture is composed of a module for each activity carried out during the operative process.

- *Scientific Papers Selection.* As we noted in the previous section, thousands of scientific papers related to the HCM are published each year. Therefore, the scientific committee has to select the most prominent papers. This subsystem provides integration with software for publishing and managing bibliographies such as EndNote.
- *Metadata Storage.* This subsystem is in charge of the introduction and storage of the metadata for each scientific paper (title, author, year, source, described mutations, etc.). Furthermore, each paper must be assigned to the inspector user who will be in charge of analyzing it. This task is performed by the supervisor users of the system, therefore only they have access to this module.
- *Inspection Process.* This module allows inspector users to access the scientific papers previously assigned to them. They analyze these papers and type relevant data in the application.
- *Supervision Process.* It provides access to the scientific papers and analyzed data introduced using the previous module. This task is performed by the supervisor users who are in charge of performing quality controls about the previous stage.
- *Correction.* If supervisor users detect some mistakes in the *inspection process* they can correct them using this module. Moreover, supervisor users can delegate this task to the inspector.
- *Publishing.* Once the analysis process is accepted, this module is in charge of committing its contents.

– *Reports Publishing*. Reports about the mutations associated with the HCM can be generated using the data analyzed in previous stages of the workflow. Supervisor users must review and edit these reports in order to commercialize them. This subsystem provides functionalities to generate, edit, and store the reports in a database.
– *Workflow Administration Module*. This subsystem is in charge of managing the workflow between all these activities. It also provides reporting tools for monitoring purposes.

## 3.2  Data Model

Figure 3 shows our proposal for the data model that supports the architecture. This data model is organized around the entities *Paper* and *Report*. Both of them constitute the core of this architecture because the system is feeding with information from the *papers*, and the *reports* are generated by the system to be commercialized.

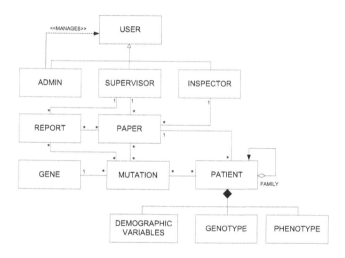

**Fig. 3.** Data Model

As we noted before, both scientific papers analyzed by the experts and the resultant information of this analysis are stored together by the Document Management System (DMS). Therefore, the entity *Paper* (and the entities associated with it such as *Mutation*, *Gene*, and *Patient*) represents both the original scientific paper (scanned PDF file, title, authors, and other metadata) and the data obtained from the analysis of this paper (mutations associated with each paper, number and type of control cases, information about described patients, etc.).

Furthermore, the workflow process described in the previous section is contemplated, and therefore, there are several entities in the data model to support it. *Admin*, *Supervisor*, and *Inspector* represent the three *User* profiles involved in the workflow process. All the entities, and modules that manage these entities, are associated with a set of user profiles. Therefore, only these user profiles have access to the information stored in such entities.

The most important characteristic of a DMS is the information that can be managed with it. As we noted, *Paper* is one of the most important entities of the data model. Metadata about original scientific papers are directly stored in this entity. Data obtained from the analysis of the papers are organized centred around the entity *Patient*. Therefore, there is a relationship between the entities *Paper* and *Patient* (a paper describes N patients and a patient is described in 1 paper). Thousands of scientific papers about the mutations associated with the HCM have been published in journals, conferences, etc. Moreover there are many internal documents in cardiology departments of the hospitals that describe patients with these mutations. Both types of data sources describe patients and information about their relatives to a greater or lesser extent. Information about these families can be entered in the system in a very intuitive, progressive, and simple way (the carefully designed user interfaces can be seen in Section 4). Information about patients and their relatives can be categorized in the following types:

- *Identification Data.* Both application internal identifiers (patient identifier, family identifier, etc.) and external domain identifiers (position in the *pedigree*) are included in this category.
- *Demographic Variables.* Data about the sex, ethnicity, age at the diagnosis of the disease, etc. are included here.
- *Genotype.* This category includes the results of the genotype study. These results present the relationship between the patients and the mutations associated with the paper where the patient is described. *Obligate carrier, homozygous carrier, normal carrier,* and *not carrier* are the possible values for this relationship. However, not all the scientific papers describe this relationship for all the mutations and patients. Therefore, an *unknown* value is available for the relationship. This philosophy is applicable for most of the variables managed by the system.
- *Phenotype.* This category includes the results of the different clinical tests that can be done in order to determine the appearance of a patient resulting from the interaction of the genotype and the environment. There are several subcategories in accordance with its nature. First, results about the *clinical diagnosis* are collected. These results determine whether the patient is affected or not by some phenotypes, which phenotypes, etc. The second group includes *environmental factors or triggers* (alcohol, hypertension, tobacco, obesity, etc.). There are many variables that can be determined in a *echocardiography, MRI, or autopsy.* These variables constitute the third group. Hypertrophy, dilatation, systolic and diastolic dysfunctions are some examples of these variables. The fourth group includes *symptoms and risk factors* (dyspnea, chest pain, abnormal blood pressure response, etc.). Variables of the *ECG* (rhythm, pre-excitation, abnormal voltage or repolarization) constitute the fifth group. The sixth group includes data about the *electrophysiological study* (inducibility of malignant arrhythmias, conduction disturbance, etc.). Finally, the last two groups include data about the *treatment* (medical treatment, surgery, etc.) and the *events* (death, cerebrovascular accident, etc.).

In brief, more than 200 variables are currently collected about each patient. However, new variables of interest can be easily introduced in the system.

### 3.3   Technology

This section briefly describes the most important technologies used in the development of the system. First, *Java 2 Platform, Enterprise Edition* (J2EE) [7][8] was the selected development platform. J2EE is a widely-used platform for server programming. This platform allows developers to create portable and scalable applications. J2EE provides a set of technologies that make the development process easier. JDBC (an API to access relational databases), JavaServer Pages (JSP, a technology to dynamically generate HTML), or JavaServer Pages Standard Tag Library (JSTL, a tag library for JSP) are several examples of such technologies provided by J2EE and used in this project. Furthermore, other technologies can be easily integrated with this platform. For example, *Jakarta Struts* [9], a framework that allows software engineers to develop applications following the architectural patterns *Model-View-Controller* and *Layers*, has been used. CSS [10] is the technology used to enhance the user interface. Finally, we have widely used JavaScript [11] to improve the dynamism and interaction of the user interface.

Technologies employed in the development of the *reports generation module* deserve special mention. *eXtensible Stylesheet Language Formatting Objects* (XSL-FO) [12] is the most important technology used in this module. XSL-FO is a mark-up language for XML document formatting which is most often used to generate reports. An XSL-FO document is an XML document where the format of a dataset is defined. This format defines the presentation of these data in a paper, screen, or other media. The XSL-FO document does not describe the layout of the text on various pages. Instead, it describes what the pages look like and where the various contents go. However, the developed system does not write XSL-FO documents. An XSLT transformation is used to convert the semantic XML, generated from data in the knowledge base of the system, into XSL-FO documents. This issue is very important because it provides independence between data and their output format. Finally, Apache FOP [13] is used to render the XSL-FO document to a specified output format. Output formats currently supported by Apache FOP include PDF, PS, PCL, AFP, XML, Print, AWT and PNG, RTF and TXT. The primary output target is PDF. However, our system generates RTF documents because the reports must be editable by the experts.

## 4   User Interfaces

The usability of the system is a key factor to guarantee its acceptance. In this context, this term denotes the ease with which users employ the application [14]. Therefore, in application domains where users do not have the required expertise level about computers, it is very important to design the user interfaces in accordance with the user preferences. For example, this system is used by experts in HCM but they do not have to know the way typical web applications work. The main issues of the user interface design are presented in this section.

Typing in the application the data resultant from the analysis of the scientific papers is the most expensive task. An average time of 3 hours has been estimated by the experts in charge of this task. Therefore, the user interface has to allow doing this task in several steps. Furthermore, scientific papers about the HCM usually present a common format. Patients and their relatives are described to a greater or lesser extent. However, sometimes these descriptions are organized by family, other times they are organized by type of data, etc. Therefore, the user interface has to take this issue into account. Figure 4 shows the developed user interface for typing data resultant from this process in the system.

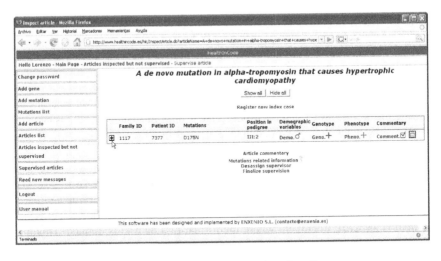

**Fig. 4.** Screenshot of the user interface (I)

The most important feature of this user interface is that it can manage a lot of information in each screen. As we noted before, each article can describe several families and each of them could have tens of studied cases. Moreover, more than 200 variables about each studied case are collected in the system. The designed interface provides the users with a centralized access point where they can introduce, consult and update all the data about the studied cases described in an article. Furthermore, a requirement of flexibility has been considered in the design of this interface. The experts that have to enter the data in the application can do it following the same organization presented in the original scientific paper. For example, they can introduce the data that identify all the patients and their relatives and, after that, they can complete other information such as genotype, phenotype, etc. But, they could type all the data about a patient before introducing other one. Some graphical icons help the user to know the state of each data category.

This design of the user interface implies that there is a lot of information in the same screen. All this information is organized and categorized in a natural way for the experts that have to use the system. First, data about each patient are organized in categories (demographic variables, genotype, phenotype, etc.). Moreover, these patients are grouped in families. A dynamic technique has been used in the implementation of the interface that allows the experts to fold and unfold the families. Figure 5 presents the same family of the previous screenshot unfolded.

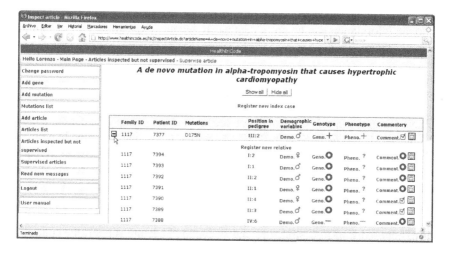

**Fig. 5.** Screenshot of the user interface (II)

These screenshots belong to a small example, with just a few cases, to improve the quality of the figures. However, there are scientific papers where more than one hundred cases are described. These papers would be unapproachable without the techniques presented in this work.

## 5   Use Interest

The main goal of this section is to emphasize the usefulness of the developed system. As we noted in previous sections, thousands of scientific papers about the mutations related to the HCM are published each year. Therefore, the task of generating reports that summarize each of these mutations would be unapproachable without a system like the one presented in this paper. There are two key factors in the design of the system. First, our system defines a workflow process. The scientific document repository building requires the ordered execution of a set of activities on the documents with several people participating in each of them. In such a complex process, the lack of control on the workflow can result in dead times, errors in the obtained results and loss of data. Second, our system is web-based. This allows the users to collaborate all over the world in order to keep the information constantly up to date.

Nowadays, there are 57 users registered in the system. One of this users has the role of *administrator*, 24 users have the role of *supervisor*, and 32 users have the role of *inspector*. These users have collaborated in the creation of a database with more than 1 GB of information. More than 3280 mutations, categorized in 85 genes, and more than 3405 scientific papers are registered in the system database. Furthermore, data about more than 14000 patients (or relatives) described in these scientific papers are available to generate high quality reports.

On the other hand, carefully designed user interfaces have a significant reduction in the time that experts need to analyze a scientific paper and to type relevant data in the application. This means a reduction of cost for the organization. But also, and much important, information can be up to date easily without expending a lot of money to hire more experts.

## 6  Conclusions and Future Work

We have presented in this paper the architecture and some implementation details of a Document Management System (DMS) to help at the diagnosis of the hypertrophic cardiomyopathy. The creation of the DMS repository is not a simple process. It requires the coordination of people and tools to carry out every activity that is part of the process. For all these process to be correctly and efficiently made, it is necessary the use of support tools that facilitate the work of each participant and ensure the quality of the obtained results.

The proposed workflow strategies and system architecture support the control and coordination of people and tasks involved in the whole process. The use of this architecture automates the completion of prone to error activities and optimizes the performance of the process and the quality of the obtained results. This architecture was defined following the recommendations of the Workflow Reference Model. Furthermore, several architectural and design patterns were used in order to obtain a modular, robust, and easy to extend system. This system was built as a web application which provides an integrated environment for the execution of all the tasks.

As lines of future work, the developed system is going to be applied to new projects of similar characteristics involving other Cardiomyopathies (Dilated, Restrictive, etc.) and Channelopathies (Brugrada syndrome, Long QT syndrome, Short QT syndrome, etc.). In addition, we are working on different implementations of the activities considered to optimize the performance of the overall process. Another future development could be the extension of the report generation module in order to support reports about several mutations. It could be very useful to analyze some mutations that occur in nearby areas of the protein. Finally, a statistical module could be developed to provide research capabilities inside the system. For example, a study about the correlation between variables could improve the quality of the final reports.

## References

1. Hollingsworth, D.: Workflow management coalition - the workflow reference model. Technical report, Workflow Management Coalition (1995)
2. van der Aalst, W., van Hee, K.: Workflow management: Models, methods, and systems (2002)
3. Fischer, L.: Workflow handbook 2003. Future Strategies Inc., USA (2003)
4. Gamma, E., Helm, R., Johnson, R., Vlissides, J.: Design Patterns: Elments of Reusable Object-oriented Sofware. Addison-Wesley, Reading (1996)
5. Grand, M.: Patterns in Java, vol. 1. John Wiley & Sons, Chichester (1998)
6. Alur, D., Crupi, J., Malks, D.: Core J2EE Patterns. Prentice-Hall, Englewood Cliffs (2003)
7. Perrone, P.J., Chaganti, K.: J2EE Developer's Handbook. Sam's Publishing (2003)
8. Bodoff, S.: The J2EE Tutorial. Addison-Wesley, Reading (2004)

9. Holmes, J.: Struts: The Complete Reference, 2nd edn. McGraw-Hill Osborne Media, New York (2006)
10. Shafer, D.: HTML Utopia: Designing Without Tables Using CSS. Sitepoint Pty Ltd. (2003)
11. Flanagan, D.: JavaScript: The Definitive Guide, 5th edn. O'Reilly, Sebastopol (2006)
12. W3C Recommendation: Extensible stylesheet language (XSL) version 1.1. Technical report, W3C (2006)
13. Apache FOP (October 2007), http://xmlgraphics.apache.org/fop/
14. Shneiderman, B.: Designing The user interface, Strategies for effective Human-computer interaction. Addison-Wesley, Reading (1998)

# BredeQuery: Coordinate-Based Meta-analytic Search of Neuroscientific Literature from the SPM Environment

Bartłomiej Wilkowski, Marcin Szewczyk, Peter Mondrup Rasmussen,
Lars Kai Hansen, and Finn Årup Nielsen

Informatics and Mathematical Modelling, Technical University of Denmark
Kongens Lyngby, Denmark
bw@imm.dtu.dk
http://neuroinf.imm.dtu.dk/

**Abstract.** Large amounts of neuroimaging studies are collected and have changed our view on human brain function. By integrating multiple studies in meta-
analysis a more complete picture is emerging. Brain locations are usually reported
as coordinates with reference to a specific brain atlas, thus some of the databases
offer so-called coordinate-based searching to the users (e.g. Brede, BrainMap).
For such search, the publications, which relate to the brain locations represented
by the user coordinates, are retrieved. We present BredeQuery – a plugin for the
widely used SPM data analytic pipeline. BredeQuery offers a direct link from
SPM to the Brede Database coordinate-based search engine. BredeQuery is able
to 'grab' brain location coordinates from the SPM windows and enter them as a
query for the Brede Database. Moreover, results of the query can be displayed in
a MATLAB window and/or exported directly to some popular bibliographic file
formats (BibTeX, Reference Manager, etc.).

## 1 Introduction

The growing number of functional neuroimaging studies of increasingly sophisticated
human brain activity brings the demand for new tools/services for integration of research findings, wider exchange of information between laboratories from the same
research area and efficient searching of related articles, reviews and other literature [1].

The dominant paradigm in current neuroimaging is that of *functional localization*.
Functional localization hypothesizes that a given human behavior is established by a
change in brain activity in a relatively limited number of spatially segregated processing units. Thus the result of an experiment under this paradigm consists of a Statistical
Parametric Map (SPM) indicating the local involvement. Often the SPM is summarized
as a list of regions, see e.g., [2,3], in which the SPM has been judged to be significantly different from zero (regions were the null hypothesis is rejected). As the typical
neuroimaging experiment investigates a highly controlled behavior and often involves a
relatively limited number of subjects, there is strong need for tools to integrate multiple
experiments in order to increase the robustness to the experiment specific implementation of the given behavior and to statistical fluctuation due to limited sample sizes.

Several methods have been proposed for neuroimaging meta-analysis and for estimation of associations between the brain locations and textual representations of behavior, for a recent review, see e.g., [1]. A set of methods are based on the so-called

A. Fred, J. Filipe, and H. Gamboa (Eds.): BIOSTEC 2009, CCIS 52, pp. 314–324, 2010.
© Springer-Verlag Berlin Heidelberg 2010

Brede Database [4]. Methods for integration include estimation of conditional probability density functions representing the localized probability of activation in response to a given behavior 'word' [5,6] and multivariate methods based on non-negative matrix factorization that aim to represent global dependencies between brain activation and semantic text labels from neuroscience publications [7].

Brain locations are reported as region coordinates relative to a specific brain atlas (usually MNI or Talairach spaces), hence, there is an interest for effective search for experiments, hence, scientific papers, which report similar coordinate sets in brain. Brain-Map [8] and Brede [4] are the databases which offer the coordinate-based searching. For Brede it is available on both the webpage and in a standalone application. A more extensive classification of the databases for fMRI coordinates can be found in [9].

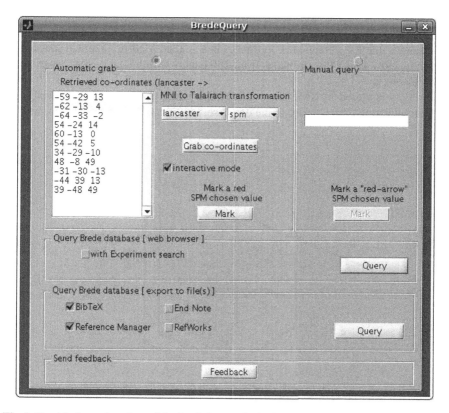

**Fig. 1.** Graphical user interface of the BredeQuery plugin for SPM. Firstly, the user can choose if the coordinates used for querying will be grabbed from an SPM's results window or will be typed manually. The grabbed (retrieved) coordinates are shown on the list. The user can switch on an interactive mode – the coordinate selected in the SPM window will be automatically selected in the plugin on the coordinates list. Moreover, the coordinates are grabbed using the chosen MNI to Talairach transformation (Brett or Lancaster MTT affine transformations). Afterwards, the user is able to display the query results in the Matlab web browser or to import them into the specified bibliographic format.

In order to enable a neuroimaging scientists to perform meta-analysis in the context of a specific ongoing study we here propose a tool that integrates retrieval of related research within the data analysis pipeline. The dominant tool for human brain mapping is undisputable the SPM[1] set of tools developed and distributed by the Functional Imaging Laboratory (London, [2]). For an analysis of the usage of imaging pipelines see e.g., [10]. Thus we have initiated the development of a *plugin* for SPM, which offers high integration with the Brede Database.

The *BredeQuery plugin* (see Fig. 1) provides the opportunity to perform coordinate-based query and retrieval of the related articles references directly from the SPM (Matlab) environment.

## 2  Brede Database

The Brede Database available through the webpage[2] records published neuroimaging experiments that list stereotaxic coordinates in so-called MNI or Talairach space [11]. Presently, close to 4000 coordinates from 186 papers with a total of 586 experiments are available.

The data is stored in XML files, and Matlab functions generate static webpages with visualization of the entries in the database, see Figure 2. Web-based searching on

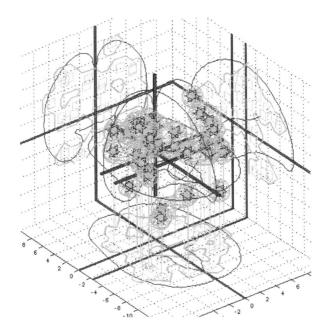

**Fig. 2.** Screenshot from one of the pages in the Brede Database showing coordinates in Talairach space. This is one of presently 586 experiments recorded in the database – an fMRI experiment resulting in 29 reported coordinates.

---

[1] Statistical Parametric Mapping.

[2] http://neuro.imm.dtu.dk/services/brededatabase/

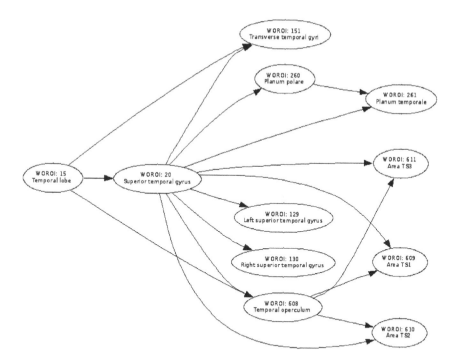

**Fig. 3.** Relationships and taxonomy of the regions in brain associated with superior temporal gyrus. The entire ontologies for brain regions and brain functions are available together with the Brede Database.

coordinates is possible from the homepage, but up till now it has required that the researcher manually typed in the query or extracted results from the image analysis program.

The Brede Database web service provides also links to other neuroscientific resources. While querying the database with a specified coordinate in brain, the user is also able to visualize the location in INC Talairach Atlas. Each publication relates by ID number to other databases like PubMed or BrainMap. Brain regions from each of the experiments are mapped to the services like MeSH, BrainInfo, CoCoMac database or Wikipedia. As the Brede webpages are public, the ordinary Web search engines enable text based search of the Brede Database. Furthermore, the researcher may navigate the database via several hyperlinked webpages including brain region, brain function and author ontologies, see Figure 3.

## 3   Related Tools

There are a few available tools with similar functionality and aims as the BredeQuery plugin.

The AMAT SPM toolbox was developed by Antonia Hamilton for the Matlab environment. It provides coordinate-based search for over 5000 coordinates from 213 published papers of which some were derived from the Brede Database. The coordinates are in MNI or Talairach space. The toolbox can locate neighboring coordinates to a given coordinate, as well as publications for a given author or year. The tool was last updated in 2005 and is available in the internet[3].

Another related toolbox, xjView, offers the SPM user, apart from viewing the images in glass view, section view or 3D render view, search of selected brain regions in databases in order to elucidate their function. It performs the searching among others in Google Scholar[4] and PubMed[5] database. This toolbox was created by Xu Cui and Jian Li and is publicly available[6].

The XCEDE SPM Toolbox [12] enables the users to capture activation data for PET/fMRI analysis and save them to the XML file in a XCEDE XML schema. Moreover, it is extending the exported XML file by automatically adding the anatomical labeling of the region in the brain for the given activity coordinates. It is achieved through two other toolboxes: Talairach Daemon[7] and Automated Anatomical Labeling[8].

## 4  Software Description

The recent version of the *BredeQuery plugin*, together with the User's Guide, can be downloaded from the webpage:

```
http://neuroinf.imm.dtu.dk/BredeQuery/
```

A graphical user interface of the BredeQuery plugin is divided into five areas where different user-actions can be performed. Firstly, the activation coordinates can be 'grabbed' from the SPM results figure into the plugin. Since the coordinates can be presented in MNI or Talairach spaces, transformations are introduced for interoperability. The coordinate-based search in the Brede Database is based on Talairach space coordinates, thus the BredeQuery plugin offers two MNI to Talairach transformations, which can be chosen by the user. The piece-wise affine transformation proposed by Matthew Brett is one of the available transformations [13]. Also included is the affine transformation MNI-to-Talairach (MTT), suggested by Jack Lancaster et al. [14]. Three separate transformations were suggested by his group: one for SPM, one for FSL and a combined 'pooled' transformation. The $MTT_{SPM}$ transformation is set as default in the BredeQuery plugin.

When the coordinates have been 'grabbed' and shown in the BredeQuery plugin, the coordinate-based querying with Brede Database can be done. One or more coordinates can be selected for querying and the results from the Brede Database (publications related to the given activity coordinate) are displayed by the plugin in a web browser (see Figure 4), exported to an XML file or saved in the bibliographic file format (BibTeX, Reference Manager, RefWorks or EndNote). We mention that the coordinates need not

[3] http://www.antoniahamilton.com/amat.html
[4] http://scholar.google.com/
[5] http://www.ncbi.nlm.nih.gov/pubmed/
[6] http://people.hnl.bcm.tmc.edu/cuixu/xjView/
[7] http://www.talairach.org/
[8] http://www.cyceron.fr/freeware/

# Brede Database - Talairach coordinate search

brede_loc_query — Search after locations (Talairach coordinates) in the Brede Database

| -62 -13 4 | Location search (one coordinate) | e.g., 14 -9 -15 |
|-----------|----------------------------------|-----------------|
| -62 -13 4 | Experiment search (several coordinates) | |
| -62 -13 4 | Visualize in INC Talairach atlas | |

| # | Distance | x | y | z | WOBIB | Description |
|---|----------|---|---|---|-------|-------------|
| 1 | 3.8 | -59 | -15 | 2 | 9 | Left medial temporal gyrus — Phonemes (WOEXP: 21) |
| 2 | 4.1 | -62 | -14 | 0 | 42 | Middle superior temporal sulcus — Voice versus non-vocal sounds (WOEXP: 141) |
| 3 | 4.1 | -62 | -14 | 0 | 42 | Middle superior temporal sulcus — Vocal stimuli versus bells, human non-vocal sounds, enveloped white noise and scrambled voices (WOEXP: 142) |
| 4 | 4.1 | -62 | -14 | 0 | 42 | Middle superior temporal sulcus — Frequency-filtered vocal versus non-vocal (WOEXP: 143) |
| 5 | 5.5 | -58 | -13 | 0 | 88 | Left superior temporal gyrus — Activation in sadness film viewing versus neutral film viewing (WOEXP: 282) |
| 6 | 5.9 | -59 | -11 | -1 | 123 | — Language perception during free viewing (WOEXP: 383) |
| 7 | 5.9 | -61 | -13 | -1 | 90 | Lateral temporal — Cued recall of familiar people. Group analysis (WOEXP: 290) |
| 8 | 7.6 | -61 | -9 | -2 | 64 | Left superior temporal gyrus — Listening to voices (WOEXP: 199) |
| 9 | 7.9 | -56 | -15 | 0 | 88 | Left superior temporal gyrus — Activation in sadness film viewing versus amusement film viewing (WOEXP: 284) |

**Fig. 4.** Brede Database query – result displayed in a web browser. List of nearby coordinates to a queried coordinate, displaying distance, the three-dimensional coordinates, the paper identifier, the anatomical label for the retrieved coordinates and short description of the experiment.

necessarily be grabbed from SPM in order to make a query. The coordinates can also be entered manually in a manner similar to the functionality of the Brede Database web service.

The user is also able to perform an 'experiment search' (available in the Brede Database service) via the BredeQuery. It has previously been suggested how a similarity can be computed between one set of coordinates and a volume or another set of coordinates [6]. This procedure required the conversion of the set of coordinates to a volume by kernel density estimation. It is, however, not necessary to convert the coordinates to a volume if only the similarity between two coordinates sets are to be compared. It will then generally be faster to compute the similarity based on all coordinate-coordinate pair-wise similarities and perform a weighted summation. There are multiple ways to compute the similarity. Presently, the web-service for the Brede Database uses the following Gaussian/Euclidean form:

$$s_{q,e} = \frac{1}{\sqrt{N}} \sum_{m=1}^{M} \sum_{n=1}^{N} \exp\left(-\frac{(x_{m,q} - x_{n,e})^2 + (y_{m,q} - y_{n,e})^2 + (z_{m,q} - z_{n,e})^2}{2\sigma^2}\right),$$

where $\sigma$ is set to 10 millimeters, $(x_{m,q}, y_{m,q}, z_{m,q})$ is the $m$th of $M$ three-dimensional query coordinates, while $(x_{n,e}, y_{n,e}, z_{n,e})$ are the $n$th of $N$ three-dimensional coordinates in the Brede Database. The factor $1/\sqrt{N}$ aims to regularize for the number of coordinates in each set so that sets with many coordinates do not dominate the search result. A corresponding weight for the query coordinates is not necessary, since this factor will be equal for all queried sets of coordinates of the database.

Following the terminology of BrainMap, a set of coordinates is in the Brede Database called an 'experiment' [15], thus the name 'experiment search'.

The Perl function that presently provides the search functionality to the Brede Database web service is part of the Brede Toolbox, and this toolbox is available on the Internet[9].

## 5   Example Session

In this section we demonstrate the use of the BredeQuery plugin on data from a block-designed auditory fMRI experiment. The experiment was conducted by Geriant Rees, University College London, and the data set was obtained from the SPM webpage[10].

Stimuli were bi-syllabic words, that were presented binaurally. The experimental condition comprised blocks of six scans alternating between rest and auditory stimulation. Data were preprocessed using the standard SPM pipeline including realignment, spatial normalisation and spatial smoothing. Following preprocessing a conventional univariate statistical analysis was conducted. In the general linear model (GLM) the design matrix comprised a box-car function convolved with the hemodynamic response function (HRF). Figure 5 presents the analysis result. The statistical parametric map was based on a t-contrast (stimulation>rest), with $p < 0.05$ corrected for multiple comparisons using family-wise-error (FWE) correction. A prominent activation was observed in the auditory cortex in the bilateral temporal lobes. An example SPM-BredeQuery user's session leading to abovementioned results may proceed with the following steps:

1. The BredeQuery plugin was loaded by choosing the BredeQuery entry in the SPM's toolbox pop-up menu. All coordinates for significant clusters from the statistical table in the *SPM Graphics windows* were grabbed by the plugin and shown in the coordinates list. They were transformed according to the chosen MNI-to-Talairach transformation. In our example, the coordinates were transformed using the Lancaster's $MTT_{SPM}$ affine transformation. The user was interested in the activation, represented by the coordinate in the MNI space as (-66,-12,2) which was selected in the statistical table – see Figure 5. In the BredeQuery window the user pressed the *Mark red SPM chosen value* button and the previously selected coordinate (-66,-12,2) in MNI space, transformed by the plugin to (-62,-13,4) in Talairach space, was marked in the plugin's coordinates list – see Figure 1.

[9] http://neuro.imm.dtu.dk/software/brede/
[10] http://www.fil.ion.ucl.ac.uk/spm/data/auditory/

2. The user has pressed the *Query* button in the *Query Brede database [web browser]* panel (shown on Figure 1) and the webpage with the query results (related articles) has appeared. The user was now able to compare the present results and conclusions with those from the retrieved articles. The webpage results from our example are displayed on Figure 4. Among the first matches from the Brede Database there are coordinates found in the *superior temporal sulcus/gyrus* from experiments with auditory stimulation. A link to the taxonomy of the regions in brain associated with superior temporal gyrus (see Figure 3) was also available. Furthermore, the abstracts of articles related to the experiment were available. Figure 6 represents the detailed, online description of one of the matched experiments.

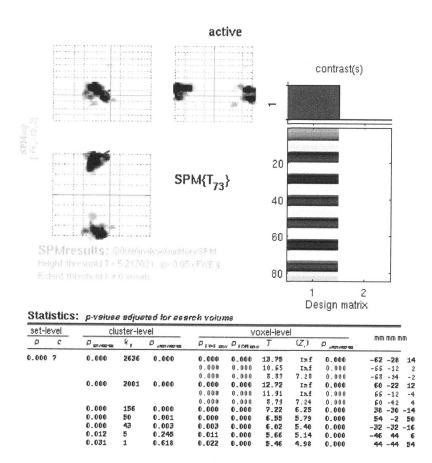

**Fig. 5.** The demonstration of example results window in SPM. The user can see regions with significant brain activation. The region of activation represented by the coordinate (-66,-12,2) in MNI space is selected. The same coordinate, transformed to the Talairach space using Lancaster's $MTT_{SPM}$ transformation is marked on the BredeQuery's coordinates list as (-62,-13,4) (see Figure 1). Afterwards, the user is able to submit coordinate-based queries to the Brede Database and get the articles related to the same (or nearby) brain regions.

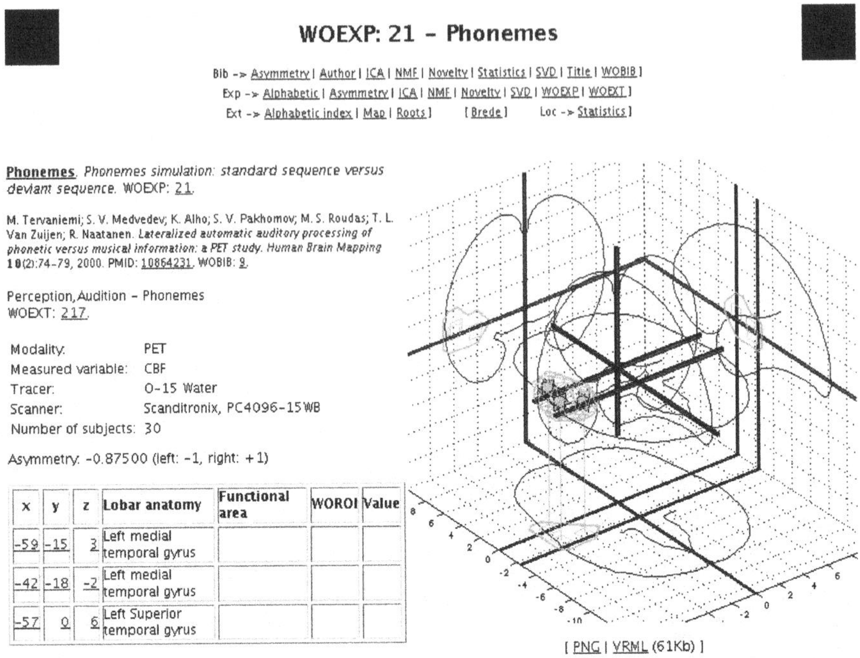

**Fig. 6.** The online description of an experiment found in the Brede Database. The detailed information about experiment called *Phonemes*, the reference to the article where it was presented, settings of the experiment (modality, type of scanner, tracer used, number of subject, etc.) is displayed. Moreover, all the reported activation coordinates are shown with the indication of the brain location. Finally, the visualization of the coordinates in Talairach space is displayed to the right of the table with coordinates' values.

3. The user wanted to reference some of the articles from the retrieved results in a manuscript. He selected an appropriate bibliographic format,(in this example case 'BibTeX'), pressed *Query* button in the *Query Brede database [export to file(s)]* panel (shown on Figure 1) and the BibTeX file with the references was obtained.
4. The user has discovered a missing feature in the BredeQuery plugin. He thus has pressed the *Feedback* button (Figure 1) and sent a comment to the develop team.

## 6   Future Work

The first official BredeQuery plugin's version was released to the SPM community on 12th March 2009. Since the plugin is still under development, all incoming feedback comments from the plugin's users are going to be taken into account while releasing updates and further versions, thus more features should be expected.

It was recently emphasized that there are many separated research communities in neuroscience, which do not want to share or exchange the experimental data [16]. Researchers have expressed concerns that sharing of data can lead to unfair use [17]. However, data sharing is important to create trusted collaboration community and is a current topic in debate on future of neuroscience [18,19] as it is believed that broad data sharing could lead to breaktroughs in our understanding of brain function [20]. Invoking online social networks and computer-based communication can support closer relationships and trust [21] hence, reduce the resistance to data sharing.

Consequently, an interesting extension of the functionality of the plugin can be a direct connection from the SPM environment to a neuroscientific research community, web service or social network. The user would be able to upload the coordinates, results of the analysis, to his own account and save in the assigned server disk space in order to process them later. He can decide whether he wants to keep it private, share only with his research group or alternately release it as a public resource to all users of the service.

It is also possible to employ the BredeQuery plugin to expand the Brede Database. The increment in number of the articles stored in the database could cause bigger interest from the neuroscientists. They could then be encouraged to register their published or unpublished publications in the database via the BredeQuery plugin together with the reported coordinates and keywords.

Finally, we are planning to employ SKEEPMED (Semantic KEyword Extraction Pipeline for MEdical Documents) which is now under development [22]. This pipeline can be used for automatic keyword extraction from abstracts and/or whole papers retrieved by Brede Database. The obtained keywords can be used to query bigger and up-to-date medical databases like PubMed, what consequently could improve the BredeQuery plugin's search results by returning to the user more recent publications related to the respective area in brain and experiments.

# 7   Conclusions

We have presented herein the BredeQuery plugin for SPM - an application which offers a direct link from the SPM environment to the Brede Database. It provides a mechanism which allows the SPM user to find references to articles which relate to the similar brain activation areas through so-called coordinate-based searching. Moreover, the BredeQuery plugin facilitates the creation of the bibliography files in popular formats.

**Acknowledgements.** We would like to thank Torben Lund and Julian Macoveanu for very constructive comments and feedback. This work is supported by Lundbeckfonden through the Center for Integrated Molecular Brain Imaging (CIMBI) – www.cimbi.org.

# References

1. Wager, T.D., Lindquist, M., Kaplan, L.: Meta-analysis of functional neuroimaging data: current and future directions. Social Cognitive and Affective Neuroscience 2, 150–158 (2007)
2. Friston, K., Ashburner, J., Kiebel, S., Nichols, T., Penny, W.: Statistical Parametric Mapping: The Analysis of Functional Brain Images. Academic Press, London (2007)
3. Pekar, J.: A brief introduction to functional MRI. IEEE Engineering in Medicine and Biology Magazine 25, 24–26 (2006)

4. Nielsen, F.Å.: The Brede database: a small database for functional neuroimaging. NeuroImage 19 (2003); Presented at the 9th International Conference on Functional Mapping of the Human Brain, June 19-22, New York (2003)
5. Nielsen, F.Å., Hansen, L.K.: Modeling of activation data in the BrainMap(TM) database: Detection of outliers. Human Brain Mapping 15, 146–156 (2002)
6. Nielsen, F.Å., Hansen, L.K.: Finding related functional neuroimaging volumes. Artificial Intelligence in Medicine 30, 141–151 (2004)
7. Nielsen, F.Å., Hansen, L.K., Balslev, D.: Mining for associations between text and brain activation in a functional neuroimaging database. Neuroinformatics 2, 369–380 (2004)
8. Laird, A.R., Lancaster, J.L., Fox, P.T.: BrainMap: The Social Evolution of a Human Brain Mapping Database. Neuroinformatics 3, 65–78 (2005)
9. Derrfuss, J., Mar, R.A.: Lost in localization: The need for a universal coordinate database. NeuroImage, doi 10 (2009)
10. Nielsen, F.Å., Christensen, M.S., Madsen, K.M., Lund, T.E., Hansen, L.K.: fMRI Neuroinformatics. IEEE Engineering in Medicine and Biology Magazine 25, 112–119 (2006)
11. Talairach, J., Tournoux, P.: Co-planar Stereotaxic Atlas of the Human Brain. Thieme Medical Publisher Inc., New York (1988)
12. Keator, D.B., Gadde, S., Grethe, J.S., Taylor, D.V., Potkin, S.G.a.: A general XML schema and SPM toolbox for storage of neuro-imaging results and anatomical labels. Neuroinformatics 4, 199–212 (2006)
13. Brett, M.: The MNI brain and the Talairach atlas. MRC Cognition and Brain Sciences Unit (1999)
14. Lancaster, J.L., Tordesillas-Gutiérrez, D., Martinez, M., Salinas, F., Evans, A., Zilles, K., Mazziotta, J.C., Fox, P.T.: Bias between MNI and Talairach coordinates analyzed using the ICBM-152 brain template. Human Brain Mapping (2007)
15. Fox, P.T., Mikiten, S., Davis, G., Lancaster, J.L.: BrainMap: A database of human function brain mapping. In: Thatcher, R.W., Hallett, M., Zeffiro, T., John, E.R., Huerta, M. (eds.) Functional Neuroimaging: Technical Foundations, pp. 95–105. Academic Press, San Diego (1994)
16. Ascoli, G.A.: The Ups and Downs of Neuroscience Shares. Neuroinformatics 4, 213–216 (2006)
17. Teeters, J.L., Harris, K.D., Millman, K.J., Olshausen, B.A., Sommer, F.T.: Data Sharing for Computational Neuroscience. Neuroinformatics (2008)
18. Kennedy, D.N.: Neuroinformatics and the Society for Neuroscience. Neuroinformatics 5, 141–142 (2007)
19. Liu, Y., Ascoli, G.A.: Value Added by Data Sharing: Long-Term Potentiation of Neuroscience Research: A Commentary on the 2007 SfN Satellite Symposium on Data Sharing. Neuroinformatics 5, 143–145 (2007)
20. Van Horn, J.D., Ball, C.A.: Domain-Specific Data Sharing in Neuroscience: What Do We Have to Learn from Each Other?. Neuroinformatics 6, 117–121 (2008)
21. Lampe, C., Ellison, N., Steinfield, C.: A face(book) in the crowd: social searching vs. social browsing. In: CSCW 2006: Proceedings of the 2006 20th anniversary conference on Computer supported cooperative work, pp. 167–170. ACM Press, New York (2006)
22. Wilkowski, B., Szewczyk, M.M., Hansen, L.K.: Bridging the gap between coordinate- and keyword- based search of neuroscientific databases by UMLS-assisted semantic keyword extraction (2009)

# Simulation of ECG Repolarization Phase with Improved Model of Cell Action Potentials

Roman Trobec, Matjaž Depolli, and Viktor Avbelj

Jožef Stefan Institute, Department of Communication Systems
Jamova cesta 39, 1000 Ljubljana, Slovenia

**Abstract.** An improved model of action potentials (AP) is proposed to increase the accuracy of simulated electrocardiograms (ECGs). ECG simulator is based on a spatial model of a left ventricle, composed of cubic cells. Three distinct APs, modeled with functions proposed by Wohlfard, have been assigned to the cells, forming epicardial, mid, and endocardial layers. Identification of exact parameter values for AP models has been done through optimization of the simulated ECGs. Results have shown that only through an introduction of a minor extension to the AP model, simulator is able to produce more realistic ECGs. The same extension also proves essential for achieving a better fit between the measured and modeled APs.

**Keywords:** ECG, Action potential, Repolarization, Myocardium, Computer simulation.

## 1   Introduction

The standard 12–lead electrocardiogram (ECG) is a diagnostic tool in cardiology for more than 60 years. It is a view on the electrical heart activity from the body surface that results from differences between potentials of myocardium cells. These in turn are a consequence of different times in which cells excite (excitation sequence) and the physiological differences in cells themselves. The mechanisms for ECG generation are still not fully understood. In our research, we tackle the problem of ECG genesis on the cellular level, more precisely, through the shape of the action potentials (APs). Focus of this work is on the shape of the repolarization phase of modeled APs and on the increase of its fidelity.

Modeling of APs on the cell level is quite complex, because a system of non-linear time-dependant differential equations has to be solved [1]. Simpler method, proposed by Wohlfart [2], models APs as a product of two sigmoidal functions ($A$ and $C$) and one exponential function ($B$):

$$
\begin{aligned}
A(t) &= \frac{1}{1+e^{-k_1 t}} \\
B(t) &= k_2((1 - k_3)e^{-k_4 t} + k_3)e^{-k_5 t} \\
C(t) &= \frac{1}{1+e^{k_6(t - k_7)}} \\
AP(t) &= A(t) \times B(t) \times C(t) \, ,
\end{aligned}
\tag{1}
$$

A. Fred, J. Filipe, and H. Gamboa (Eds.): BIOSTEC 2009, CCIS 52, pp. 325–332, 2010.

where component $A(t)$ controls the initial upstroke (phase 0), $B(t)$ the immediate fast repolarization and the AP plateau (phases 1 and 2, respectively), and $C(t)$ the repolarization part (phase 3). Because of the characteristics of exponential functions, however, the whole AP curve is influenced to some extent by all the components.

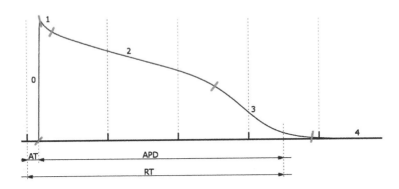

**Fig. 1.** An example of AP with corresponding phases 0 – 4. AT, APD, and RT are shown at 90% repolarization.

A model of AP curve with four phases is shown in Figure 1. Activation time (AT) represents delay between the myocardium excitation start and individual cell activation, which is defined by the excitation sequence (both are shown on the spatial model in Figure 2). Repolarization time (RT) is a sum of AT and action potential duration (APD) and is a measure of the delay between the myocardium excitation start and individual cell repolarization end.

The mechanisms forming the repolarization phase of ECG are still under investigations. For example, the shape of the T wave, its width and its slopes have not been elucidated yet in all details. Even more mysterious is the genesis of the U wave [3,4]. Several hypotheses are frequently quoted: U wave genesis is a consequence of the late repolarization of Purkinje fibers [5]; U wave is generated because of mechanoelectrical feedback [6,7]; U wave is a residual of the late repolarization of cells in midmyocardium [8].

A new view of the mid-myocardium hypothesis was presented in [9] that suggests that the end of the T wave is taken as the residual of cancellation of opposing potential contributions throughout the myocardium during the repolarization while the U wave arises because of imbalance of potentials in the late repolarization because of the prolonged repolarization of mid-myocardium. Recently, it was shown that there are other alternatives for T and U wave genesis [10]. U waves can be generated even if the repolarization of the mid-myocardium is not prolonged.

Leaving physiological causes aside and looking on the problem purely mathematically, we experienced drawbacks of the Wohlfart's AP model in two ways. First, we were unable to get a good fit between the Wohlfart model and the measured APs from the intact heart. Second, we identified the inability to control AP phases 2 and 3 independently, to be the limiting factor in our simulator's abilities to produce properly

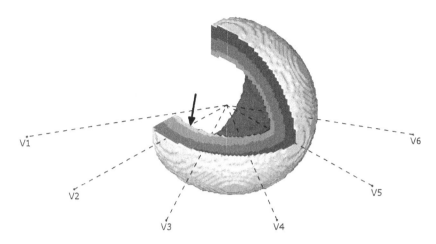

**Fig. 2.** Spatial model of the left ventricle. ECG lead positions are shown with dashed lines. Arrow points to the excitation trigger area. The excitation time is shown on the cutout of the myocardial wall; each level of gray representing 10 ms.

shaped ECGs. This comes about because the slopes in these phases determine the shape and duration of the T wave and the time of its appearance. We propose an extension of the Wohlfart model in terms of changing the repolarization part $C(t)$ in Equation (1), as a solution for the problems mentioned above.

The rest of the paper is organized as follows. In Methods, the spatial model of the left ventricle and the simulation procedure are described. The simulation results obtained with the Wohlfart model are compared with a measured ECG. In section 3, the proposed model extension is introduced together with an example of an improved AP curve fit and measured ECG fidelity. The paper concludes with an overview of the obtained results and further work.

## 2   Methods

### 2.1   Model of the Left Ventricle

We constructed a three-dimensional model from 65628 cubic cells with a volume of $1 \text{ mm}^3$, stylized in a cup-like shape, shown in Figure 2. The model is onion-like composition of twelve layers, which enables different APs to be assigned to each layer. Results presented in this paper were obtained by constructing three thicker layers: epi, mid and endo, each of them composed of four identical thinner layers.

Implementation of faster longitudinal conduction between cells of the same layer (along the wall) than transversal conduction between cells of different layers (across the wall) emulates faster conduction paths of the Purkinje fibers.

### 2.2   Simulation Method

ECG is simulated by fist calculating the excitation sequence for all the cells and then projecting the sum of differences in cell potentials on approximate positions of ECG leads. Simulation procedure is integrated into a simulator, that takes parameters of Equation (1) as input and generates ECGs on predefined positions as output. This simulator is then used in simulation based optimization that solves the inverse problem of identification of AP parameters.

The ECG simulator works with cell APs and a simple rule for each cell. Excited cells behave as sources of electrical potential determined by their AP functions. Every excited cell stimulates its neighboring non-excited cells to become excited with a small delay, which depends on the layer of the neighboring cell and its position relative to the excited cell. Because of the onion-like layering, cell neighbors along the myocardium wall will be of the same layer while neighbors perpendicular to the wall will be of different layers. If neighbors are from different layers, the delay of 2 ms results in transversal conduction velocity of 0.5 m/s. On the other hand, if neighbors belong to the same layer, the delay of 1/3 ms results in longitudinal conduction velocity of 3 m/s. Both velocities are in accordance with measured values on myocardial tissue [11]. For the purposes of cell excitation, a cell neighborhood is comprised of the 26 cells that share at least one vertex with the cell, as shown in Figure 3b). This type of neighborhood is selected over the simpler one depicted in Figure 3c), because it allows an uninterrupted excitation along a single layer even when layers are so thin that in some parts of the model, cells of the same layer only touch in their corners.

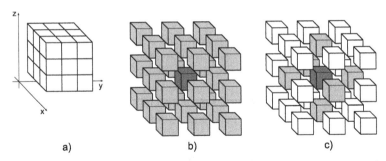

a)                          b)                          c)

**Fig. 3.** a) Block of 3x3x3 cells as it would appear in the model. b) and c) show exploded views of this block on two types of neighborhoods, used in cell excitation and the dipole formation, respectively. Cells colored light gray belong to the neighborhood of the dark gray cell, while the white cells are not a part of the neighborhood.

Six observation points were selected around the model, 4 cm away from the epicardial layer, at angles $-120°$ (V1) to $30°$ (V6), in increments of $30°$, as in a real ECG precordial leads placing (see Figure 2). For each observation point, an ECG is simulated with the following procedure. We assume formation of a dipole between cell $i$ and its immediate neighborhood $\Omega$ (caused by cells having different prescribed APs and different ATs), in the same way as in [12]. This includes only neighbors with coincident faces, i.e., 6 neighbors for the spatial model, as shown in Figure 3c). Dipole moment $\mathbf{D}_i$ is proportional to the vector sum of differences in potentials $V$:

$$\mathbf{D}_i(t) \propto \sum_{j \in \Omega(i)} (V_i(t) - V_j(t)) . \qquad (2)$$

ECG leads are simulated as a sum of dipole potential contributions at the observation point $P$ from all $N$ cells:

$$V_P(t) \propto \sum_{i=1}^{N} \frac{|\mathbf{D}_i(t)| \cdot \cos \phi}{|\mathbf{R}_{i,P}|^2} , \qquad (3)$$

where $\mathbf{R}_{i,P}$ is a directional vector from the cell $i$ to $P$, and $\phi$ is the angle between $\mathbf{D}_i$ and $\mathbf{R}_{i,P}$.

Simulator based optimization with evolutionary algorithm works on top of the above simulation procedure. It deduces optimal parameters for three AP groups from a predefined target ECG on a predefined location. Currently, a measured ECG on V2, which has been published in [13], is used as the target. The evaluation algorithm starts with a number of random inputs for the simulator and generates ECGs on target location for each input. Then it combines and modifies inputs in an evolution-like procedure, resulting in inputs that produce ECGs very similar to target ECG. Finally, the result of the optimization is the input that produces the most similar ECG.

### 2.3   Simulation Results – Wohlfart AP Model

APs for epicardium, mid, and endocardium have been generated with Wohlfart model, using coefficients from Table 1. Resulting APs are shown in the upper part of Figure 4. The repolarization phase of the simulated and measured ECGs on $V_2$ are shown in the

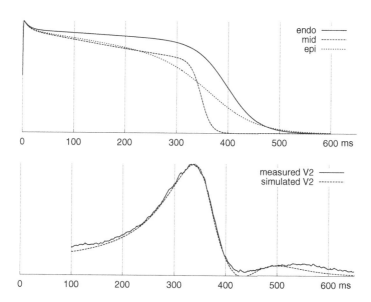

**Fig. 4.** Wohlfart APs for epicardial, mid and endocardial layer (top), simulated and measured ECG repolarization phase on the lead $V_2$ (bottom). Note that the ECG on the lower part of the figure is generated through Equation (3), where besides the APs (upper part of the figure), also the shape of the heart model plays an important role.

**Table 1.** Coefficients of Equation (1) used for modeling APs from Figure 4

| LAYER | $k_1$ | $k_2$ | $k_3$ | $k_4$ | $k_5$ | $k_6$ | $k_7$ |
|-------|-------|-------|-------|-------|---------|--------|-----|
| endo  | 2.5   | 100   | 0.9   | 0.1   | 0.00194 | 0.0755 | 326 |
| mid   | 2.5   | 100   | 0.9   | 0.1   | 0.00260 | 0.0345 | 339 |
| epi   | 2.5   | 100   | 0.9   | 0.1   | 0.00228 | 0.0376 | 296 |

lower part of Figure 4. The simulated ECG fits well the measured signal, however, some details around the T and U waves are still inadequate.

## 3  Modified AP Model

In the Wohlfart model, used in the previous section, AP phases 2 and 3, and the transition between them cannot be independently controlled. Shape of the transition between phases 2 and 3 is determined by the shape of the transition between phases 3 and 4, and to some extent by the shape of phase 2. This dependence constraints possible shapes of resulting T and U waves and consequently limits the usability of our simulator. Therefore, we propose a modification of the factor $C(t)$, which controls the repolarization part of the ECG. Instead of using a simple sigmoid we introduce an asymmetric sigmoidal function, which requires an additional parameter $k_8$:

$$C(t) = 1 - \left(1 + e^{-k_6(t-k_7) + \ln(2^{\frac{k_6}{k_8}} - 1)}\right)^{\frac{-k_8}{k_6}} \qquad (4)$$

Incorporating the extended AP model, the simulator immediately shows improvements. The results of simulation based optimization on the same target ECG as before are shown on Figure 5 and coefficients for the APs are presented in Table 2. The ECG fidelity is increased while the APs retain their typical shape. The newly introduced asymmetry (4) of phase 3 is barely noticeable.

The confirmation that proposed modification of the AP model does not reflect a quirk of our simulator but is an actual improvement of the model can be found through examination of the measured APs. Trying to fit both modified and unmodified Wohlfart AP model (searching for parameters that would result in the most similar shape) to measured APs published by Druin et al. [8], difference in model fidelity can be observed. An

**Table 2.** Coefficients of Equation (4) used for modeling APs from Figure 5

| LAYER | $k_1$ | $k_2$ | $k_3$ | $k_4$ | $k_5$     | $k_6$  | $k_7$ | $k_8$  |
|-------|-------|-------|-------|-------|-----------|--------|-------|--------|
| endo  | 2.5   | 100   | 0.9   | 0.1   | 0.000744  | 0.0873 | 369   | 0.0291 |
| mid   | 2.5   | 100   | 0.9   | 0.1   | 0.000643  | 0.0991 | 399   | 0.0354 |
| epi   | 2.5   | 100   | 0.9   | 0.1   | 0.000992  | 0.0284 | 376   | 0.0147 |

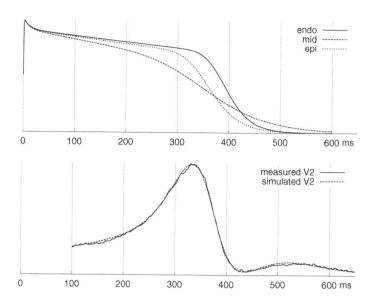

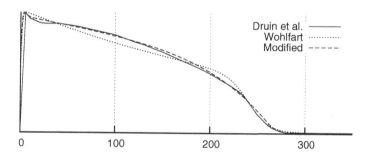

**Fig. 5.** Extended Wohlfart APs for epicardial, mid and endocardial layer (top), simulated and measured ECG repolarization phase on the lead $V_2$ (bottom)

**Fig. 6.** Endo AP published by Druin et al. [8] (solid line), fitted with the Wohlfart model (dotted line) and proposed modified model (dashed line)

example is shown in Figure 6, where the modified model, using coefficients (1.0, 145.6, 0.8, 0.374, 0.00130, 0.0160, 225, 0.244) for $k_1$ to $k_8$, respectively, fits the target AP more accurately than does the unmodified model. Both models were fitted to measured APs using the same optimization method, based on an evolutionary algorithm.

## 4    Conclusions

We have created a simple three-dimensional model of a left ventricle for a computer simulation of ECGs. The simulation is based on a variety of different AP sets based on the Wohlfart AP model, which were shown to have some limitations in the simulation of known phenomena in the myocardial wall. Examining the AP model closer, problem with its fidelity was discovered and identified as the most probable cause of the mentioned simulator limitations. The problem was solved through addition of another degree of freedom to the AP model.

We are preparing new simulations of ECGs with the modified AP model and improvements of the optimization. Currently, only one ECG lead can be targeted at a time, which leads to inaccuracies of other ECG leads. Although the modified AP model both increases fidelity of simulated ECGs and enables better approximation of measured APs, there are still differences between measured APs and APs acquired through our simulation based optimization. If we succeed in reconciling these differences, we expect that the simulator will provide helpful in explaining some of the complex phenomena of the repolarization phase.

**Acknowledgements.** The authors acknowledge financial support from the Slovenian Research Agency under grant P2-0095.

# References

1. Ten Tusscher, K., Noble, D., Noble, P., Panfilov, A.: A model for human ventricular tissue. Am. J. Physiol. Heart Circ. Physiol. 286, H1573–H1589 (2004)
2. Wohlfart, B.: A simple model for demonstration of SST-changes in ECG. Eur. Heart J. 8, 409–416 (1987)
3. Surawicz, B.: U wave: Facts, hypotheses, misconceptions, and misnomers. J. Cardiovasc. Electrophysiol. 9, 1117–1128 (1998)
4. Pérez Riera, A., Ferreira, C., Filho, C., Ferreira, M., Meneghini, A., Uchida, A., Schapachnik, E., Dubner, S.L.Z.: The enigmatic sixth wave of the electrocardiogram: the U wave. Cardiol. J. 15, 408–421 (2008)
5. Watanabe, Y.: Purkinje repolarization as a possible cause of the U wave in the electrocardiogram. Circulation 51, 1030–1037 (1975)
6. Franz, M.: Mechano-electrical feedback in ventricular myocardium. Cardiovasc Res. 32, 15–24 (1996)
7. Schimpf, R., Antzelevitch, C., Haghi, D., Giustetto, C., Pizzuti, A., Gaita, F., Veltmann, C., Wolpert, C., Borggrefe, M.: Electromechanical coupling in patients with the short QT syndrome: further insights into the mechanoelectrical hypothesis of the U wave. Heart Rhythm 5, 241–245 (2008)
8. Druin, E., Charpentier, F., Gauthier, C., Laurent, K., Le Marec, H.: Electrophysiologic characteristics of cells spanning the left ventricular wall of human heart: Evidence for presence of M cells. J. Am. Coll. Cardiol. 26, 185–192 (1995)
9. van Ritsema Eck, H., Kors, J., van Herpen, G.: The U wave in the electrocardiogram: A solution for a 100-year-old riddle. Cardiovasc Res. 67, 256–262 (2005)
10. Depolli, M., Avbelj, V., Trobec, V.: Computer-simulated alternative modes of U-wave genesis. J. Cardiovasc. Electrophysiol. 19, 84–89 (2008)
11. Macfarlane, P., Lawrie, T. (eds.): Comprehensive Electrocardiology: Theory and Practice in Health and Disease, 1st edn., vol. 1. Pergamon Press, Oxford (1989)
12. Miller, W., Geselowitz, D.: Simulation studies of the electrocardiogram. I. The normal heart. Circ. Res. 43, 301–315 (1978)
13. Avbelj, V., Trobec, R., Gersak, B.: Beat-to-beat repolarisation variability in body surface electrocardiograms. Med. Biol. Eng. Comput. 41(5), 556–560 (2003)

# Advances in Computer-Based Autoantibodies Analysis

Paolo Soda and Giulio Iannello

Integrated Research Centre, University Campus Bio-Medico of Rome, Italy
{p.soda,g.iannello}@unicampus.it

**Abstract.** Indirect Immunofluorescence (IIF) imaging is the recommended method to detect autoantibodies in patient serum, whose common markers are antinuclear autoantibodies (ANA) and autoantibodies directed against double strand DNA (anti-dsDNA). Since the availability of accurately performed and correctly reported laboratory determinations is crucial for the clinicians, an evident medical demand is the development of Computer Aided Diagnosis (CAD) tools supporting physicians' decisions.

In this paper we present a comprehensive system that helps in recognising the presence of ANA and anti-dsDNA autoantibodies. The analysis of CAD performance shows its potential in lowering the method variability, in increasing the level of standardization and in serving as a second reader reducing the physicians' workload. The system has been successfully tested on annotated datasets.

## 1 Introduction

Indirect Immunofluorescence (IFI) imaging is the recommended laboratory technique for the detection of autoantibodies in patient serum, which are at the basis of autoimmune disorders, a chronic inflammatory process involving connective tissues [1, 2]. To this aim, antinuclear autoantibodies (ANA) and autoantibodies directed against double strand DNA (anti-dsDNA) are markers very commonly used. The former are used in patients with suspected connective tissue diseases (CTD) [1], whereas the latter are employed to diagnose Systemic Lupus Erythematosus (SLE). SLE is a chronic inflammatory disease of unknown aetiology affecting multiple organ systems, [3], which is considered a very serious sickness further to be regarded as an invalidating chronic disease.

In IIF the availability of accurately performed and correctly reported laboratory determinations is crucial for the clinicians, demanding for highly specialized personnel that are not always available. Moreover, the readings in IIF are subjected to interobserver variability limiting the reproducibility of the method [4, 5]. To date, the highest level of automation in IIF tests is the preparation of slides with robotic devices performing dilution, dispensation and washing operations [6, 7]. Although such instruments helps in speeding up the slide preparation step, the development of Computer-Aided-Diagnosis (CAD) systems supporting IIF diagnostic procedure would be beneficial in many respects. Indeed, CAD systems have proven their effectiveness in other medical contexts, such as mammography [8] or chest radiography [9].

Interest in autoimmune diseases is motivated by the increase of their reported incidence, partly due to the improved diagnostic capabilities as well as the growing

A. Fred, J. Filipe, and H. Gamboa (Eds.): BIOSTEC 2009, CCIS 52, pp. 333–346, 2010.

awareness of this clinical problem in general medicine. Recent research efforts in IIF have focused on the classification of ANA slides with promising results [10, 11, 12, 13, 14, 15]. However, to our knowledge, no works have discussed a CAD systems that should help the physicians in the detection of different autoantibodies.

In this paper we present a recognition tool supporting the detection of ANA and anti-dsDNA autoantibodies. The analysis of CAD performance shows its potential in lowering the method variability, in increasing the level of standardization and in serving as a second reader. The system has been successfully tested on annotated datasets.

## 2  Background and Motivations

### 2.1  ANA Tests

Current guidelines for ANA tests recommend the use of HEp-2 substrate diluted at 1:80 titer [16] and require to classify both the fluorescence intensity and the staining pattern. The same guidelines suggest scoring the former semi-quantitatively and independently by two physicians. Since technical problems can affect test sensitivity and specificity, they suggest using both positive and negative controls. The former allows the physician to check the correctness of the preparation process; the latter represents the auto-fluorescence level of the slide under examination. To diagnose the sample, specialists compare the sample with the corresponding positive and negative controls. This is a problematic task affecting the reliability of the diagnosis [4, 5].

To reduce the variability of multiple readings of the same sample, recently it has been recently proposed to classify the fluorescence intensity into three classes, named *negative*, *intermediate* and *positive*. On the one hand, these three classes maintain the clinical significance of the IIF test and, on the other hand, this class revision gets ground truth robust enough to develop a classification system [2].

Using HEp-2 cells as a substrate, the positive samples may reveal different patterns of fluorescent staining that are relevant to diagnostic purposes Fig. 1. Although more than thirty different nuclear and cytoplasm patterns should be identified [17], in the literature they are classified into one of the following groups [12]:

- *Homogeneous* (HO): diffuse staining of the interphase nuclei and staining of the chromatin of mitotic cells;
- *Peripheral Nuclear* (PN): solid staining, primarily around the outer region of the nucleus, with weaker staining toward the center of the nucleus;
- *Speckled* (SP): a fine or coarse granular nuclear staining of the interphase cell nuclei;
- *Nucleolar* (NU): large coarse speckled staining within the nucleus, less than six in number per cell;
- *No Pattern* (NP): unclassifiable pattern.

It is worth noting that sometimes two concomitant staining patterns can be observed in the same well. In these cases, further dilution and/or better focusing may help to recognize different overlapping staining.

Some papers have recently applied pattern recognition and data mining techniques to classify one of the two sides of ANA classification, i.e. fluorescence intensity [10, 11] and staining pattern [12, 13, 14, 15]. However, none of them presented an overall CAD supporting both recognition tasks. In this respect, we presented our fluorescence intensity classification system in [10, 11], while we discussed a system that classifies the staining pattern in [15]. It differs from [12], [13] and [14] for two main reasons. First, they aim only at classifying the pattern of individual cells. Second, their data sets differ from ours since we use images acquired from the real patients sera diluted at 1:80, which therefore exhibits positive fluorescence intensity at various grading. Indeed, in [14] the authors employed only sera of positive controls, whereas in [13] and [12] the authors used a different data set, which is constituted by samples diluted at 1:160 and also containing cells that were negative, i.e. they did not exhibit a detectable fluorescence intensity.

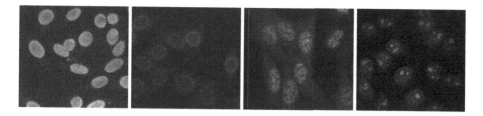

**Fig. 1.** Examples of the homogenous, rim, speckled and nucleolar staining patterns (left to right)

## 2.2   Anti-dsDNA Tests

Crithidia Luciliae (CL) substrate diluted at 1:10 titer is used to diagnose SLE [18]. CL is an unicellular organism containing both the nucleus and a strongly tangled mass of circular dsDNA, named as kinetoplast. The fluorescence of kinetoplast is the fundamental parameter to define the positiveness of a well, whereas the fluorescence of other parts of CL cells, e.g. the basal body or the flagellum, is not a marker of anti-dsDNA autoantibodies and, consequently, of SLE. Fig. 2 shows five stylized cells representative of different cases.

Panels A and B depict positive cases. The former shows a cell where only the kinetoplast exhibits a fluorescence staining higher than cell body, while in the latter also the nucleus is highly fluorescent. The other three panels show negative cells. In panel C, the cell is clearly negative since all cell body has a weak and quite uniform fluorescence. Panels D and E depict cells where regions different from the kinetoplast exhibit strong fluorescence staining. Indeed, in panel D the basal body, which is similar to kinetoplast in size and type of fluorescence staining, is lighter than cell body. In panel E, one or more parts different from the kinetoplast has a strong staining. Finally, notice also that occasionally fluorescence objects, i.e. artefacts, may be observed outside cell body.

These observations suggest that several and different reasons of uncertainty can be observed in CL images, making the right determination of kinetoplast staining a demanding task. Such motivations are at the basis of CL tests inter-observer variability. Other reasons of variability, common also to ANA tests, are the photo-bleaching effect and the lack of quantitative information supplied to physicians [2].

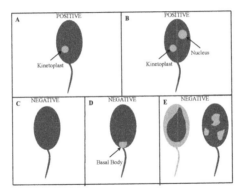

**Fig. 2.** Positive and negative cases depicted through stylized cells. Dark and light grey represent low and high fluorescence, respectively.

## 3    System Architecture

A typical CAD system is composed of several blocks that control data acquisition and storage, interact with users and support the diagnosis. Since in this paper we focus on strategies to help the physicians in the classification task, next subsection presents the architecture of the system that classifies ANA images, whereas subsection 3.2 describes our approach to recognise anti-dsDNA slides.

### 3.1    ANA Image Classification

ANA image classification system is based on a cascade of two steps: the first classifies the fluorescence intensity, whereas the second recognizes the staining pattern of positive wells (Fig. 3). Details of the corresponding classification systems are discussed in previous papers of the same authors [10, 11, 15]. The interested readers may refer to them for further details.

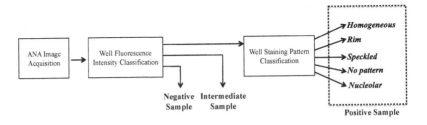

**Fig. 3.** Flow-chart of classification procedure for ANA tests

*Fluorescence Intensity Classification.* The system that classifies the fluorescence intensity uses a decomposition approach, which reduces the multiclass task into multiple and less complex binary problems [19, 20]. The rationale of our choice is given by the results coming out from the feature selection phase, based on measures related to first and second order histograms, which enforce the evidence that the classification could be reliably faced by introducing one specialized expert per each class that the system should recognize. Different dichotomizers, i.e. the discriminating functions that subdivide the input patterns in two separated classes, perform the corresponding recognition task. To provide the final classification, their outputs are combined according to a given rule, usually referred to as *selection rule*.

Among the different decomposition approaches presented in the literature [19, 20, 21, 22], we used the one usually named as *one-per-class* where the binary learning functions separate a single class from all the others. Hence, each module is specialized on the classification of positive, negative and intermediate samples, respectively.

Given the set of binary decisions, we apply two different selection rules .

The first consists of a binary combination of the outputs, referred to as Binary Selection (BS). Let us denote $O(x)$ the MES output and $Y_j(x)$ the output on sample $x$ of the $j$th classifier devised to recognized the class $C_j$ from the others, with $j = [1, ..., L]$. Since each module has a binary output, i.e. 1 or 0, possible input combinations to the selection module can be grouped into three categories: (i) those for which only one module $j$ classifies the sample in its class $C_j$, (ii) those for which more modules classify the sample in its own class, (iii) those for which no module classifies the sample in its class.

According to these considerations, the following conservative selection rule is adopted. In case (i) the class of the module whose output is 1 is chosen as a final output, since all the classifiers agree in their decision. In case (ii) the sample is rejected since two or more modules indicate that the sample belongs to their own class. In case (iii) the sample is rejected since no module indicates that the sample belongs to its class. It is worth noting that this approach does not require any reliability estimation.

Alternatively, a strategy based on reliability estimation that chooses an output in any of the possible combinations of outputs may be introduced, referred to as Reliability-based Selection (RbS). Let us then denote $\psi_j(x)$ the reliability parameter of the $j$th module when it classifies the sample $x$. Since in case (i) all the modules agree in their decision, we choose as before the class of the module whose output is 1 as a final output. Conversely, in cases (ii) and (iii) the final decision is performed looking at the accuracy of each modules' classifications. More specifically, in case (ii), $m$ modules vote for their own class, with $2 < m \leq L$, whereas the others $(L - m)$ ones indicate that $x$ does not belong to their own class (i.e. their outputs are 1 and 0, respectively). To solve the dichotomy between the $m$ conflicting modules we look at the reliability of their classification and choose the more reliable one. Formally:

$$O(x) = C_j, \text{where } j = \arg \max_{i:Y_i(x)=1} (\psi_i(x)). \tag{1}$$

In case (iii), all modules classify $x$ as belonging to another class than the one they are specialized in (i.e. their outputs are 0). In this case, the bigger the reliability parameter $\psi_j(x)$, the less the probability that $x$ belongs to $C_j$, and the bigger the probability that it belongs to the other classes. These observations suggest selecting the following selection rule:

$$O(x) = C_j, \text{where } j = \arg \min_{i:Y_i(x)=0} (\psi_i(x)). \tag{2}$$

In other words, we first find out which module has the minimum reliability and then we choose the class associated to it as a final output.

*Staining Pattern Classification.* To classify the well staining pattern into one of the groups reported in Section 2.1 (i.e. HO, PN, SP, NU and NP), we adopt the approach depicted in Fig. 4. First, we segment the image to locate the cells; second, we classify the staining pattern of several cells and, third, we classify the staining pattern of the whole well on the strength of the classification of its cells.

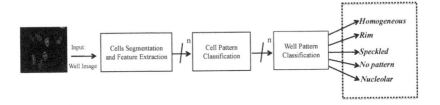

**Fig. 4.** Representation of the approach applied to classify the well staining pattern

In our opinion, such an approach addresses some key points of IIF staining pattern classification. Indeed, a recognition approach based on the classification of individual cells has the potential for detecting the occurrence of multiple patterns, i.e. the predominant and the minor ones. Furthermore, this approach is tolerant with respect to misclassifications in cells recognition, since the final label of the well is computed by using several pieces of information, i.e. the classifications of individual cells. Indeed, if enough cells per well are available, it is reasonable that cells misclassification, if limited, does not affect the well pattern classification.

The first step of the approach depicted in Fig. 4 requires to locate the cells: in this respect we use some morphological filters and global thresholding techniques [10]. In the second step, which asks for staining pattern classification of individual cells, we adopt once more a decomposition approach, where each dichotomizer is composed of an ensemble of classifiers. Both BS and RbS rules have been adopted again to select the final label of input cells [15]. Each module uses different feature sets related to texture components, computed on the basis of both statistical and spectral measures. The former are associated to properties of the first and the second order histogram, whereas the latter are calculated by partitioning the spectrum of the Fourier Transform into angular and radial bins. Furthermore features related to Wavelet Transform and Zernike Moments have been computed. Results of discriminant analysis show again that all the extracted features have limited discriminant strength over the classes, but different feature subsets discriminate better each class from the others, enforcing the rationale of a classifier selection approach.

On the basis of this system that recognizes individual cells, we determine the staining pattern of the whole well. To this end, we tested different voting rules, such as absolute and relative majority as well as the Weighted Sum (WS) rule, which is based on weighting the classifications of individual cells of the well. Formally, for each well we define $WS_i$ as:

$$WS_i = \sum_X \phi(x) \cdot I_i(x) \tag{3}$$

where the summation is over the set $X$ of cells that belong to the well under consideration, $\phi(x)$ is the reliability of each cell classification and $I_i(x)$ denotes an indicator variable defined as follows:

$$I_i(x) = \begin{cases} 1 \text{ if the cell } x \text{ is classified to class } C_i \\ 0 \text{ otherwise.} \end{cases} \tag{4}$$

The index of the final class of well staining pattern is $\upsilon = \arg\max_i(WS_i)$, i.e. the class for which $WS_i$ is maximum. The experimental results show that the WS rule outperforms the others in whole well recognition.

## 3.2 Anti-dsDNA Image Classification

To recognize the staining of each well we initially classify the single cells and then label it on the strength of their classification. Fig. 5 details the corresponding steps. The first relies upon the observation that kinetoplast fluorescence implies the presence in the image of a compact set of pixels brighter than other regions. Conversely, the absence of such a set suggests that the well is negative. For instance, the cell shown in panel C of Fig. 2 is classified as negative while the cells in the other panels contain regions candidate to be a kinetoplast and proceed in the next classification steps.

As a second step, we are interested in locating and classifying single cells. Hence, we segment only the images containing at least one set of pixels candidate to be a kinetoplast and, then, the segmented cells are classified. Finally, the staining of the well is computed on the basis of the labels of all cells.

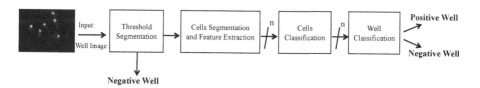

**Fig. 5.** Description of the approach applied to classify Crithidia Luciliae wells

We deem that such an approach provides two benefits. First, the initial threshold segmentation allows a rapid classification of several images ($\approx 0.1s$ per image). Second, it provides us a certain degree of redundancy, integrating together the information extracted from different cells of the same well. This permits to lower the effect of either erroneously segmented cells or artefacts.

Following paragraphs focus on cell and well classification blocks, respectively.

*Cell Classification.* The cells have been segmented using Otsu's algorithm [23]. Then, morphological operations, such as filling and connection analysis, output a binary mask for cutting out the cells from the image. Cells connected with the image border have been suppressed. Further to features associated to statistical and spectral measures (as in section 3.1), from each segmented cell we extract also standard, circular and circular rotation invariant Local Binary Pattern [24]. The interested readers may refer to [25] for further details. Discriminant analysis permits us to identify the subset of most discriminant features to be used with the expert system.

*Well Classification.* Each image contains several cells whose kinetoplasts can be positive or negative. According to previous considerations, as a first step we are interested in identifying clearly negative images, i.e. images that do not contain any compact fluorescent set of pixels. To this aim, we perform a threshold-based segmentation that looks for connected regions satisfying conditions on both intensity and dimension.

When the image contains regions candidate to be a kinetoplast, we have to determine if it is a true positive image. For instance, false positive images occur when there is a fluorescent mass inside cell body, as shown in panel D and E of Fig. 2. To this end, we classify single cells as reported above and then, on their basis, we apply the Majority Voting (MV) rule which assigns to the image the label corresponding to the class receiving the relative majority of votes. Ties lead to reject the image.

## 4    Data Set

Since, to our knowledge, there are not reference databases of IIF images publicly available, we populated an image repository of both ANA and anti-dsDNA images. To this aim, two IIF specialists independently classified each sample according to classes introduced in Section 2.

With reference to ANA images, we have used the 600 images of HEp-2 slides to test the fluorescence intensity classification system. The a priori probabilities of positive, negative and intermediate classes are 36.0%, 32.5% and 31.5%, respectively. To classify the staining pattern according to the approach depicted in Fig. 4, we first populated a referring data set of fluorescent cells by randomly selecting 37 images of positive wells from our database. The a priori probabilities of HO, PN, SP, NU and NP class for such wells are 24.3%, 21.6%, 35.1%, 18.9% and 0.0%, respectively. Two third of segmented cells from each of those images are chosen at random, cropped to a rectangular region, stored in TIFF format and singly classified by two specialists. Notice that the classes introduced in section 2.1 do not cover all the possibilities since some cells, identifiable with an irregular shape, can be corrupted during the slide preparation process. For this reason at the level of single cell pattern recognition we introduce the *artefact* (AR) class, which complements those presented above. The cells data set consists of 573 labelled cells, therefore subdivided: 23.9% HO, 21.8% PN, 37.0% SP, 8.2% NU and 9.1% AR. With regard to anti-dsDNA images, we acquired and annotated 342 wells , 74 positive (21.6%) and 268 negative (78.4%).Furthermore, specialists have labelled a set of cells only on images where the threshold-based segmentation identifies fluorescent regions, realising a cells dataset composed of 1487 cells belonging to 34 different wells. The cells are therefore labelled: 928 as positive (62.4%) and 559 as negative (37.6%).

# 5   Experimental Evaluation

In order to classify IIF images we have investigated the performance that could be achieved appealing to popular classifiers belonging to different paradigms, such as neural network (Multi-Layer Perceptron or MLP), statistical classifier (Nearest Neighbour or NN), bayesian classifier ( Naïve Bayes) and Support Vector Machines (SVMs). For space reason we report in the following only the best performance.

In case of fluorescence intensity classification dichotomizers are of NN classifiers, in case of staining pattern recognition they are an ensemble of classifiers combining NN and MLP, whereas for CL classification we use SVM. In all cases, preliminary tests have permitted to determine the best configuration of classifiers' parameters.

The error rate has been evaluated according to a eight-fold cross validation method; the rates reported in the following are the mean of the tests.

In section 5.1 we first present the performance of the system that labels the fluorescence intensity, second we report the results of well staining pattern classification and, third, we discuss the global performance achieved in ANA tests. In section 5.2 we report the results attained in the classification of anti-dsDNA slides.

## 5.1   Results of ANA Classification

With reference to fluorescence intensity classification, the second and third columns of Table 1 report the absolute performance attained employing both the selection rules. As to the binary selection rule (BS), the overall miss rate is quite low. At a deeper analysis, the selection scheme does not exhibit a false negative rate. Hence, the positive samples erroneously classified are assigned to the intermediate class, whereas intermediate samples wrongly recognized are assigned to the positive class. Furthermore, no negative samples are misclassified and occasionally they are rejected. The selection module rejects approximately 11% of samples, which is the counterpart we have to pay for such low error rates. Therefore, with reference to not rejected samples, the hit rate is 98.5%.

It is worth noting that in medical application, the two kinds of errors, i.e. false positive and false negative, have very different relevance. In order to increase the test sensitivity, the former misclassification can be tolerated to a larger extent since false positive leads to non-necessary analysis, whereas the latter should be as low as possible.

Turning attention to the RbS rule (third column of Table 1), i.e. the zero-reject strategy based on reliability estimation of each classification act, note that the hit rate increases from 87% up to more than 94%. Hence, some of the samples that are rejected by

**Table 1.** Recognition rate of both the fluorescence intensity and single cell staining pattern classifiers, adopting the two selection rules

|  | Fluorescence Intensity | | Single Cell Staining Pattern | |
|---|---|---|---|---|
|  | BS | RbS | BS | RbS |
| Hit (%) | 87.4% | 94.3% | 60.8% | 75.9% |
| Miss (%) | 1.3% | 5.6% | 10.4% | 24.6% |
| Reject (%) | 11.3% | - | 28.8% | - |

the previous approach are now correctly classified. Nevertheless, there are also samples previously rejected that are now misclassified, increasing the overall miss rate of the recognition system up to 5.67%. Moreover the performance on negative samples is still fine, since 99% of them are correctly recognized.

The fourth and fifth columns of Table 1 report the performance of the system that classifies the staining pattern of individual cells.

On the one hand, applying the BS rule, the classification accuracy of HO, PN and NU classes ranges from 51% to 60%, whereas the best and worst recognition performance are attained for cells of SP and AR classes, i.e. 75% and 29%, respectively. However, as introduced in Section 3, such a rule introduces a fixed reject rate that aims at lowering the misclassifications. Indeed, the hit rate on the classified samples for HO, PN, SP, NU and AR classes is 81.3%, 84.6%, 93.0%, 89.0% and 50.1%, respectively.

On the other hand, applying the RbS rule, the classification accuracy of HO, PN and NU classes ranges from 71% to 74%, whereas the best and worst recognition performance is attained for cells of SP and AR classes, i.e. 88% and 44%, respectively.

Whatever the selection rule, we deem that misclassifications of HO, PN and SP samples are related to their similarities of staining pattern and texture. Indeed, the discrimination between such classes is a burdensome issue also for well-trained specialists. Furthermore, errors on NU and NP classes are related to the small cardinality of such sets. The variability among AR samples is high, since this class contains those cells corrupted during the slide preparation that exhibit irregular shape and texture.

It is worth noting a direct comparison of these results with respect to previous works on the same topic [14], [13] and [12] is not possible, since their recognition tasks differ from ours. Indeed, in [14] the authors employed only sera of positive controls, whereas in [13] and [12] the authors used a different data set, which is not only constituted by samples diluted at 1:160, but also containing cells that were negative, i.e. they did not exhibit a detectable fluorescence intensity.

On the strength of cells classification, we then determine the whole well staining pattern applying the Weighted Sum (WS) rule (see formula 3). To evaluate the corresponding recognition performance, we proceed similarly to a leave one out approach working at the well level rather than at the cells one: at each iteration one well (and therefore all its cells) constitutes the test set, while the others populate the training set. Using this approach, we achieve a hit rate of 85.3%. This performance, although promising, shows an error rate that could be still too high to make the system usable in the medical practice. To overcome such a limitation, in an operating scenario we may apply the reject option to the decision taken by the WS criterion. In this respect, we have to estimate the reliability of the decision provided by this rule, and then to compare it with respect to a threshold, similarly to what we did to reject individual cells. It looks reasonable to adopt as a reliability estimator the quantity:

$$\rho = \frac{\max_i (WS_i)}{\sum_i WS_i} = \frac{WS_v}{\sum_i WS_i} \qquad (5)$$

where $v$ is the index of the final class of well staining pattern and $i$ varies over the four classes homogeneous, rim, speckled and nucleolar (see the background). Indeed, the rationale of this choice is that the final classification is as much reliable as a larger number of cells are classified in the final class of the well. Applying such an option, with a threshold equal to 0.57 we get an error rate of 5.8%. Notice that this value is smaller than the estimated intra-laboratory variability, which it has been measured equal to 7.4% in [4]. The corresponding reject rate is 17.6% which looks fairly limited. This performance seems very good and makes the system usable in practice, especially as a second reader to support the specialists' decisions.

**Table 2.** Performance rate of the overall system that classifies ANA slides, achieved when the the most liberal and conservative setup are applied

|  | Liberal Setup | | | Conservative Setup | | |
|---|---|---|---|---|---|---|
|  | Hit | Miss | Reject | Hit | Miss | Reject |
| Positive Samples | 78.6% | 21.4% | 0.0% | 67.2% | 5.5% | 27.3% |
| Negative Samples | 98.9% | 1.1% | 0.0% | 89.4% | 0.0% | 10.6% |
| Intermediate Samples | 92.3% | 7.7% | 0.0% | 85.0% | 3.4% | 11.6% |
| Total | 89.5% | 10.5% | 0.0% | 80.0% | 3.1% | 16.9% |

On the basis of the previous results concerning both fluorescence intensity and well staining pattern recognition, we discuss now the overall perspective performance attainable by a CAD composed by these systems.

Since they can apply two selection rules, different setups can be used. Among all, we focus on the two extreme available arrangements, which are referred to as liberal and conservative. On the one hand, a classification system may be thought as "liberal" when it makes positive classifications with weak evidence so it classifies nearly all positives correctly, but it often has high false positive rates. On the other hand, it may be defined as "conservative" when it makes positive classifications only with strong evidence so it makes few false positive errors, but it often has low true positive rates as well.

In our case, the most liberal configuration is realized as follows. Both the fluorescence intensity and the staining pattern classification systems apply the RbS criterion.

A conservative setup is carried out as follows. The fluorescence intensity classification system employs the BS criterion, whereas the system that recognizes the single cell staining pattern is based on the RbS rule. To label the staining pattern of the whole well, the weighted voting criterion works with the reject option presented above (equation 5).

The results of the liberal and conservative setup are shown in Table 2. In case of liberal configuration, the overall recognition rate is 90%, approximately, whereas in the conservative one it is 80%. Such a variation is essentially due to the introduction of reject options both at the stage of fluorescence intensity and staining pattern classification, respectively. Their use aims at lowering the misclassifications: indeed, the miss rate of the conservative configuration is one third of the liberal setup corresponding one, i.e. 3.1% vs. 10.5%. The side effect is that the 16.9% of samples are rejected. It is worth noting that the staining pattern classification influences only the recognition rate of positive samples. Therefore, the employment of a two stages recognition approach (Fig. 3) permits to achieve low false negative rate in both setups, as discussed for fluorescence intensity recognition results.

Besides the two configurations presented above, others should be used. However, these two arrangements represent the most conservative and liberal ones that can be set on the basis of the systems discussed in this work. The other setups present intermediate performance between such extrema.

### 5.2   Results of Anti-dsDNA Classification

In this subsection we present the results achieved in the classification of individual cells and wells of antids-DNA slides.

With reference to cell classification, the second and third columns of Table 3 report the confusion matrix achieved when SVM classifiers are applied to recognise positive and negative CL cells. The cells have been randomly divided into 34 subsets, one for each well, to perform a 34-folds cross validation. In this way we are sure that cells belonging to the same well are in the training or test sets only. The results point out that the accuracy is balanced over the two classes and confirm previous papers reporting that SVMs have very good performance on binary classification task, since they have been originally defined for this type of problem [26]. The overall cells classification accuracy is 94.2%.

**Table 3.** Confusion matrix for cell and well classification in case of anti-dsDNA images. *Pos* and *Neg* stand for Positive and Negative classes.

|  |  | Single cell classification | | Threshold-based segmentation | | Well classification | |
|---|---|---|---|---|---|---|---|
|  |  | Hypothesised class | | Hypothesised class | | Hypothesised class | |
|  |  | Pos | Neg | Pos | Neg | Pos | Neg |
| True | Pos | 95.5% | 4.5% | 100.0% | 0.0% | 98.6% | 1.4% |
| Class | Neg | 7.9% | 92.1% | 17.5% | 82.5% | 2.1% | 97.9% |

With regards to well classification, the results reported in the fourth and fifth columns of Table 3 show that the threshold-based segmentation does not exhibits false negatives. Hence, the remaining 35.4% of images are then labelled having recourse to cell labels provided by 34-folds classification. The overall performance are reported in the last two columns of Table 3, showing that the 99.4% of wells have been correctly classified.

## 6   Conclusions

In this paper we have proposed a system supporting the classification of different kinds of IIF images, which permit to detect ANA and anti-dsDNA autoantibodies. Such a CAD system could be used: (i) to reduce the interobserver variability, (ii) to increase the level of standardization of the reading procedures, (iii) to act as a second reader to reduce the workload of senior IIF experts. In particular, according to the preliminary rates measured on our prototype, all the images can be read by a junior (e.g. resident) IIF expert and his/her diagnoses compared with those automatically provided by the system. If the two classifications match, the diagnosis is confirmed and the specialist needs no more work. A senior IIF expert should read mismatched and rejected samples to confirm or not the diagnosis of the junior expert. According to this scenario the system would allow a remarkable workload reduction for a senior IIF expert (more than 80%).

In the end, we are currently engaged in populating a larger annotated database, improving the developed tools, as well as in performing extensive clinical trials.

**Acknowledgements.** The authors gratefully acknowledge the contribution of DAS s.r.l of Palombara Sabina (*www.dasitaly.com*), the comments of Dario Malosti, as well as Antonella Afeltra, Leonardo Onofri and Amelia Rigon for their collaboration in IIF images annotation and evaluation.

# References

1. Kavanaugh, A., Tomar, R., Reveille, J., Solomon, D.H., Homburger, H.A.: Guidelines for clinical use of the antinuclear antibody test and tests for specific autoantibodies to nuclear antigens. American College of Pathologists, Archives of Pathology and Laboratory Medicine 124, 71–81 (2000)
2. Rigon, A., Soda, P., Zennaro, D., Iannello, G., Afeltra, A.: Indirect immunofluorescence in autoimmune diseases: Assessment of digital images for diagnostic purpose. Cytometry B (Clinical Cytometry) 72, 472–477 (2007)
3. Lee, P.P.W., Lee, T.L., Ho, M.H.K., Wong, W.H.S., Lau, Y.L.: Recurrent major infections in juvenile-onset systemic lupus erythematosus–a close link with long-term disease damage. Rheumatology 46, 1290–1296 (2007)
4. Bizzaro, N., Tozzoli, R., Tonutti, E., Piazza, A., Manoni, F., Ghirardello, A., Bassetti, D., Villalta, D., Pradella, M., Rizzotti, P.: Variability between methods to determine ANA, anti-dsDNA and anti-ENA autoantibodies: a collaborative study with the biomedical industry. Journal of Immunological Methods 219, 99–107 (1998)
5. Feltkamp, T.E.W., Klein, F., Janssens, M.: Standardisation of the quantitative determination of antinuclear antibodies (ANAs) with a homogeneous pattern. Annals of the Rheumatic Diseases 47, 906–909 (1988)
6. Das: Service Manual AP16 IF Plus. Palombara Sabina, RI (2004)
7. Bio-Rad Laboratories Inc.: PhD System (2004), http://www.bio-rad.com
8. Cheng, H.D., Cai, X., Chen, X., Hu, L., Lou, X.: Computer-aided detection and classification of microcalcifications in mammograms: a survey. Pattern Recognition 36, 2967–2991 (2003)
9. Van Ginneken, B., Ter Haar Romeny, B.M., Viergever, M.A.: Computer-aided diagnosis in chest radiography: a survey. IEEE Transactions on Medical Imaging 20, 1228–1241 (2001)
10. Soda, P., Iannello, G.: Experiences in ANN-based classification of immunofluorescence images. International Journal of Applied Science, Engineering and Technology 2, 102–107 (2006)
11. Soda, P., Iannello, G., Vento, M.: A multiple experts system for classifying fluorescence intensity in antinuclear autoantibodies analysis. Pattern Analysis & Applications 12(3), 215–226 (2009)
12. Sack, U., Knoechner, S., Warschkau, H., Pigla, U., Emmerich, F., Kamprad, M.: Computer-assisted classification of HEp-2 immunofluorescence patterns in autoimmune diagnostics. Autoimmunity Reviews 2, 298–304 (2003)
13. Perner, P., Perner, H., Muller, B.: Mining knowledge for HEp-2 cell image classification. Journal Artificial Intelligence in Medicine 26, 161–173 (2002)
14. Hiemann, R., Hilger, N., Michel, J., Nitscke, J., Bohm, A., Anderer, U., Weigert, M., Sack, U.: Automatic analysis of immunofluorescence patterns of HEp-2 cells. Annals of the New York Academy of Sciences 1109, 358–371 (2007)

15. Soda, P., Iannello, G.: Aggregation of classifiers for staining pattern recognition in antinuclear autoantibodies analysis. IEEE Transactions on Information Technology in Biomedicine 13, 322–329 (2009)

16. Center for Disease Control: Quality assurance for the indirect immunofluorescence test for autoantibodies to nuclear antigen (IF-ANA): approved guideline. NCCLS I/LA2-A 16 (1996)

17. Solomon, D.H., Kavanaugh, A.J., Schur, P.H.: Evidence-based guidelines for the use of immunologic tests: Antinuclear antibody testing. Arthritis Care & Research 47, 434–444 (2002)

18. Tozzoli, R., Bizzaro, N., Tonutti, E., Villalta, D., Bassetti, D., Manoni, F., Piazza, A., Pradella, M., Rizzotti, P.: Guidelines for the laboratory use of autoantibody tests in the diagnosis and monitoring of autoimmune rheumatic diseases. American Journal of Clinical Pathology 117, 316–324 (2002)

19. Dietterich, T.G., Bakiri, G.: Solving multiclass learning problems via error-correcting output codes. Journal of Artificial Intelligence Research 2, 263–286 (1995)

20. Mayoraz, E., Moreira, M.: On the decomposition of polychotomies into dichotomies. In: ICML 1997: Proceedings of the Fourteenth International Conference on Machine Learning, pp. 219–226 (1997)

21. Jelonek, J., Stefanowski, J.: Experiments on solving multiclass learning problems by $n^2$ classifier. In: Nédellec, C., Rouveirol, C. (eds.) ECML 1998. LNCS, vol. 1398, pp. 172–177. Springer, Heidelberg (1998)

22. Soda, P.: Facing polychotomies through classification by decomposition: Applications in the bio-medical domain. Biomedical Engineering Systems and Technologies, Communications in Computer and Information Science, 291–304 (2009)

23. Otsu, N.: A threshold selection method from gray-level histograms. IEEE Transactions on Systems, Man, and Cybernetics 9, 62–66 (1970)

24. Ojala, T., Pietikäinen, M., Mäenpää, T.: Multiresolution gray-scale and rotation invariant texture classification with local binary pattern. IEEE Transactions on Pattern Analysis and Machine Intelligence 24, 971–987 (2002)

25. Soda, P., Onofri, L., Rigon, A., Iannello, G.: Analysis and Classification of Crithidia Luciliae fluorescent images. In: Foggia, P., Sansone, C., Vento, M. (eds.) ICIAP 2009. LNCS, vol. 5716, pp. 558–566. Springer, Heidelberg (2009)

26. Vapnik, V.N.: The nature of statistical learning theory. Springer, NY (1995)

# Support Vector Machine Diagnosis of Acute Abdominal Pain

Malin Björnsdotter[1], Kajsa Nalin[2], Lars-Erik Hansson[3], and Helge Malmgren[4]

[1] Institute of Neuroscience and Physiology, University of Gothenburg
SE-405 30 Göteborg, Sweden
malin.aberg@neuro.gu.se
[2] Centre of Interdisciplinary Research/Cognition/Information, University of Gothenburg
SE-405 30 Göteborg, Sweden
[3] Department of Surgery, Sahlgrenska University Hospital/Östra, SE-416 85 Göteborg, Sweden
[4] Department of Philosophy, Linguistics and Theory of Science, University of Gothenburg
SE-405 30 Göteborg, Sweden

**Abstract.** This study explores the feasibility of a decision-support system for patients seeking care for acute abdominal pain, and, specifically the diagnosis of acute diverticulitis. We used a linear support vector machine (SVM) to separate diverticulitis from all other reported cases of abdominal pain and from the important differential diagnosis non-specific abdominal pain (NSAP). On a database containing 3337 patients, the SVM obtained results comparable to those of the doctors in separating diverticulitis or NSAP from the remaining diseases. The distinction between diverticulitis and NSAP was, however, substantially improved by the SVM. For this patient group, the doctors achieved a sensitivity of 0.714 and a specificity of 0.963. When adjusted to the physicians' results, the SVM sensitivity/specificity was higher at 0.714/0.985 and 0.786/0.963 respectively. Age was found as the most important discriminative variable, closely followed by C-reactive protein level and lower left side pain.

## 1 Introduction

Medical diagnosis and decision-making, that is, the classification of patients into disease groups based on various symptoms, is a highly complex problem – as evidenced by the near decade-long training required by specialist physicians. The possibility of computer-based decision support systems is, therefore, highly appealing as an assistive tool in medical diagnostics. This study investigates one such decision-making process, namely the diagnosis of patients seeking care for acute abdominal pain at emergency wards, and a potential method for automatic diagnosis of these patients.

Emergency ward doctors face a highly demanding situation, where medical decisions with substantial impact on patients must be made under time pressure. The large number of potentially relevant physical measurements, from blood factors to face color, in combination with a stressful situation yield a challenging decision-making process. Moreover, physicians have reported lack of relevant experience and continuous feedback among other factors that affect the decision-making process negatively [1]. In addition, disease symptoms are highly variable between individuals, leaving the doctor to rely heavily on experience [2].

A. Fred, J. Filipe, and H. Gamboa (Eds.): BIOSTEC 2009, CCIS 52, pp. 347–355, 2010.

Standardized, computer-based decision support systems, automatically identifying typical disease patterns in patient data, are thus attractive as a complement to the trained physician. These systems generally consist of statistical models (classifiers) which are trained to identify patterns related to a given disease in supplied patient data where the final diagnosis is known. The classifiers are then applied to new patients, where an instantaneous diagnosis is made in order to assist the doctor.

Results on computer-aided diagnosis of abdominal pain were reported as early as in 1972, where de Dombal reported a surprisingly high diagnostic accuracy (91.8% vs. 79.6%) using decision support compared with the unaided examination [3]. Moreover, a large British multi-center study with more than 16.000 patients confirmed the utility of computer aided diagnostic with accuracies of 65% vs. 46% [4].

Computer-aided diagnostic systems for acute abdominal pain have not, however, achieved general acceptance – in part due to bureaucratic obstacles, in part to methodological concerns. The latter include the construction of extensive patient databases, representative for the complexity of the specific field of interest. To assure high quality of data, disease symptoms and related variables require collection and recording by trained specialists using standardized forms. Careful follow-ups must be performed to ascertain a definitive diagnosis. Substitution for missing data – an unavoidable issue in medical data collection – require a method that is statistically sound as well as applicable in practice. The high number of potential input variables, combined with a realistic number of patient cases, makes feature selection necessary to alleviate the curse of dimensionality for accurate and efficient automatic diagnosis [5]. The study reported here fulfils all the mentioned desiderata with the exception that not all data were collected by specialists.

A common acute abdominal disease, often a reason for emergency hospital admission – especially in elderly patients — is diverticulitis of the colon [6,7,8,9]. The diagnosis is typically made at the emergency department, and is based on medical history, clinical indications and laboratory data. The clinical presentation of acute diverticulitis was recently described by Laurell and colleagues, as well as the natural short-term development of the disease [10]. Primary diagnosis sensitivity, by the physician, was reported to be 64%, with a specificity of 97%. Moreover, Laurell et. al. identified nonspecific abdominal pain (NSAP) as one of the most important differential diagnoses. For NSAP, the primary diagnosis sensitivity was reported to be 43%, with a specificity of 90%.

In the current paper, we investigate the feasibility of using a decision support system for the automatic diagnosis of acute diverticulitis, contrasted with all other reported cases of abdominal pain and from the diagnosis category NSAP. Using a state-of-the-art classifier, namely (linear) support vector machines, and feature selection, we also attempt to understand the underlying factors that are key to identifying diverticulitis.

## 2    Methods

### 2.1    Data Acquisition

Mora Hospital in northern Sweden is a district hospital serving a population of 87 000 individuals, providing full emergency services. During the period of February 1997 to

June 2000, all patients older than one year of age admitted to the hospital with abdominal pain of duration of up to 7 days were registered in a database. Details were recorded according to a standardized form for history, clinical indications and laboratory results. The attending physician suggested a diagnosis, and a final diagnosis was given when the patient left the hospital. A definitive diagnosis was later established by a follow-up study of the patient's journal around one year later. Data on 3 337 patients were thus acquired.

## 2.2 Variable Transformation

The original data (formulated as 52 items on the standardized form) were reformatted to create logically independent input variables. All data on pain localization were transformed to binary form. All "Yes"/"Don't know"/"No" responses on clinical questions and investigations were coded numerically as 2/1/0 in order to keep the implied ordinal information.

Non-reported values were assumed irrelevant and within the normal range, and missing data were therefore substituted by estimated normal (healthy) values. An added benefit of this approach to missing data substitution is that normal values can be used in practical applications of decision support systems without any knowledge of the statistics of the present sample. Importantly, the variable substitution included gender specific adjustments where required, for example assuming "normal" values for males in variables such as time from last menstruation.

A final dataset consisting of 3 337 patients, each with 117 variables and an initial diagnosis by a trained physician, was thus obtained. Out of the 3 337 patients, 148 obtained diverticulitis as a definitive (retrospective) diagnosis, whereas 1340 were diagnosed as having NSAP. The investigated decision-support system was subsequently trained on the dataset to predict the final diagnosis and differentiate the diseases.

## 2.3 Data Partitioning

The patients were divided into a training (90%) and a validation dataset (10%). All reported results refer to the validation dataset, unless otherwise specified. The automatic decision support system was trained on the training data, and subsequently applied to the validation data. The performance of the physicians' initial diagnosis were, similarly, computed on the validation dataset.

## 2.4 Support Vector Machines

Support vector machines (SVMs) is a type of classification algorithm which maximizes the geometric margin between the data classes and the separating hyperplane [11].

Given our training data:

$$\mathcal{D} = \{(\mathbf{x}_i, y_i) | \mathbf{x}_i \in \mathbb{R}^p, y_i \in \{-1, 1\}\}_{i=1}^n \qquad (1)$$

where $y_i$ is the disease category (-1 or 1) to which patient $\mathbf{x}_i$ belongs, the hyperplane that maximally separates the data points must fulfill the following inequalities:

$$\omega \cdot \mathbf{x}_i + b \geq d \text{ for all } i \text{ where } y_i = 1 \qquad (2)$$

$$\omega \cdot \mathbf{x}_i + b \leq -d \text{ for all } i \text{ where } y_i = -1 \qquad (3)$$

where $\omega$ is the weight vector, $b$ is the bias and $d$ is the separating margin.

The SVM model is trained by adapting the weights to the given data such that the margin $d$ is maximized. To this end, the matlab toolbox LS-SVMlab, developed by the group SCD/sista in the department ESAT at the KULeuven, Belgium [11], available at http://www.esat.kuleuven.be/sista/lssvmlab/, was used.

### 2.5  Variable Ranking

Although the database is substantial and there is a reasonable number of instances (patients) compared to the number of available variables, feature selection, that is, the identification of a lower number of highly discriminatory features, can be expected to boost classification performance [5,12].

A simple method was implemented for variable ranking and subsequent selection as follows:

$$v_i = \text{abs}(\frac{\mu_0 - \mu_1}{\sigma_0 + \sigma_1}) \qquad (4)$$

where $\mu_0$ and $\mu_1$ represent the mean value of variable $i$ over the patterns (patient data) belonging to class 0 and 1 respectively, and $\sigma_0$ and $\sigma_1$ are the standard deviations within each class. The variable ranking value is thus a measure of variable stability, over the patterns, as well as how well each variable taken by itself separates the data classes.

Diagnostic performances were evaluated for variable subsets formed by applying the classifier to a $N$ of the highest ranked variables, where N ranged from one to all 117 variables.

### 2.6  Class Imbalance Correction

The diagnosis diverticulitis is heavily under-represented compared to the remaining diseases (148 instances out of 3 337) as well as to NSAP (1340). In order to reduce the effect of the imbalance during classifier training, a number of randomly selected instances (patients) of the over-represented category was removed until both categories were equally represented.

### 2.7  Performance Measure

The measure of diagnostic performance was based on the sensitivity (e.g. the proportion of the patients diagnosed with diverticulitis correctly identified as such) and specificity (e.g. the proportion of patients who did not have diverticulitis and were correctly identified as such). For the SVM classifier, the receiver operating characteristic curve (ROC; a plot of the sensitivity versus [1-specificity] for varying classifier thresholds) was computed for a summarized performance estimation. The area under the curve (AUC) was subsequently obtained, where larger AUC values indicate better diagnostic performance.

The estimation of the physicians' performance cannot be summarized in a ROC-curve, however. In order to directly compare the resulting performance between the SVM and the doctors, therefore, the specificity and sensitivity for the final SVMs (including the optimal number of variables) were computed at the corresponding levels of the physicians.

## 3   Results

### 3.1   Classification Performance

First, the SVM was applied to attempt discrimination between diverticulitis and the pool of all other diseases for a varying number of included variables (see table 1 for a summary of the classification performance results). A maximum AUC of 0.95 was found for 64 variables (see figure 2), and a histogram illustrating the resulting discrimination ability of the classifier is shown in figure 1A. At this optimal point, the SVM obtained a substantially higher sensitivity of 1.00 – that is, correctly identified all of the diverticulitis cases – than the physicians at 0.571. However, as is clear from figure 1A, given this high sensitivity the specificity suffers, and a high amount of false positives are inevitable: the SVM obtained a specificity of 0.823, as opposed to the doctors who produced a much higher specificity of 0.987. When adjusted to the physicians' values (of 0.571 and 0.987, respectively), the SVM achieved a comparable (yet slightly lower) sensitivity/specificity score of 0.571/0.981 (at 105 variables) and 0.5/0.987 (at 111 variables).

Non-specific abdominal pain (NSAP) appears more difficult to distinguish from the pool of all other diseases than diverticulitis. At its highest discriminative power, the SVM validation sensitivity and specificity are low at 0.687 and 0.721 respectively. Again, the doctors' diagnosis achieves a low sensitivity of 0.455 but a high specificity of 0.909. Also here, when adjusted to the physicians' results the sensitivity/specificity obtained was comparable (but lower) for the SVM than the physicians at 0.455/0.878 (at 16 variables) and 0.41/0.909 (at 16 variables).

The discrimination between diverticulitis and NSAP, however, was considerably improved. The doctors achieved a sensitivity of 0.714 and a specificity of 0.963, whereas the best SVM resulted in a substantially higher sensitivity of 1 and a satisfactory specificity of 0.858 (AUC: 0.959). As can been seen in figure 1C, the validation data were distinctly separable. Also, when adjusted to the physicians sensitivity/specificity, the SVM scored higher in both cases at 0.714/0.985 (at 44 variables) and 0.786/0.963 (at 19 variables), respectively.

**Table 1.** Summary of performance results for the diagnosis of diverticulitis (diver.) and non-specific abdominal pain (NSAP) in terms of sensitivity/specificity.

|  | diver. vs. others | NSAP vs. others | diver. vs. NSAP |
|---|---|---|---|
| Physicians | 0.571/0.987 | 0.455/0.909 | 0.714/0.963 |
| SVM at maximum AUC | 1/0.823 | 0.687/0.721 | 1/0.858 |
| SVM at physicians' sensitivity | 0.571/0.981 | 0.455/0.878 | 0.714/0.985 |
| SVM at physicians' specificity | 0.5/0.987 | 0.41/0.909 | 0.786/0.963 |

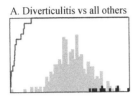 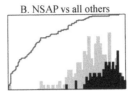 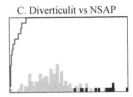

**Fig. 1.** Histogram illustrating the separability of A) Diverticulitis (dark) vs. all other diseases (light), B) Non-specific abdominal pain (dark) vs. all other diseases (light) C) Diverticulitis (dark) vs. non-specific abdominal pain (light). The blue line represents the receiver operating characteristic (ROC) curve.

## 3.2   Variable Selection

The variable selection proved to have substantial impact on all datasets (figure 2). For the discrimination between diverticulitis and all other diseases, the addition of variables from one through 30 had a large effect on classifier performance in the validation dataset, after which the performance declined. Similarly, on the NSAP vs. all other diseases task, going from one to two features shows a dramatic increase in performance, whereas further addition does not have a large effect. On the other hand, for the more specific case of diverticulitis vs. NSAP, it is clear that several variables contain large amounts of information regarding the categories – there is a sharp increase in performance up to nine variables, and an equally sharp decrease after the addition of 15 more variables.

A closer inspection of the highly rated features (see table 2-4) reveals that, for any dataset combination, age is the most important discriminating factor, closely followed by C-reactive protein level. As can be expected, similar variables are important for the discrimination between diverticulitis and all other diseases as for that between diverticulitis and NSAP, namely: initial pain localization (both in the left lower quadrant and the right upper quadrant), current pain localization (in the lower left quadrant) and tenderness on palpation (in the left lower quadrant). The focus on left side pain in the diagnosis

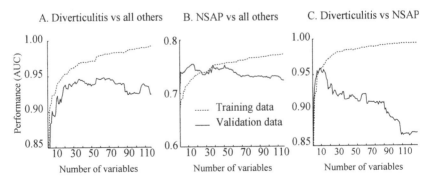

**Fig. 2.** Performance as a function of the number of included variables for A) Diverticulitis vs. all other diseases, B) Non-specific abdominal pain vs. all other diseases C) Diverticulitis vs. non-specific abdominal pain.

**Table 2.** The top ten ranked variables for diverticulitis vs. all other diseases

| Variable | Weight |
|---|---|
| Age | 1.22099 |
| C-reactive protein level | 1.05363 |
| Initial pain localization; left lower quadrant | 0.917118 |
| Tenderness on palpation | 0.901306 |
| Current pain localization; lower left quadrant | 0.89903 |
| Current pain localization; right upper quadrant | 0.634565 |
| Initial pain localization; right upper quadrant | 0.604262 |
| Vomiting | 0.588099 |
| Previous abdominal surgery | 0.562905 |
| Abdominal scars | 0.551697 |

**Table 3.** The top ten ranked variables for non-specific abdominal pain (NSAP) vs. all other diseases

| Variable | Weight |
|---|---|
| Age | 0.605346 |
| C-reactive protein level | 0.369473 |
| Serum bilirubin level | 0.350293 |
| Systolic blood pressure | 0.303903 |
| Decrease/absence of bowel movements | 0.286566 |
| Visible bowel movements | 0.28654 |
| Development of pain intensity; increase | 0.277085 |
| Localized swelling | 0.267 |
| S-amylase level | 0.264904 |
| Serum alanine aminotransferase level | 0.251346 |

**Table 4.** The top ten ranked variables for diverticulitis vs. non-specific abdominal pain

| Variable | Weight |
|---|---|
| Age | 1.51921 |
| C-reactive protein level | 1.12222 |
| Initial pain localization; left lower quadrant | 1.03731 |
| Current pain localization; lower left quadrant | 0.944773 |
| Tenderness on palpation; left lower quadrant | 0.885332 |
| Development of pain intensity; increase | 0.762057 |
| Current pain localization; right upper quadrant | 0.677884 |
| Local muscular defence | 0.674493 |
| Leukocyte level | 0.664661 |
| Initial pain localization; right upper quadrant | 0.657553 |

of diverticulitis agrees with clinical knowledge and previous research [9]. However, in the case of NSAP vs. all other diseases, other variables (predominantly various fluid measurements) are highly rated. Moreover, the resulting rating coefficients are much smaller, thus indicating lower differentiation between variables.

## 4 Discussion

We have investigated the utility of using a decision support system for the computer aided diagnosis of acute diverticulitis and non-specific abdominal pain (NSAP), as well as for the discrimination between the two, using the Mora acute abdominal pain database.

The general performance of the SVM was comparable to that of the doctor. Moreover, both sensitivity and specificity were higher than those of the physicians in the distinction between diverticulitis and NSAP.

The discrimination between diverticulitis and the pool of other diseases, as well as that between NSAP and the other diseases, was substantially worse than the differentiation between the two disease categories. This suggests that incorporating known information about the other disease categories, including their respective distribution, in the training of the classifier model can aid in the subsequent diagnosis of new cases. This can, for example, be achieved using an ensemble of classifiers. For the case of multi-class data, where the diagnosis of all patients and thus all diseases is desired, this is, moreover, required for inherently binary classifiers such as SVMs. Standard schemes for ensemble encoding include the one-against-all and one-against-one approach. The latter is more computer-intensive than the former, but typically yields better results. Importantly, it also provides insight into the distinction between diseases. Moreover, the one-against-one approach is a more well-defined problem, contrasting data categories with inherent similarities and differences, as was evidenced by the findings in our study. The one-against-all scheme, on the other hand, is more relevant to the problem of diagnosis.

The class imbalance was adjusted by simple under-sampling of the majority class. More sophisticate methods could be employed to this end, such as the SMOTE algorithm [13].

The simple feature ranking and subsequent selection utilized in this study proved to be effective in boosting classifier performance. However, more suitable approaches can be used to obtain optimal variable subsets, such as evolutionary algorithms [14,15,16]. Moreover, there is ample reason to believe that non-linear relationships pertaining to the disease category exist between some parameters, and reducing the complexity of the data structures can potentially allow for better performance with non-linear classifiers [17]. We are currently pursuing this strategy.

## 5 Conclusions

Automatic computer-based disease classification is a promising tool for the diagnosis of acute abdominal pain, but requires substantial research before a clinical implementation is feasible. The support vector machine is highly suitable for the discrimination between binary disease categories, and achieved results comparable to the medical doctor. Moreover, the classifier obtained higher sensitivity and specificity than the physicians in the distinction between diverticulitis and NSAP. Age and C-reactive protein level, as well as left side pain sensations, were identified as important factors for the identification of diverticulitis.

# References

1. Nalin, K.: Den ideala kliniska beslutsprocessen. en studie av arbetsprocessen på en kirurgisk akutmottagning/The ideal clinical decison process. A study of the work process in an acute surgical ward. Masters thesis in Cognitive Science, University of Gothenburg (2006)
2. Hansson, L.E.: Akut Buk. Studentlitteratur, Lund (2002)
3. de Dombal, F., Leaper, D., Staniland, J., McCann, A., Horrocks, J.: Computer-aided diagnosis of acute abdominal pain. British Medical Journal 2(5804), 9–13 (1972)
4. Adams, I.D., Chan, M., Clifford, P.C., Cooke, W.M., Dallos, V., de Dombal, F.T., Edwards, M.H., Hancock, D.M., Hewett, D.J., McIntyre, N.: Computer aided diagnosis of acute abdominal pain: a multicentre study. British Medical Journal (Clinical research ed.) 293(6550), 800–804 (1986)
5. Bellman, R.E.: Adaptive Control Processes. Princeton University Press, Princeton (1961)
6. Ambrosetti, P., Robert, J., Witzig, J., Mirescu, D., Mathey, P., Borst, F., Rohner, A.: Acute left colonic diverticulitis: a prospective analysis of 226 consecutive cases. Surgery 115(5), 546–550 (1994)
7. Ferzoco, L., Raptopoulos, V., Silen, W.: Acute diverticulitis. The New England Journal of Medicine 338(21), 1521–1526 (1998)
8. Young-Fadok, T., Roberts, P., Spencer, M., Wolff, B.G.: Colonic diverticular disease. Current Problems in Surgery (37), 459–514 (2000)
9. Laurell, H., Hansson, L., Gunnarsson, U.: Acute abdominal pain among elderly patients. Gerontology 52(6), 339–344 (2006)
10. Laurell, H., Hansson, L., Gunnarsson, U.: Acute diverticulitis – clinical presentation and differential diagnostics. Colorectal Disease 6(9), 496–501 (2007)
11. Suykens, J., Gestel, T.V., Brabanter, J.D., Moor, B.D., Vandewalle, J.: Least Squares Support Vector Machines. World Scientific, Singapore (2002)
12. Blum, A., Langley, P.: Selection of relevant features and examples in machine learning. Artificial Intelligence 97(1-2), 245–271 (1997)
13. Chawla, N., Bowyer, K., Hall, L., Kegelmeyer, W.: Smote: synthetic minority over-sampling technique. Journal of Artificial Intelligence Research 16, 321–357 (2002)
14. Hernandez Hernandez, J.C., Duval, B., Hao, J.-K.: A genetic embedded approach for gene selection and classification of microarray data. In: Marchiori, E., Moore, J.H., Rajapakse, J.C. (eds.) EvoBIO 2007. LNCS, vol. 4447, pp. 90–101. Springer, Heidelberg (2007)
15. Åberg, M., Löken, L., Wessberg, J.: An evolutionary approach to multivariate feature selection for fMRI pattern analysis. In: Encarnação, P., Veloso, A. (eds.) Proceedings of the First International Conference on Bio-inspired Systems and Signal Processing (BIOSIGNALS), Funchal, Madeira, Portugal, pp. 302–307. INSTICC Press (2008)
16. Björnsdotter Åberg, M., Wessberg, J.: An evolutionary approach to the identification of informative voxel clusters for brain state discrimination. IEEE Journal of Selected Topics in Signal Processing 2(6), 919–928 (2008)
17. Åberg, M.C., Wessberg, J.: Evolutionary optimization of classifiers and features for single trial EEG discrimination. BioMedical Engineering Online 6(32) (2007)

# Near Field Communication and Health: Turning a Mobile Phone into an Interactive Multipurpose Assistant in Healthcare Scenarios

Giuliano Benelli and Alessandro Pozzebon

University of Siena, Department of Information Engineering
Via Roma 56, 53100 Siena, Italy
{benelli,alessandro.pozzebon}@unisi.it
http://www.dii.unisi.it

**Abstract.** In this paper we discuss the introduction of the Near Field Communication (NFC) technology in the management of the assistance operations in the hospitals. NFC is a new short range communication system based on RFID technology.

NFC systems can work like traditional RFID systems, where a master device reads some information from a slave device, but they can also set up a two-way communication between two items. In particular, NFC devices can be integrated on mobile phones, widely enhancing the intercommunication capabilities of the users.

The introduction of NFC in sanitary environments can help to make safer all the assistance operations. Next to the realization of a NFC electronic case history, we also studied the realization of electronic medical prescription and the use of this technology for the exchange of patient data between doctors and between nurses, in order to avoid errors in the attendance operations.

The final idea is to change a mobile phone into an interactive multipurpose assistant for people working in hospitals or in harness with patients.

**Keywords:** NFC, RFID, Informative Systems, Electronic case history, Electronic medical prescription, Wireless Communication.

## 1 Introduction

Near Field Communication (NFC) is a new short range telecommunication technology directly deriving from RFID identification systems.

Like many RFID systems, NFC works at the frequency of 13.56MHz and is based on the physical principle of inductive coupling.

The main difference between the two systems derives from the fact that, while RFID is strictly an identification technology, NFC has been studied to be properly used as a wireless communication technology between devices brought to a short distance between them.

Nowadays many mobile devices producers are beginning to realize phones and PDAs equipped with NFC circuitry, also providing the software to realize applications using this technology.

A. Fred, J. Filipe, and H. Gamboa (Eds.): BIOSTEC 2009, CCIS 52, pp. 356–368, 2010.

The main fields of application of NFC include proximity payments, peer-to-peer communication and obviously, strictly deriving from RFID, access control.

In sanitary environments RFID has found many applications: one of these has been the managing and the identification of patients.

In particular an electronic case history located on passive RFID bracelets has been studied and realized in a previous work [5], showing the benefits deriving from the chance to get vital information directly form the electronic support, with a reduction of the times of assistance and of the risks deriving from human errors.

The idea of this article is to show how much the performances of RFID systems can be widen moving to NFC technology, which can be used to execute the same operations made by mobile RFID devices, but adding many new functions once unfeasible.

## 2    NFC Technology

Before talking about how to introduce NFC in sanitary environments it's important to describe its main technological features.

It's obviously impossible to speak about NFC without briefly introducing RFID, which represents a fundamental technological background.

In the second subsection are then described the main characteristics of NFC communication protocol, in order to understand the various ways of interaction between different devices.

### 2.1    The Origins: RFID

The technological structure of NFC systems is quite the same as the one of RFID systems. NFC uses the same physical principles and partly the same kind of devices.

It's therefore difficult to understand the structure and the possible uses of NFC with investigating a little the main features of Radio Frequency Identification technology.

RFID is an automatic identification technology which uses the electromagnetic field as the mean of identification [3]. Usually RFID systems are composed by two devices: a Reader, which generates the interrogating electromagnetic field, and the Transponder, which is located on the item to be identified and returns back to the reader the ID (*Identification*) code and the additional information.

When the Transponder comes inside the EM (*Electromagnetic*) field of the reader it can be interrogated and it can send back the data using the same field.

There are many kinds of RFID systems, working at different operative frequencies. In particular we can find Low Frequency (125-135kHz), High Frequency (13.56MHz), Ultra High Frequency (868-915MHz) and Microwave ($> 2GHz$) systems.

Every different RFID application needs a particular care in the choice of the right technological solution: for example even if Ultra High Frequency systems can provide large read ranges, they have a lot of problems od electromagnetic compatibility.

The same happens for the powering methods of the transponders: in fact we can find passive, active or semipassive transponders, offering very different features. While an active transponder can be read from a distance ten times wider than a passive one, his higher price can make it unsuitable in applications in which the number of items to be identified is very high.

## 2.2 NFC Technological Features

NFC belongs to the family of RFID, but it has specific technological features [7].

It only works at the frequency of 13.56MHz, that is an unregulated band. This means there aren't any licenses required and restrictions concern only the electromagnetic emissions, in order to limit the impact of the system on human body.

Differently from traditional RFID technology, passive and active devices can be integrated into the same system.

NFC can reach a maximum read range of around 20cm but common devices are not able to read from distances larger than 4 or 5cm. The decision to create products with low read ranges comes not only from the physic limitations of the technology but also from the fact that short ranges ensure a bigger protection from outside intrusions. These requirements mainly come from the aim to use NFC to implement proximity wireless payment systems.

NFC devices can currently communicate at three different speeds, 106kbit/s, 212kbit/s and 424kbit/s but in the future higher rates will be probably achieved.

NFC protocol differentiates the device initiating and controlling the communication, called *Initiator* and the device answering the request from the initiator, called *Target* [6].

NFC protocol also presents two different operative modalities: a passive mode, with a single device generating the field and the other one using this field to exchange the data, and an active mode, in which the two devices generate their own EM field.

As a consequence NFC devices have studied to integrate on the same support the functions covered by the Reader an the Transponder. This means that we can have three different types of communication:

– The traditional communication protocol of RFID systems, in which the NFC equipped device acts as a Reader and it can get the information stored onto a Transponder and can also write on it.
– A bidirectional communication, in which two NFC devices exchange data between themselves. This case is particularly interesting because, even if the bit rates currently available are not too high, the particular protocol implemented makes the establishing of the communication very easy.
– A communication between a turned on device and a turned off one. In this case the second item is seen by the first one simply as a Transponder, making it suitable for identification and access control purposes.

All these different methodologies of data exchange have brought to the realization of many kinds of systems covering a wide range of applications not only in the field of identification but also in the one of personal communication.

Moreover, the short reading range of the devices makes NFC systems considerably safe, because intruders should arrive too much close to the devices to steal the data.

The large number of possible applications has finally led to the integration of the NFC technology onto the most common communication devices currently available on the market: the mobile phones. NFC phones can then be used as keys, credit cards or business cards and the number of possible applications is virtually infinite.

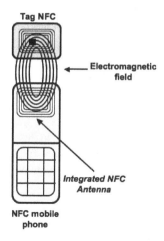

**Fig. 1.** An NFC system

## 3   NFC in the Real World

Even if quite new, NFC is a technology whose introduction is supposed to be very fast. In this sense the number of possible devices and applicative fields is growing day by day, making necessary a deep knowledge of the possible technological solutions.

In the following sections we describe briefly the most important typologies of NFC devices and the main applications.

The last subsection describes all the work that has been made in order to create worldwide accepted standards for NFC systems.

### 3.1   NFC Devices

NFC systems can integrate three different kinds of devices:

1. Fixed Read/Write or Read Only terminals;
2. Mobile Read/Write or Read Only devices;
3. Read Only Tags.

The fixed terminal can be common RFID readers working at 13.56MHz ISO14443 compliant or can be specifically studied devices created to perform specific actions deriving from the particular operation they have to execute.

Some examples of ad-hoc devices can be the NFC POS systems or the electronic ticketing terminals.

Mobile devices are mainly represented by phones or PDA, even if in this case too specific platforms can be studied to satisfy particular requirements.

Many mobile phones producers, including big companies like Nokia, Samsung and Motorola are studying and realizing particular phones with NFC technology integrated, and specific studies assert that by the year 2010 half of all mobile devices will support NFC.

In particular, while many companies have begun to sell specific versions of common phones equipped with NFC infrastructure, Nokia has been the first brand to put on the market a totally NFC phone.

The most important feature that has to be underlined in NFC phones is that on this kind of devices are collected both the functions of transponder and reader. In fact the phone presents an antenna with its circuitry to perform communication operations, but it also has an internal 4kbyte MIFARE card turning the device working in passive modality into a transponder.

The last kind of devices is represented by Tags. In this case common ISO14443 RFID tags can be used, even if specific products have nevertheless been realized.

Currently four different types of tag can be used in NFC systems:

- Type 1 is based on ISO14443A, is produced only by Innovision Research & Technology and has a 96-byte memory. This kind of tags are very cheap;
- Type 2 is based on ISO14443A, is produced only by Philips (MIFARE UltraLight), and has only a 48-byte memory;
- Type 3 is based on FeLica (a specification compatible with the ISO18092 standard for passive communication mode) and produced only by Sony. These tags have higher memory capacities (up to 2kbytes) and reach the 212kbit/s rate;
- Type 4 is compatible with the ISO14443A/B standard and is produced by several manufacturers. It has a large addressing-memory capability and reaches rates up to 424kbit/s.

These four kinds of tags represent four strongly different products, and every time that an NFC application has to be realized, the choice of the right kind of tag has to been made extremely carefully.

## 3.2  NFC Applications

NFC can be used in several different fields. Three different categories of possible applications have been identified [7]:

- Service initiation;
- Peer-to-Peer;
- Payment and Ticketing.

In the 'Service Initiation' scenario NFC is used in a way similar to RFID. The NFC device reads the ID code or the saved data from a tag and uses them in many different ways.

In this case the NFC reader can be a fixed terminal or a mobile device, while the Identification device can obviously be a transponder, but can also be a turned-off mobile device. In fact the ID code of the internal tag of NFC phones can be read even if the device is off, allowing for example the use of the phone as an electronic key.

The information retrieved from the transponder (stored data or UID code) can be simply read and displayed, can be used to set up a connection (in this case data can be a URL or a phone number) or can be used for access control in the same way as RFID keys.

One example of this kind of applications can be the 'smart poster'. In this application an NFC tag is located near an informative point: the user brings an NFC phone near the tag, reads an URL stored on it and uses it to connect to the Internet site providing the information requested.

The 'Peer-to-Peer' category is something totally different from RFID systems. In this kind of applications a two-way communication is set up between two devices working in Active mode.

If the amount of data to be exchanged is not too large this can be done using directly the NFC channel. If the amount of data is too big (for example an image), NFC channel can be used to set up another wireless connection (Bluetooth, Wi-Fi) in a way totally invisible to the user, and then send the data through this connection.

In this case NFC is used exclusively to set up the connection. For example in an Internet Point the user can get the Wi-Fi settings touching a specific hot-spot with the NFC terminal and then transfer them, also with NFC, to the device to be connected to Internet.

The last scenario 'Payment and ticketing' is currently the most studied due to high the interest of many banking companies in this technology [1].

The idea is to turn a mobile phone in an electronic wallet or in an electronic credit card. While nowadays a card can be used for a single payment function, with NFC will be possible to collect many different functions on a single multimodal platform.

As we told before NFC is implicitly safe due to its short ranges.

The possible payment operations can be divided in two main groups: micro-payments and macro-payments.

Micro-payments are represented by the electronic wallet. An amount of virtual money is loaded onto the phone and the user can pay various services like tickets or car parks simply bringing the phone next to payment terminal.

Macro-payments can be a little more complicated because they necessarily involve the collaboration with banks. In this case the phone will replace the Credit Cards or Bancomats in payment operations working with POS system. In macro-payments it's mainly used the identification capability of NFC.

### 3.3   NFC Standards and Organizations

The high interest in NFC technology has brought many companies, coming from very different business areas, to join together into an organization called NFC Forum [4].

In particular the Forum is composed by manufacturers of devices, developers of applications and financial institutions. Among the most important we can cite Hewlett-Packard, Microsoft, Sony, Texas Instruments, Nokia, Motorola, Samsung, IBM, MasterCard, Visa and AT&T, but the most important fact is the the Forum has been joined by companies from all over the world.

The forum has the main purpose to promote the introduction of NFC technology in common applications on a worldwide scale and tries to do this by proposing standard-based specifications, interoperable solutions, and providing stable frameworks for application development.

Being basically an RFID technology, NFC systems are compliant with the ISO14443 standard for proximity cards used for identification.

Next to this specific standards for NFC have been developed. In particular the following standards have been published:

- ISO18092: *Near Field Communication - Interface and Protocol - 1 (NFCIP-1)*: this standard basically specifies the modulation schemes, coding, transfer speeds and frame format of the RF interface of NFC systems;
- ISO21481: *Near Field Communication - Interface and Protocol - 2 (NFCIP-2)*: this standard specifies the mechanism to detect and select the communication mode out of the possible NFC communication modes;
- ISO22536: *Near Field Communication Interface and Protocol (NFCIP-1) - RF Interface Test Methods*: this standard defines the test methods for the RF-interface of NFC systems;
- ISO23917: *NFCIP-1 - Protocol Test Methods*: this standard complements the previous one and specifies the protocol tests;
- ISO28361: *Near Field Communication Wired Interface (NFC-WI)*, this standard specifies the digital wired interface between two components, a Transceiver and a Front-End, including the signal wires, binary signals, the state diagrams and the bit encodings for three data rates.

## 4   NFC in Sanitary Environments

Even if currently the most studied application field regards the payment scenarios, the versatility of NFC technology encourages its use in many other fields not directly involved in commercial operations.

In particular the availability of common mobile phones equipped with NFC hardware encourages its wide use in applications with high interactivity and security requirements.

In sanitary environments the assistance operations usually involve a high number of actors, including doctors, nurses and obviously patients. Moreover all these operations require a high level of reliability in the interaction among different people because a wrong medical prescription can cause big problems to the patients while in the assistance operations in Emergency Rooms 5 minutes can make the difference between life of death.

The informative and communicative capabilities of NFC can be therefore used to reduce all the errors or misunderstandings deriving from wrong data exchanges or from slow information access.

If the NFC Forum predictions on NFC devices diffusion will demonstrate to be true in the next years many people all over the world will be provided with NFC devices and, due to the standardization of the technology, simply downloading a specific software on their phone they will be able to access at some services modeled on the needs of their particular working environment.

The final idea is to turn the personal mobile phone of doctors and nurses into a multipurpose device joining the personal uses with the common working activities of these categories of people. The phone may therefore become a key, a pen, a sheet of paper, an organizer and, obviously, a communication device.

In the specific case of sanitary environments we studied and developed a set of applications deriving from the first two applicative fields, i.e. *Service initiation* an *Peer-to-Peer* because the *Payment and Ticketing* field is evidently more distant than these two from the specific needs of an hospital.

Next to this, with a vast diffusion of NFC phones, these devices could also become personal electronic case histories for common people, with vital information stored inside the internal memory. In emergency situations the fast reading of these data could speed up the assistance operations and reduce potentially fatal errors. Therefore in the last section we describe an end-user solution, moving the NFC phone from the sanitary operators to the patients.

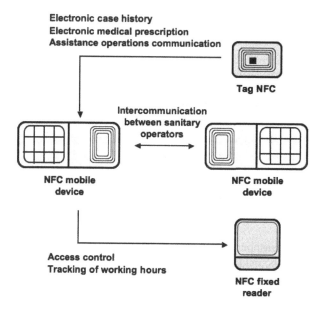

**Fig. 2.** NFC applications in sanitary environments

## 4.1  *Service Initiation* Applications in Sanitary Environments

*Service Initiation* doesn't mean only identification or access control. In this applicative field are included all the applications which use transponders as contactless memories to be read from mobile or fixed devices.

What can be read is not only the UID code of the transponder, but also the information stored on it, which can be codified in order to use in the best possible way the small amount of memory available, and can be ciphered in order to make it unreadable from external users.

Taking care of patients means the performing of a lot of different activities, from the care of wounded people in the Emergency Room to the administration of medicines. Some of these activities can be made safer an faster with the use of NFC phones as reader devices.

In particular the following applications have been studied and developed:

- Access control: entrance in reserved areas and tracking of the working hours of employees;
- Electronic case history;
- Electronic communication of assistance operations;
- Electronic medical prescriptions.

The first application is evidently the closest to the original target of RFID, the automatic Identification. In fact, as we said before, NFC phones can be identified from NFC readers exactly as transponders, even when they are turned off.

This allows to use phones as keys to obtain access to restricted areas in the same way as magnetic strip cards. Moreover, the phone can be used to record the accesses and the exits from the working place, a function currently managed in Italian hospitals with cards.

In this case the main advantage derives from the incorporation of these functions on the mobile device, preventing the employees to bring with them the requested cards.

The Electronic Case History application derives from a previously studied similar system based on RFID technology [5].

In many assistance operations the quickness of intervention is one of the most important features to be achieved. Next to this there are some vital information concerning the patient that must be provided to the doctor before performing the intervention: data like blood type, allergies or vaccines are fundamental to avoid dangerous errors.

One of the best ways to ensure a correct and fast reading of the information is to store them into an electronic device and retrieve it with a multimedia support. In this sense the idea is to provide to patients an electronic bracelet concerning mainly of an NFC transponder, where data can be stored and read quickly with a phone in case of need.

Obviously NFC transponder cannot store large amounts of data. As a consequence a severe choice has to be made between strictly vital information, which has necessarily to be provided to the doctor and will be then kept directly on the bracelet, and less important data, like for example the chronology of all the medical interventions made on a patient, which can be stored into a remote database and then retrieved only on request using Wi-Fi, GPRS or UMTS connection [2].

Even if our application has been studied and tested using MIFARE 4K transponders, which provide a 4kbyte EEPROM, we studied an organization of the information to make it storable also into 256 byte transponders.

This is the bytes subdivision chosen in our application:

- 40 bytes for first name and family name;
- 16 bytes for the tax code, whose decoding allows the recover of birth date, place of birth and gender;
- 10 bytes for the sanitary code;
- 8 bytes for allergies: every byte is a flag indicating the presence/absence of the corresponding allergy;
- 8 bytes for vaccines: every byte is a flag indicating the presence/absence of the corresponding vaccine;
- 8 bytes for infectious diseases: every byte is a flag indicating the presence/absence of the corresponding disease;

- 8 bytes for various information like blood type (1 byte encoding), HIV positivity or smoking/non smoking;
- 100 bytes with a specific codification for the 10 last hospitalizations. Every hospitalization is codified with 10 bytes where the first to bytes are a code corresponding to the specific medical ward, the third and the fourth indicate the kind of intervention and the last six bytes are the date;
- 50 bytes for textual accessory information.

Once the NFC device is brought in proximity of the bracelet, it reads the string of bytes, decodes it and shows the data into a graphic interface. It also gets the UID code of the tag in order to use it as the identification mean to retrieve the information stored inside a remote database.

The application which manages the reading, decoding, recover and reproduction of the information is a Java Midlet, also incorporating simple read and write tag functions.

These functions will be used in the third application, in which transponders are seen as the means to communicate the type of assistance operations performed on a specific patient.

The idea is that every bed in the hospital will be equipped with a transponder. Every time that an intervention is made, before performing the operations the doctor or the nurse reads on the transponder the previous treatment made to the patient, in order to avoid dangerous errors like repetitions in drug administration.

After the assistance has been made the operation performed is written on the transponder with the NFC phone, in the same way as are written SMSs, in order to inform the ones who will make the following intervention.

In the last application we studied the realization of an electronic medical prescription where sheets of paper are replaced with transponders.

The technological infrastructure is the same of the former task, allowing then to incorporate its functionalities into the same software developed for all the other applications.

Instead of writing the medical prescription on a paper, the doctor writes it into a transponder in the same SMS way of the other application. Usually the length of a prescription is less than 200 characters so usually there are no problems of shortage of memory. Anyway 4kbyte cards can be used in order to avoid any risk of incomplete descriptions.

Once the prescription has been written the transponder can be brought to the chemist who can read it with its own NFC phone.

In many cases handwritten medical prescriptions are very difficult to be read: such a kind of system will help to avoid errors in the administration of medicines, making safer the treatment of patients.

## 4.2 *Peer-to-Peer* Applications in Sanitary Environments

As described before the *Peer-to-Peer* scenario is the one in which the most interesting innovations are introduced.

The phone is in fact used as a short range communication device allowing people to exchange data simply bringing close to them their mobile phones. The absence of

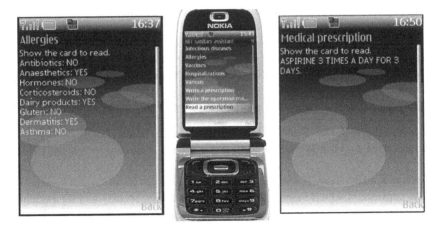

**Fig. 3.** Electronic case history - allergies, menu and electronic medical prescription

direct interaction from the users to set up the communication channel makes very fast the beginning of the process of information exchange.

This functionality can be joined with the system described in the previous section in order to make the information about patients retrieved with the reading of the electronic bracelets transferable among the people operating around a specific patient.

In fact, usually many different people attend at the assistance of people in a hospital, and in some cases the operations can last even some hours.

Every time that a doctor replaces another one he can download the electronic case history simply bringing his phone close the one of the former doctor.

This operation can be performed for every kind of data that has to be exchanged between two different employees.

For example a doctor can download from a remote server a particular information about a patient like an image of an x-ray and then he can send it to all the other doctors operating with him simply using the NFC channel.

As described before, if the amount of information to be exchanged is too big to be transferred with NFC, this technology can be used to set up a connection in a fast way with other technologies like Bluetooth which can then be used as the real transfer channel.

### 4.3  An End-User Application: The Telephone as an Electronic Case History

Next to all the applications to be implemented on the terminals belonging to sanitary operators, the chance to find NFC technology on common phones will considerably enlarge the range of possible users, opening the way to new kinds of applications.

The internal 4kbyte MIFARE card of the phones can be used to store quite a large amount of data, part of whom can be reserved to vital information to be provided in case of emergency situations. In particular, the organization and the access to data is strictly modular, with single sectors of the internal memory that can be protected and made unchangeable once written.

As we described before in the section dedicated to the electronic case history with an adequate coding a lot of information can be stored in a relatively small portion of memory, keeping enough free space for other kinds of operations (In fact the internal memory is often used to store data like electronic tickets, electronic credit, etc...). In particular, looking at the subdivision studied for the electronic case history we can see that strictly vital information (allergies, vaccines, infectious diseases and other information like blood type and HIV positivity) can fill less than 50 bytes. Next to this, in particular situations the amount of memory reserved for sanitary information can be extended to include specific data about particular phisical caracteristics or disabilities of the specific user (e. g. the presence of pacemakers, particular malformations of organs, etc...).

The versatility of such a kind of system derives also from the fact that the information stored inside the internal card can be read with every kind of NFC reader device, even if the phone is turned off or the battery is low.

In a typical emergency scenario, during the assistance operations the doctor can read the information about the patient directly with his own NFC phone (obviously equipped with a decoding software similar to the one used for the reading of the electronic bracelets), simply bringing it close to the phone of the patient, making safer and faster all the following medical aid actions.

## 5    NFC Devices Used in the Realization of the System

The devices used in the realization of this study are a mobile phone provided with an ad-hoc development kit, a set of transponders and a reader.

The mobile phone used is a Nokia 6131 NFC: it's a particular version of the common 6131 phone with the NFC circuitry embedded. Nokia sells this product in a particular 'Experimental' version, which allows the developers to use all the features of the telephone without the need to validate their applications.

Moreover Nokia provides a Java Development Kit with all the APIs needed to realize NFC Midlet and with a Simulation Environment which can be used to test the applications but can also work as an emulated phone interacting through a NFC reader, connected to the computer, with an external phone.

The Simulation Environment also provides some different simulated transponders in order to fully test several kinds of applications.

In the real testing phase we used MIFARE transponders of different types. In particular we used a MIFARE UltraLight transponder, which is very thin (like a sheet of paper), but it has a small memory (only 512 bit), and a MIFARE 4k, which is thicker but can store up to 4kbytes of data. Both these transponders are ISO 14443 compliant.

Finally we used a common RFID reader to study the access control application because, once turned off, the phone is read like a common RFID transponder and no specific hardware is requested.

## 6    Conclusions and Future Work

The aim of this work was to show how much NFC technology can increase the quality of service in situations very distant from the standard payment scenarios.

The performances of NFC systems have proved to be extremely high for what concerns the reliability of the communication channel.

During the reading of the transponder no error has been recorded, especially in the case of the electronic case history, which involved the largest amount of data to be moved.

In addiction the reading and decoding of the data took only some fractions of second, making the information immediately available when the phone was brought in proximity of the transponder.

The modularity of this system, due to the fact that single tasks can be easily integrated into the same software, makes it upgradeable simply adding the new functions to the underlying structure.

Among new applications to be studied we can find some ones deriving from already existing RFID systems, but now made simpler from the presence of mobile phones. In particular we can list the following fields:

– The identification between mother and baby with the use of transponders located onto the cradles or with electronic bracelets.
– The tracking and identification of blood sacks.
– The assistance operations inside the ambulance, which can be made safer using the traditional GPRS and UMTS connections available on mobile phones.

Next to these in many other situations NFC can be used to increase the level of reliability without the need to enlarge the number of devices to be bought. This fact is extremely important because in many cases the introduction of a new service is prevented due to the high costs.

The only expenses to be made to introduce such a system are basically the ones for the mobile phone and for the readers to control the accesses, because currently transponders can be bought with few euro cents, making this expense virtually unimportant.

If the NFC Forum valuations will demonstrate to be true, in the next years we will see a vast diffusion of NFC phones, making the applications described before simply downloadable and executable on the personal phones of doctors and nurses, without the need to buy specific devices.

## References

1. Smart Card Alliance: Proximity Mobile Payments: Leveraging NFC and the Contactless Financial Payments Infrastructure. Smart Card Alliance (2007)
2. Bing, B.: Wireless local area networks. Wiley, Chichester (2002)
3. Finkenzeller, K.: RFID handbook fundamentals and applications in contactless smart card and identification. Wiley, Chichester (2003)
4. NFC Forum: Near Field Communication and the NFC Forum: The Keys to Truly Interoperable Communications. NFC Forum (2007)
5. Benelli, G., Parrino, S., Pozzebon, A.: Rf-health: an integrated management system for a hospital based on passive rfid technology. In: EHST 2008, 2nd International Workshop on e-Health Services and Technologies. INSTICC Press (2008)
6. ECMA International: Near Field Communication White Paper. ECMA International (2007)
7. Innovision Research and Technology Near Field Communication in the real world: turning the NFC promise into profitable, everyday applications. Innovision Group (2007)

# Electronic Health Records: An Enhanced Security Paradigm to Preserve Patient's Privacy

Daniel Slamanig and Christian Stingl

Carinthia University of Applied Sciences, Austria
School of Medical Information Technology · Healthcare IT & Information Security
{d.slamanig,c.stingl}@cuas.at

**Abstract.** In recent years, demographic change and increasing treatment costs demand the adoption of more cost efficient, highly qualitative and integrated health care processes. The rapid growth and availability of the Internet facilitate the development of eHealth services and especially of electronic health records (EHRs) which are promising solutions to meet the aforementioned requirements. Considering actual web-based EHR systems, patient-centric and patient moderated approaches are widely deployed. Besides, there is an emerging market of so called personal health record platforms, e.g. Google Health. Both concepts provide a central and web-based access to highly sensitive medical data. Additionally, the fact that these systems may be hosted by not fully trustworthy providers necessitates to thoroughly consider privacy issues. In this paper we define security and privacy objectives that play an important role in context of web-based EHRs. Furthermore, we discuss deployed solutions as well as concepts proposed in the literature with respect to this objectives and point out several weaknesses. Finally, we introduce a system which overcomes the drawbacks of existing solutions by considering an holistic approach to preserve patient's privacy and discuss the applied methods.

## 1 Introduction

In recent years many countries have installed eHealth initiatives and working groups in order to develop strategies to harmonize the exchange of health related information using the Internet. A central aspect of eHealth is called the electronic health record (EHR) which integrates all relevant medical information of a person and represents a lifelong documentation of the medical history. Considering implementations of EHRs, one of the most critical factors of success is the protection of the patient's privacy, which is clearly reflected in surveys concerning such systems [1]. Additional issues are, that the EHR is patient-centered and that the patient herself moderates her EHR [2]. This means, that solely the patient is able to grant access to her medical data to other parties and to nominate delegates respectively. Moreover, the study in [2] shows, that patient access to their electronic records needs to be developed in partnership with the patients.

In this paper we are considering web-based EHR systems which enable persons to manage their EHRs by means of a web-application.

Besides the classical aspects such as standardization, interoperability, time and location independent access, legal frameworks, etc. basic requirements for web-based EHRs

A. Fred, J. Filipe, and H. Gamboa (Eds.): BIOSTEC 2009, CCIS 52, pp. 369–380, 2010.

are the possibility for a patient to freely define the structure, e.g. folders, of the EHR and to share medical data or even the entire EHR with other parties, e.g. physicians, relatives. Furthermore, we will introduce and discuss further aspects and concepts which are especially applicable for patient-moderated and web-based EHRs. This consideration for EHR architectures helps to improve the trustworthiness of an EHR system by means of privacy preserving techniques.

The organization of the rest of the paper is as follows: In the subsequent section 2 we motivate why privacy issues are important in this context and define potential attacker models. We propose and discuss basic management, security and privacy relevant objectives, which need to be considered when designing privacy enhanced EHRs in section 3. Based on these objectives in section 4 we investigate systems which are explicitly designed to provide EHRs as well as systems which are not originally developed to manage EHRs but may also be used to store health related data. In the subsequent section 5 we introduce a novel system for EHRs called PE$^2$HR (Privacy Enhanced EHR) that realizes the security and privacy objectives defined in section 3. In the remaining section 6 we will provide a conclusion and discuss some future aspects.

## 2    Motivation

Web-based EHR solutions are growing in popularity and provide mechanisms that enable people to comfortably manage their medical data online and furthermore help to improve the quality of care, by means of availability of all relevant medical data. However, in context of the Internet we are confronted with a set of attacker models and threats that need to be considered when designing systems which deal with sensitive data. For example, the lack of anonymity in Internet communication and the trustworthiness of providers hosting such EHR systems may constitute serious problems. Additionally, there exists a phenomenon often denoted as privacy myopia. This means, that people often are not aware of dangers related to the "ordinary" use of the Internet, e.g. that they reveal IP-addresses which may enable third parties to link several actions and may even enable them to identify the physical users behind their computers. Furthermore, users often give away their data very easily, without reflecting on potentially negative consequences.

### 2.1   Attacker Models

In order to identify realistic threats, we need to consider potential attackers firstly, which are listed subsequently.

- Client Intruder: This kind of attackers breaks into the client computer, e.g. they compromise a host by means of malware, e.g. trojan horses, and thus obtain access to sensitive data which should solely be available to its owner. This threat is highly realistic and indeed one of the major problems in context of the Internet and should be reduced by means of trusted computing [3] in the future.
- Eavesdropper: An eavesdropper compromises or owns a subset of nodes of the communication infrastructure and thus is able to inspect messages which are routed

over them. This attacker is usually able to link communicating parties by means of addresses used by the communication media, e.g. IP-addresses, and to fully access content data of messages in absence of transmission encryption, e.g. SSL/TLS.

- Curious insider or server intruder: This attacker is in possession of administrative privileges and thus has full access to log information as well as content data of the EHR system. Clearly, in analogy to the client intruder an external adversary may also compromise a host which provides server applications.

In this work we are especially interested in the latter two types of attackers, since in absence of trusted computing and remote attestation in particular, it is very hard to decide whether client hosts are "secure" or not.

In context of eavesdropping, even if transmission encryption is used, the attacker may be able to obtain information about users which are communicating with an EHR system. Thus, an attacker is able to derive communication patterns. In order to hide these patterns from external parties, e.g. how often a user logs into the system, we need to investigate communication anonymity (cf. section 5). Additionally, there exist potential attackers which are located at the provider of the EHR system. As a recent study [4] shows, more than 50% of attacks against information system are conducted by insiders, and hence this threat is highly realistic. Consequently, the trustworthiness of service providers is in question and we need to design and investigate concepts, which protect content as well as metadata, e.g. relationships between users and data objects, from insiders in order to improve their privacy.

## 3   Security and Privacy Objectives for EHR'S

In this section we propose and discuss security and privacy objectives which need to be considered when designing privacy enhanced EHRs. Before discussing the main objectives in detail, we want to point out that there are some basic functionalities which need to be provided by any serious EHR system. These functionalities comprise amongst others the availability of the system, the confidentiality of data transmitted between users and the system, e.g. by using SSL/TLS, and the integrity of stored data.

A major criterion for the choice of these objectives was the influence of a user on the degree of privacy. More precisely, considering a single objective, the privacy protection depends mainly on the provider or on the user, subject to the applied method. For example, confidentiality can be realized on the one hand by means of client-side encryption by the user, e.g. XML-Encryption, and on the other hand by means of server-side encryption by the provider, e.g. database encryption.

Subsequently we will provide a brief discussion regarding our objectives.

### 3.1   Anonymity

Anonymity is often referred to as the property of being not identifiable with respect to a set of actions inside a group of people, the so called anonymity set [5]. Intuitively the degree of anonymity is the higher, the larger the anonymity set is and the more uniformly the actions are distributed within this set. Considering an EHR system we can define anonymity at three different levels.

- Anonymous Communication: Anonymous communication is guaranteed, if an observer is not able to determine a communication relationship between two communicating parties by means of information revealed by the communication channel.
- Sender- and Receiver-anonymity: A communication relationship between a sender and receiver provides sender-anonymity, if the receiver is not able to identify the sender by means of received messages. The receiver-anonymity can be defined analogously.
- Data Anonymity: A system provides data anonymity, if data stored in the system of the receiver and related to a specific sender can not be linked to the sender by the receiver and any other person. This means, that even an insider of a system is not able to establish a relationship between a patient and her related data. Consequently, serious measures to provide data anonymity must be realized by the sender.

### 3.2  Authentication

If access to a system is restricted to an authorized set of users, the systems needs to establish the identity of a potentially authorized user. This is in general realized by means of authentication mechanisms. In authentication or identification protocols the holder of an identity usually claims a set of attributes including an identifier and interactively proves the possession of the claimed identity to a verifier based on these attributes. This identifier is usually unique within a specific context, e.g. within a single application, and thus enables the system to link an authentication and subsequent transactions to a specific user.

The above mentioned authentication mechanisms obviously establishes a one-to-one mapping between an authenticating user and her identity. In contrast, anonymous authentication (cf. section 5.2) provides mechanisms such that the before mentioned one-to-one mapping does not longer exist. In particular, an authenticating user proves solely her membership in a group of authorized users, whereas the verifier is not able to decide which member of this group actually conducted the authentication.

### 3.3  Authorization

Authorization is the concept of providing access to resources only to users who are permitted to do so. Usually the process of authorization takes place after a successful authentication. Mainly, authorization concepts in systems are realized by means of discretionary access control (DAC) strategies, e.g. access control lists (ACLs) or mandatory access control (MAC) strategies, e.g. role based access control (RBAC) [6]. In the former case, the access policies for objects are specified by the their owners whereas in the latter case access policies are specified by the system. The before mentioned strategies represent only a selection methods which exist in the literature and in practice today [7], however they share one important commonality. These strategies are implemented by means of application layer mechanism which can be easily bypassed by insiders of the respective system. This means, that medical data may be accessed by unauthorized insiders circumventing the access control system.

## 3.4   Confidentiality

In context of EHRs, which provide web-based access to health related data, methods to guarantee confidentiality are essential. In general, confidentiality is realized by means of encryption and relies on the protection of the respective cryptographic keys. For the key management we distinguish two widely used techniques. On the one hand cryptographic keys are solely accessible to the user and all cryptographic operations are performed by the user's client (client-side encryption). On the other hand, the system at the provider is responsible for the key management and all cryptographic operations (server-side encryption). From the security point of view, client-side encryption provides a higher level of confidentiality, since content data is not available in plaintext at any time at the provider. Thus, the number of feasible attacks can be reduced significantly. It must be mentioned, that in this paper we are not considering proprietary encryption-software, e.g. PGP, that may be applied by the user in addition to mechanism provided by the EHR system. Thus, if we speak of client-side encryption we mean that these mechanism are provided by the EHR system, however, all cryptographic operations are performed by the client.

## 3.5   Deniability

One major advantage of a web-based EHR is the time and location independent availability of health related data. However, under certain circumstances this can be disadvantageous and even result in dramatic consequences for the user. In order to demonstrate this problem, we will provide an example which is in our opinion highly realistic. Assume that a person was suffering from a cardiovascular diseases, diseases of the musculoskeletal system, drug addiction or a mental diseases like (burn out) depression. This disease is in detail documented in the EHR of the person, however does not affect the current state of health of that person significantly. It is obvious, that this potentially compromising information should solely be available for persons who are directly involved in the medical treatment process of the person and are authorized to access these data. Exactly this is in our opinion a basic requirement of an EHR. Assuming that an EHR can provide this requirement, there is absolutely no way for unauthorized persons to gain access to these data by means of the EHR system. Unfortunately, since the EHR is accessible via the Internet, the user herself may be "motivated" or even enforced to present this compromising data during a job interview or an insurance contract conclusion. This is what we call the disclosure attack (cf. [8]). It must be emphasized, that a person which has presented her EHR under such circumstances is not able to prove this involuntary disclosure to another party later on. Thus, there exists the need for mechanisms to plausibly deny the existence of highly compromising information (e.g. a cured burn out depression) from people who dot not need to know that information at all.

## 3.6   Unlinkability

Unlinkability of items of interest means that relations between items, which a priori exist, can not be identified through pure observation of the system [5,9]. A system containing $n$ users provides perfect unlinkability, if the relation of an object and a user $u_i$ exists with probability $p = 1/n$ for all objects. Hence, an insider of the system can not gain any information on links between users and objects by means of solely observing the system. In context of EHR systems we additionally need to consider static as well as dynamic aspects of unlinkability. The static aspect covers data objects which are stored in the EHR system and unlinkability is provided, if an insider at the system is not able to establish links between data objects and users significantly better than guessing. The dynamic aspect covers user's interaction with the system. In particular, we raise the question whether an eavesdropper or an insider at the system is able to link instances of authentication protocols and transactions with the system together and to a specific user. Clearly, a system which does not provide dynamic unlinkability also negatively influences static unlinkability aspects. For example, if a transactions represents an access to a specific data object and this transaction can be linked to a specific user then the data object can be linked to the user, although the system may provide static unlinkability.

### 3.7 Data Structure

The data structure defines primarily the logical structure of the EHR, e.g. a hierarchy of users, folders, subfolders and documents. In contrast to the objectives discussed above, the data structure does not contribute to the overall security of the system. However, the data structure is the main component regarding the usability and efficiency of the EHR system. Moreover, the degree of structuredness massively influences the concepts used for authorization. Especially, when considering sharing of health data between several parties the absence of any structure complicates standardized mechanisms for this task.

We want to point out that the data structure always contains information on the entire system (metadata), which may potentially reduce the degree of privacy, e.g. unlinkability, authorization, provided by the system. For example, if a system holds pseudonymized or even anonymized medical documents then authentication information and information provided by the data structure can be used to identify the holder of the respective documents.

## 4    Investigation of EHR Systems

Prior to presenting our proposed solution in detail, we investigate systems which are either explicitly designed to provide web-based EHR functionality, i.e. Personal health record platforms, PIPE, and systems which may be used by people to "build" their own web-based EHR, i.e. virtual hard disks. This investigation is based on the security objectives introduced in section 3 and summarized in Table 1. In the remainder of this section we provide a discussion of the above mentioned systems with respect to the security objectives.

### 4.1    Virtual Hard Disk

This approach provides remote storage space which is accessible via the Internet, e.g. cloud storage services. It offers the user the possibility to realize arbitrary folder structures for data management, usually by means of the WebDav protocol. Some typical representatives are iDisk (Apple), Xdrive (AOL) and Gspace (Google). Considering

**Table 1.** This table provides an analysis of the virtual hard disk concept, Google Health, the PIPE system and our system introduced in section 5 regarding the objectives defined in section 3. Thereby C denotes client-side, S server-side, SC a combination thereof, × denotes that this feature is not provided and ? denotes that it is not clear whether this feature can be provided. In context of authentication T denotes traditional and A anonymous authentication.

| Objective | Virtual Hard Disk | Google Health | PIPE | PE$^2$HR |
|---|---|---|---|---|
| Communication anonymity | O | O | O | C |
| Sender anonymity | × | × | × | C |
| Data anonymity | × | × | ? | C |
| Authentication | T | T | T | A |
| Authorization | S | S | C | C |
| Confidentiality | ? | ? | ? | C |
| Deniability | × | × | × | C |
| Unlinkability | × | × | ? | C |
| Data structure | C | SC | × | SC |

these products one can conclude that authentication is realized by means of traditional authentication methods, e.g. username/password, and authorization is realized by means of DAC or MAC strategies. The data structure is solely determined by the user. In general methods to guarantee confidentiality are not integrated into these products, however server-side encryption could be established. Methods for the remaining objectives are not yet implemented in the above mentioned products, to the best of the authors knowledge.

### 4.2   Personal Health Record Platforms

Personal health record platforms provided by major vendors such as Google (Google Health) or Microsoft (Health Vault) are growing in popularity. For example, Google provides a patient centric and patient moderated system, that offers the possibility to organize health related data of a person and moreover enables the integration of third party services offered by physicians, hospitals and pharmacies. As above, the same arguments hold for these systems, but the server takes influence on the data structure by demanding certain aspects of this structure.

### 4.3   PIPE

The architecture PIPE (pseudonymization of information for privacy in eHealth) [10], [11] focuses on the management of person related medical documents in a pseudonymized fashion. In this context pseudonymization means the replacement of personal information by a document related specifier which is not linkable to the holder of the document. The authorization for pseudonymized documents is realized by means of a hierarchical structure of cryptographic keys and encrypted document related specifiers. Both can be shared with other users. One major aspect of the architecture is the establishment of key-backup mechanisms based on threshold secret sharing schemes. The latter aspect positively influences the availability of cryptographic keys, however, has no positive impact on security and privacy properties considered in this paper.

Authentication against the system is realized by applying digital signatures, whereas it must be pointed out that the protocol used in [10,11] easily allows impersonation attacks. However, the authentication solely provides access to the encrypted master cryptographic key of the user, which will subsequently be decrypted at the client. In context of an EHR the above mentioned pseudonymization is in our opinion impractical when using different medical document types, because they always contain unstructured narrative text passages (even CDA Level 3). Consequently, the pseudonymization has to be performed manually, and hence the effort would be unacceptable in our opinion. Additionally, data anonymity can not be guaranteed when not using anonymous authentication. For example, if a patient integrates a pseudonymized medical finding into the system, then this document will be linkable to the authenticating party (the patient) by an insider. The same argument holds for the objective unlinkability. Confidentiality is not taken into the consideration in [10,11] due to the pseudonymization. The remaining objectives are not provided by this architecture. The data structure can not be analyzed seriously due to the facts that a simple conceptual model is used and further crucial details are not published. Additionally, it must be emphasized that the "pseudonymization" of more complex conceptual models requires more sophisticated methods [12].

## 5   Privacy Enhanced EHR

In this section we discuss methods that help to preserve the patient's privacy in context of EHRs with respect to the security and privacy objectives defined in section 3.

### 5.1   Anonymous Communication

Mechanisms that provide anonymity and unlinkability of messages sent over a communication channel are denoted as anonymous communication techniques and have been intensively studied in recent years, see [13] for a sound overview. There are several implementations available for low-latency services like Web browsing, e.g. Tor [14], JAP [15] as well as high-latency services like E-Mail, e.g. Mixminion [16].

These anonymous communication channels help to improve the privacy of users in context of eavesdroppers and curious communication partners. Especially, regarding the latter one anonymity can be preserved if electronic interaction does not rely on additional identifying information at higher network layers, i.e. the application layer. For example, a user who queries a public web page using an anonymous communication channel may remove all identifying information from higher network layers and thus can stay anonymous. However, if service providers offer their services only to authorized sets of users (e.g. subscription-based services, closed communities), they require identification which in general takes place at higher layers by means of authentication mechanisms. In the latter context anonymity can however be preserved by means of anonymous authentication.

### 5.2   Anonymous Authentication

Anonymous authentication aims to provide a somewhat paradoxical solution to enhance user's privacy in context of authentication. It provides mechanisms such that a user is able to prove membership in a group $U' \subseteq U$ of authorized users $U$, whereas the verifier does not obtain any information on the identity of the user. Clearly, anonymous communication systems are a prerequisite for providing anonymity in context of anonymous authentication.

A naive approach to realize anonymous authentication would be to give a copy of a secret $k$ to every user $u \in U$, which could be used in conjunction with a traditional authentication scheme. Obviously, the revocation of a single user $u_i$ would result in a reinitialization and thus in reissuing a new secret $k'$ to every user $u \in U \setminus u_i$. Hence, this approach is far from being practical. Improved techniques for anonymous authentication were explicitly treated in [17,18,19] and can additionally be derived from group signatures [20,21, etc.], ring signatures [22,23, etc.] or similar concepts as (deniable) ring authentication [24], whereas the latter class of signatures and authentication schemes is preferable to group signatures in the context of large groups, since they can be generated "ad hoc" without depending on an explicit setup phase.

### 5.3 Authorization and Confidentiality

In contrast to strategies implemented by means of application layer mechanism (see section 3.3), there exists the possibility to realize DAC based on cryptographic tokens [25]. In particular, all resources are encrypted by their owners (client-side encryption) which hold the corresponding secret keys and are stored encrypted in the system. In particular, if a user grants access to another user she provides a cryptographic token to this user. This cryptographic token represents the secret key to the respective data object, encrypted with the public key of the grantee. In other words, access control based on cryptographic tokens is realized by means of the ability of authorized persons to properly decrypt resources, e.g. content data. This access control strategy provides a serious advantage in comparison with the before mentioned strategies, namely insiders are solely able to bypass the access control by breaking the underlying cryptographic primitives (symmetric resp. asymmetric cryptosystem). Additionally, it can be used to realize fine-grained access, i.e. to single data objects, in contrast to approaches which allow to share all data or no data with other persons, e.g. physicians, [26].

### 5.4 Pseudonymization

Pseudonymization of person related data $(u, x)$ is the process of replacing every person identifier $u$ for example by the value $nym = E_k(u)$, where $E_k$ is an appropriate symmetric encryption function with a corresponding secret key $k$. Since $k$ is kept secret it is practically impossible to invert $E_k(\cdot)$ without the knowledge of $k$ and thus compute $u$ given the value $nym$. However, a person which is in possession of $k$ can easily compute $D_k(nym) = u$ using the corresponding decryption function $D_k(\cdot)$. Hence $(nym, x)$ can not be linked to $u$ anymore.

We realize pseudonymization by letting every user $u_i$ choose a second identifier $P_{u_i}$ uniformly at random, i.e. a pseudonym [27]. This pseudonym is used by her to identify data objects that are related to her. In order to prevent the linkage between a user and a pseudonym, the pseudonym is solely stored in an encrypted fashion, $E_{k_{u_i}}(P_{u_i})$, in the EHR system. The unlinkability holds, since $P_{u_i}$ is independently chosen from $u_i$

and furthermore $E_{k_{u_i}}(P_{u_i})$ can only be inverted by $u_i$, who holds the corresponding key $k_{u_i}$, which may be derived from a appropriately chosen password or passphrase defined by $u_i$. This simple example can be generalized to pseudonymize an arbitrary conceptual model [12]. The resulting pseudonymized conceptual model provides data anonymity and static unlinkability (concerning any observer of the system) and it enables highly efficient implementations. Additionally, by means of anonymous authentication the system provides dynamic unlinkability. Furthermore, it must be emphasized that the conceptual model can be defined by the EHR system and users are able to freely create their own structures with respect to the conceptual model.

### 5.5   Multiple Identities

However, there still exists the precarious disclosure attack which can lead to the disclosure of the complete EHR of a person. Therefore we need a measure to provide plausible deniability in a cryptographically provable sense. As countermeasure we propose the use of so called multiple identities [28]. In this context multiple identities can be described by means of dividing the EHR of a person into so called sub-identities (see Figure 1).

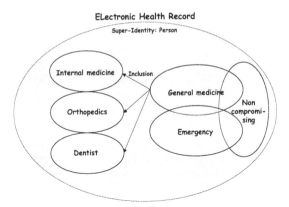

**Fig. 1.** Multiple identities

A user can assign a subset of her EHR to each of these sub-identities. Thereby, these subsets do not need to be disjoint. Subject to the person, the medical data are presented to, the user is able to choose one of her sub-identities, e.g. a special prepared, non compromising one and consequently opens the assigned subset of medical data. Hence, a user can hide sensitive data in a special sub-identity in order to prevent disclosure of medical data. However, one drawback of this approach is that passwords which are used to derive the cryptographic keys for the respective identities need to be chosen independently of each other. More precisely, there must not be any relationship between passwords which clearly could be computed by an adversary too. However, we assume that in practice the number of identities and passwords respectively will be moderate.

Furthermore, this concept additionally provides the possibility to create so called super-identities which can hold several sub-identities. Thus, super-identities can be used to comfortably manage the respective sub-identities.

## 6 Conclusions

In this paper we have discussed security and privacy aspects which are especially of relevance in context of web-based EHR systems. Following these objectives, we have investigated deployed solutions for EHR systems as well as concepts discussed in the literature. Regarding these systems we can conclude that either patient's need to fully rely on the trustworthiness of the provider of an EHR system (e.g. Google Health) or there exist methods to bypass the implemented security concepts. This is due to the fact that security concepts are focusing solely on specific aspects and not the entire EHR system. However, in our opinion it is absolutely necessary to consider all the relevant security objectives in order to provide an adequate protection of the patient's privacy. Nevertheless, by applying specific methods, e.g. anonymous authentication, one is confronted with additional challenges. For example, in context of anonymous authentication it is apparently impossible to realize user specific resource limits. Actually, we are working toward a solutions for the aforementioned problem based on blind signature techniques. Furthermore, we are investigating strategies for the choice of anonymity sets for anonymous authentication. The latter aspect is crucial, since not appropriately chosen strategies may lead to unwanted user identification [29].

## References

1. HI: Harris Interactive, Survey on Medical Privacy (2004),
   http://www.harrisinteractive.com/news/newsletters/healthnews/
   HI_HealthCareNews2004Vol4_Iss13.pdf
2. Pyper, C., Amery, J., Watson, M., Crook, C.: Access to Electronic health records in primary care – a survey of patients' views. Med. Sci. Monit. 10(11), 17–22 (2004)
3. TCG: Trusted Computing Group (2008),
   http://www.trustedcomputinggroup.org
4. CSI: Computer Crime and Security Survey 2007, Computer Security Institute (2007),
   http://www.gocsi.com/forms/csisurvey.jhtml
5. Pfitzmann, A., Köhntopp, M.: Anonymity, Unobservability, and Pseudonymity - A Proposal for Terminology. In: Workshop on Design Issues in Anonymity and Unobservability, pp. 1–9 (2000)
6. Win, K.T.: A review of security of electronic health records. HIM J. 34(1), 13–18 (2005)
7. Bishop, M.: Computer Security: Art and Science. Addison-Wesley, Reading (2002)
8. Stingl, C., Slamanig, D.: Privacy Enhancing Methods for eHealth Applications: How to Prevent Statistical Analyses and Attacks. Int. J. Business Intelligence and Data Mining 3, 236–254 (2008)
9. Steinbrecher, S., Köpsell, S.: Modelling Unlinkability. In: Dingledine, R. (ed.) PET 2003. LNCS, vol. 2760, pp. 32–47. Springer, Heidelberg (2003)
10. Riedl, B., Neubauer, T., Goluch, G., Boehm, O., Reinauer, G., Krumboeck, A.: A Secure Architecture for the Pseudonymization of Medical Data. In: Proceedings of the The Second International Conference on Availability, Reliability and Security (ARES 2007), pp. 318–324. IEEE Computer Society, Los Alamitos (2007)

11. Riedl, B., Grascher, V., Neubauer, T.: A Secure e-Health Architecture based on the Appliance of Pseudonymization. Journal of Software 3, 23–32 (2008)
12. Slamanig, D., Stingl, C., Lackner, G., Payer, U.: Preserving Privacy in a Web-based Multiuser-System (German). In: Horster, P. (ed.) Proceedings of DACH-Security 2007, pp. 98–110. IT-Verlag (2007)
13. Danezis, G., Diaz, C.: A Survey of Anonymous Communication Channels. Technical Report MSR-TR-2008-35, Microsoft Research (2008)
14. Dingledine, R., Mathewson, N., Syverson, P.: Tor: The Second-Generation Onion Router. In: Proceedings of the 13th USENIX Security Symposium, p. 21 (2004)
15. Federrath, H.: Privacy Enhanced Technologies: Methods, Markets, Misuse. In: Katsikas, S.K., López, J., Pernul, G. (eds.) TrustBus 2005. LNCS, vol. 3592, pp. 1–9. Springer, Heidelberg (2005)
16. Danezis, G., Dingledine, R., Mathewson, N.: Mixminion: Design of a Type III Anonymous Remailer Protocol. In: SP 2003: Proceedings of the 2003 IEEE Symposium on Security and Privacy, Washington, DC, USA, pp. 2–15. IEEE Computer Society, Los Alamitos (2003)
17. Boneh, D., Franklin, M.: Anonymous authentication with subset queries. In: Proc. of the 6th ACM conference on Computer and communications security, pp. 113–119 (1999)
18. Lindell, Y.: Anonymous Authenticaion. Whitepaper Aladdin Knowledge Systems, 2007 (2007), http://www.aladdin.com/blog/pdf/AnonymousAuthentication.pdf
19. Schechter, S., Parnell, T., Hartemink, A.: Anonymous Authentication of Membership in Dynamic Groups. In: Franklin, M.K. (ed.) FC 1999. LNCS, vol. 1648, pp. 184–195. Springer, Heidelberg (1999)
20. Ateniese, G., Camenisch, J., Joye, M., Tsudik, G.: A Practical and Provably Secure Coalition-Resistant Group Signature Scheme. In: Bellare, M. (ed.) CRYPTO 2000. LNCS, vol. 1880, pp. 255–270. Springer, Heidelberg (2000)
21. Chaum, D., van Heyst, E.: Group signatures. In: Davies, D.W. (ed.) EUROCRYPT 1991. LNCS, vol. 547, pp. 257–265. Springer, Heidelberg (1991)
22. Rivest, R.L., Shamir, A., Tauman, Y.: How to Leak a Secret. In: Boyd, C. (ed.) ASIACRYPT 2001. LNCS, vol. 2248, pp. 552–565. Springer, Heidelberg (2001)
23. Dodis, Y., Kiayias, A., Nicolosi, A., Shoup, V.: Anonymous Identification in Ad Hoc Groups. In: Cachin, C., Camenisch, J.L. (eds.) EUROCRYPT 2004. LNCS, vol. 3027, pp. 609–626. Springer, Heidelberg (2004)
24. Naor, M.: Deniable Ring Authentication. In: Yung, M. (ed.) CRYPTO 2002. LNCS, vol. 2442, pp. 481–498. Springer, Heidelberg (2002)
25. Stingl, C., Slamanig, D., Rauner-Reithmayer, D., Fischer, H.: Realization of a Secure and Centralized Data Repository (German). In: Horster, P. (ed.) Proceedings of DACH Security 2006, pp. 32–45. IT-Verlag (2006)
26. Demuynck, L., Decker, B.D.: Privacy-Preserving Electronic Health Records. In: Dittmann, J., Katzenbeisser, S., Uhl, A. (eds.) CMS 2005. LNCS, vol. 3677, pp. 150–159. Springer, Heidelberg (2005)
27. Chaum, D.: Untraceable electronic mail, return addresses, and digital pseudonyms. Commun. ACM 24, 84–90 (1981)
28. Slamanig, D., Stingl, C.: Privacy Aspects of eHealth. In: Proceedings of the The Third International Conference on Availability, Reliability and Security (ARES 2008), pp. 1226–1233. IEEE Computer Society, Los Alamitos (2008)
29. Slamanig, D., Stingl, C.: Investigating Anonymity in Group Based Anonymous Authentication. In: Svenda, P. (ed.) The Future of Identity in the Informaton Society - Challenges for Privacy and Security. IFIP International Federation for Information Processing. Springer, Boston (2009)

# Augmented Feedback System to Support Physical Therapy of Non-specific Low Back Pain

Dominique Brodbeck[1], Markus Degen[1], Michael Stanimirov[1], Jan Kool[2], Mandy Scheermesser[2], Peter Oesch[3], and Cornelia Neuhaus[4]

[1] School of Life Sciences, University of Applied Sciences, Northwestern Switzerland, Muttenz, Switzerland
{dominique.brodbeck,markus.degen,michael.stanimirov}@fhnw.ch
http://www.fhnw.ch/hls
[2] Department of Health, Zurich University of Applied Sciences, Winterthur, Switzerland
{jan.kool,shem}@zhaw.ch
[3] Clinic Valens, Center for rehabilitation, Valens, Switzerland
p.oesch@klinik-valens.ch
[4] University Children's Hospital UKBB Basel, Basel, Switzerland
cornelia.neuhaus@ukbb.ch

**Abstract.** Low back pain is an important problem in industrialized countries. Two key factors limit the effectiveness of physiotherapy: low compliance of patients with repetitive movement exercises, and inadequate awareness of patients of their own posture. The Backtrainer system addresses these problems by real-time monitoring of the spine position, by providing a framework for most common physiotherapy exercises for the low back, and by providing feedback to patients in a motivating way. A minimal sensor configuration was identified as two inertial sensors that measure the orientation of the lower back at two points with three degrees of freedom. The software was designed as a flexible platform to experiment with different hardware, and with various feedback modalities. Basic exercises for two types of movements are provided: mobilizing and stabilizing. We developed visual feedback - abstract as well as in the form of a virtual reality game - and complemented the on-screen graphics with an ambient feedback device. The system was evaluated during five weeks in a rehabilitation clinic with 26 patients and 15 physiotherapists. Subjective satisfaction of subjects was good, and we interpret the results as encouraging indication for the adoption of such a therapy support system by both patients and therapists.

## 1 Introduction

Low back pain (LBP) is a very frequent condition in industrialized countries leading to high burden to the health care system [1,2]. In the majority of cases, no patho-anatomical causes for the complaints are present and they are classified as non-specific LBP [3]. Active and supervised movement exercises are effective in reducing pain and restoring function in non-specific LBP [4,5]. However, there are several factors that limit the effectiveness of such exercises and lead to poor therapy outcomes:

- proprioception is inadequate for the lumbar spine
- movement exercises are difficult to learn

A. Fred, J. Filipe, and H. Gamboa (Eds.): BIOSTEC 2009, CCIS 52, pp. 381–393, 2010.

– exercises require a lot of repetition to be effective
– appropriate feedback requires continuous presence of a physical therapist

Thus, the main problems are insufficient patient motivation to comply with exercise regimes, as well as the inability of the patients to exercise independently. The aim of the Backtrainer project is to address these limitations of current conservative therapy, by automatically monitoring movement exercises in real time, generating a motivating, game-like visual feedback, and storing patients' performance data for later assessment.

The system has to be easy to use and simple enough, in order for it to be adopted by physical therapists, and so that it can be used by patients at home. Although monitoring of back movements and postures in laboratory settings, but also at the workplace, has gained a lot of attention in the past, there are no technical solutions available that would fulfill these requirements.

This leads to the following research questions:

– What is a minimal sensor configuration that still produces enough data, in order to generate valid feedback for movement exercises?
– What basic set of exercises needs to be supported by the software to cover most common therapy needs?
– In what form can the software provide intuitive feedback about posture to focus patients' attention on the aspects that are relevant for the task?
– How can the software motivate patients to use the system?

The problem was approached in three steps: (i) identify the minimal sensor configuration, (ii) build a prototype system, and (iii) evaluate the system in a controlled clinical setting.

## 2  Background

There have been numerous research projects addressing the measurement of the kinematics of the spine. High-end systems usually use optical sensors, with either passive or active markers that are glued onto the skin. These systems have a very high precision, often in the sub-millimeter range, and are used by the video-game and movie industry for motion capturing and animation generation. Other measurement setups consist of sensors based on ultrasonic waves to measure distances, electro magnetic tracking systems [6,7], resistance strain gauges, optical fibres [8], or sensors that use a combination of accelerometer, compass, and gyroscope to determine the orientation of the sensor [9].

Each sensor technology has its advantages but also its drawbacks. Optical systems can often capture only in a small area, and there must be an unobstructed line of sight from the camera to the markers. Systems based on electromagnetism can be influenced by the environment (i.e. training machines built of iron or steel).

Research has also been done to compare the accuracy of skin mounted (glued) sensors with radiographs or magnetic resonance imaging used to identify the actual positions of vertebrae [10,11]. The results show that positions and motions of the skin markers can be used as an estimate for the calculation of the position and orientation of the underlying vertebrae. It should also be noted that the goal of our system is not diagnosis, but to support physiotherapists whose work is also based on surface observations.

Some recent publications focus on camera based systems. In [12] a system based on low cost cameras has been described. This system works with infrared cameras and a headband, with mounted reflectors that can be used to recognize and identify different sitting postures. Although the low cost camera approach seems promising, it is not applicable in our situation because of the line of sight problem, which applies to other optical systems as well.

The therapy system that was proposed in [13] also uses cameras to track gestures from stroke patients and, in addition, provides augmented feedback in a game-like setup. Another promising approach, using multi-modal feedback in neural rehabilitation, is described in [14]. This system uses visual as well as auditory means for feedback on functional, task-oriented exercises.

## 3   Identification of Sensor Configuration

In order to satisfy the requirement of a simple and easy to handle system, we had to identify the minimal configuration of sensors that would still produce enough data in order to classify movement quality and performance, and to generate valid feedback. The hypothesis was that it is sufficient to measure the orientation of the lumbar spine at two points with three degrees of freedom.

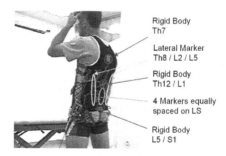

Rigid Body
Th7

Lateral Marker
Th8 / L2 / L5

Rigid Body
Th12 / L1

4 Markers equally
spaced on LS

Rigid Body
L5 / S1

**Fig. 1.** 22 Markers were positioned in the region of the lumbar spine (LS), and tracked with a high-precision optical motion capturing system, in order to analyze the motion of the lower back during standard movement tasks

We used an optical motion capturing system (Optotrak Certus from NDI) with 22 infrared LEDs positioned on the subjects' back (see Figure 1). The precision of this system is in the sub-millimeter range, and data was collected with a sampling rate of 30 Hertz. In addition to the telemetry data, we captured the movement of the subjects on video at a rate of 15 frames per second. This provided us with a setup that is sufficiently overdetermined, to allow us to simulate and virtually evaluate many different potential sensor configurations.

Then we identified a set of movement tasks that are commonly used for the management or in the diagnosis of low back pain:

1. Posture Correction (while seated)
2. Range of Motion (flexion/extension)

3. Range of Motion (lateral flexion)
4. Range of Motion (rotation)
5. Stabilization of lumbar spine during knee extension
6. Lifting test (light load: 7.5 kg)
7. Lifting test (heavy load: 7.5-45 kg individually)

We recruited 22 healthy subjects between the ages of 18 to 55 years, and asked them to perform the above sequence of tasks. This produced a total of 2.4 hours of video synchronized with 5.6 million 3D positions for all of the markers. Standard statistical tools are too limited to explore and analyze this large body of information. We therefore developed a highly interactive visualization application that would allow us to visually explore the data, and experiment with different scenarios.

The visualization application consists of multiple coordinated views that simultaneously show:

– x, y, z positions of all the markers projected onto the three planes of the body (sagittal, coronal, transversal)
– 3D view of the marker positions in space
– synchronized video frames
– missing values (due to line of sight problems)
– derived values (e.g. spatial angles between pairs of markers, distances)

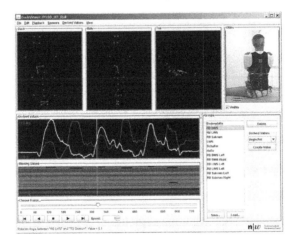

**Fig. 2.** The interactive visualization application uses multiple coordinated views to show different aspects, in order to explore and analyze all the data that was recorded

The software allows to plug in any number of algorithms that produce derived values from selected marker positions. We developed algorithms to measure angles and distances, as well as various projection methods onto the body planes. The videos were examined visually by physiotherapy experts, and marked up at points in time where subjects lost their ability to stabilize the lumbar spine during the exercises.

This analysis revealed the following results:

- The shape of the lumbar spine can be quantified by measuring the angle between vertebrae Th12/L1 and L5/S1.
- The difference between correct, stable movements and unstable, potentially dangerous movements can be identified in the data, and corresponds to visual assessment by physical therapists.
- The necessary data to evaluate the quality of movement can be acquired by the use of skin surface sensors.

Based on these results we confirmed our hypothesis that for our intended goal to support therapy of low back pain, it is sufficient to measure the orientation of the lumbar spine at two points, with three degrees of freedom.

## 4   The Backtrainer System

### 4.1   System Overview

The Backtrainer prototype consists of two inertial sensor modules, capable of measuring the three rotational degrees of freedom. Each of the two modules is positioned on the patients' back using an elastic band (Figure 8). Motion data from the two sensor modules are transmitted to a PC.

On the PC, a therapy software receives the signals and reconstructs the movements of the lumbar spine. The software further consists of a patient database, and a set of movement exercises that can be configured for the individual patient. The software supports the therapist in instructing complex movements, and allows the patient to exercise independently in a motivating, game-like environment, and document therapy activities and progress.

Figure 3 shows an architectural overview of all components of the system.

### 4.2   Inertial Sensors

Of the sensing technologies discussed in section 2, we chose to use inertial sensors, because of their low cost and simplicity of use. We first considered to develop a sensor based on accelerometers, gyroscopes, and magnetometers on our own, but recently, many commercial sensors of this type that match our specifications have become available, and therefore we decided against it.

We used sensors from two different manufacturers. One was a wireless system (InertiaCube3 from Intersense), and another one was a system where the sensors are connected to the PC with USB cables (MotionNode from GLI Interactive). We abstracted the interfacing of the sensors in the software, so that we are able to switch systems easily.

### 4.3   Software

The software is separated into several layers to guarantee high flexibility and extensibility (Figure 3). The core application layer is responsible for the hardware abstraction,

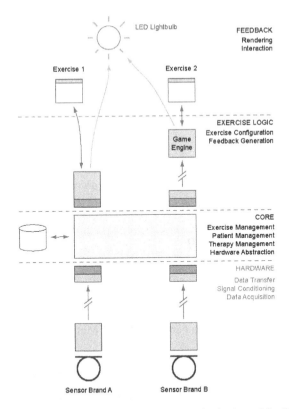

**Fig. 3.** Overview of the Backtrainer system. Hardware, exercise logic, and feedback are abstracted into separate layers to guarantee high flexibility and extensibility.

as well as basic patient, exercise and therapy management. The exercises are abstracted into a separate layer, which makes it possible to easily add any number of exercises to the system. This exercise logic layer takes care of configuration, generation of feedback, and determination of the exercise success level. Rendering and interaction are done in the feedback layer, to support the use of different feedback modalities.

In order to design the user interaction, a task analysis was performed based on scenarios of future therapy sessions. The resulting use cases are summarized in Fig. 4.

Analysis of the use cases and iterative prototyping together with therapists, lead to three areas of interaction that were then implemented in the following main screens:

**Device.** Management of the hardware (Initialization, calibration, monitoring).
**Personae.** Management and selection of therapist and patients, monitor therapy (exercise performances, access to historical data).
**Exercises.** Selection and performance of exercises (configuration, personalization, feedback).

These three screens reproduce the main work flow executed in a therapy session:

1. Initialize the device and make sure the sensors are mounted correctly and deliver signals.

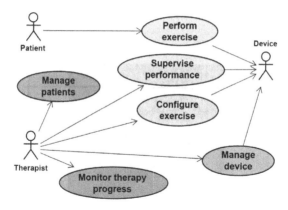

**Fig. 4.** Use cases resulting from the task analysis

2. Select therapist and patient, review previous sessions and decide on the exercises to be performed in the current session.
3. Select, perform and evaluate exercises.

## 4.4 Exercises

The exercises are the main concept within the Backtrainer system, they target the training area and can be divided into the two groups of mobilizing and stabilizing exercises.

**Mobilization Exercises.** The aim of mobilization exercises is to restore the range of motion of the patient. The physiotherapist, together with the patient, set range limits which enclose the required movement range to achieve the treatment goal. This range is visualized to the patient by a white ball moving within the predefined range limits. The ball turns its color to green if the wanted limits are reached and to red if these are exceeded. This information assures him/her that motions within this range will be most effective. This is important, because it prevents patients to be overcautious or overambitious, which would result in a lower success rate of the therapy. The success level of a particular performance of the mobilizing exercise is defined as the ratio of number of times that the limit was reached to the total number of attempts.

Figure 5 shows the application window with the mobilizing exercise selected. The slider and buttons on the right allow the adjustment of exercise parameters.

**Stabilization Exercises.** The goal of stabilization exercises is to hold the lumbar spine in a given stable position, while performing movements such as "squats", lifting weight, or changing from sitting to standing. For this type of exercise, it is important to provide the patient with an augmented feedback of the posture of the lumbar spine, as proprioception of this region is typically low.

The metaphor of a "green range" was introduced, meaning that motions within this range are perfectly tolerable. This range is adjustable and allows to define the level of difficulty for the exercise.

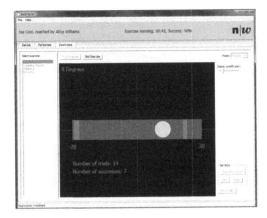

**Fig. 5.** Mobilizing feedback

Our first approach for the visualization of the lumbar spine posture used a comic like stick figure that had a bendable spine. Tests showed however that this approach worked only, if the subject was positioned exactly as the figure on the screen (i.e. standing upright). In other situations (e.g. sitting, kneeling), this concrete depiction turned out to be more confusing than helpful.

We therefore replaced the figure by a more abstract visualization of a sphere balancing on a curved convex surface. Figure 6 shows these two approaches side by side. When the patient leaves the "green range", then the ball slides down on one side of the surface, and changes its color from yellow gradually to red, depending on how much the current measured angle is away from the green angle.

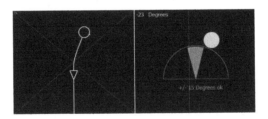

**Fig. 6.** Stabilizing feedback: The first version (left) used a figure like feedback but was then replaced by a more abstract visualization using a ball balanced on a bump

**Game Exercise.** As described in section 1, movement exercises with many repetitions are key to therapy effectiveness. Patient compliance with repetitive movement exercise regimes is problematic though. We developed a simple game with the aim of enhancing motivation and compliance.

In the game, the patient controls a bat at the bottom of the window (Figure 7). The bat can be moved from left to right, according to the difference in rotational angles between the upper and lower sensor modules. Which of the three rotational planes should be used for the mapping, can be freely chosen by the therapist, depending on the therapy goal.

The task in this game is to catch the balls that are rolling from the back toward the front at randomly chosen offsets from the center line. The ratio of caught vs. missed balls is displayed as a score, and the final score is recorded as the success level of this exercise in the patients' therapy history.

**Fig. 7.** Game feedback: A ball (grey) is rolling from the back toward the player and has to be caught by a bat (blue)

This simple game is implemented in the Backtrainer software using basic "OpenGL" commands, but the software design explicitly addresses the possibility to integrate more sophisticated games that can be based on so called game-engines (Figure 3). Game engines facilitate the implementation of 3D games providing elaborate functions for the realization of virtual worlds, avatars and leveling systems.

**Lightbulb Feedback.** The movements that a patient executes while performing a movement exercise often involve a rotation of the line of vision of the patient (e.g. rotation of the upper body in the coronal plane), or the line of vision is not oriented horizontally (e.g. patient lying on the chest). In such situations, it is unpractical to provide a visual feedback on a computer screen with a fixed position. It would be helpful to provide feedback that is not directional in nature, but embedded in the environment in an ambient way.

Auditory feedback is one possibility. However, physiotherapy for low back patients is often performed in clinical therapy settings, where many patients exercise in the same room at the same time. In such a setting, auditory feedback can be distracting and confusing.

Research in the field of human computer interaction suggests the use of "ambient displays". Ambient displays are abstract, peripheral displays that visualize information on the *periphery of a user's attention* [15]. This approach can also be adapted to physiotherapy, to let the patient concentrate on the exercises and nevertheless perceive feedback about the movements.

For the Backtrainer system, we built a very simple but effective component to generate an "ambient feedback" using a modified color changing LED Lightbulb (Figure 8). The Lightbulb was instrumented with a USB-interface in order to be able to change the color from software for any combination of the two primary colors red and green.

The Lightbulb matches the ball metaphor that is used for the feedback shown in the software on the computer screen. Some exercises can be performed with only the Lightbulb feedback (e.g. stabilizing exercise) while others (e.g. game exercise) use the

Lightbulb as an additional feedback modality. Apart from being an ambient feedback system for the patients, the bulb can also be used by physiotherapists coaching several patients at the same time. The therapist can observe the emitted color from a distance and intervene to adapt exercise settings, i.e. if there are too many "misses" while playing the game exercise.

Figure 8 shows a therapy situation with the Backtrainer mounted on the patients back, the LED Lightbulb and the physiotherapist. The patient is performing a stabilizing exercise (weight lifting while stabilizing the lumbar spine).

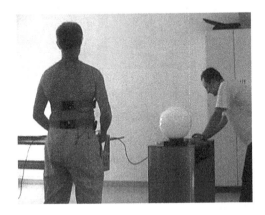

**Fig. 8.** The Backtrainer system in a therapy situation

## 5    Clinical Evaluation

In order to evaluate the Backtrainer system, we performed an exploratory study in a controlled clinical setting. The goal of this study was to use the system under realistic conditions, in order to evaluate the practicality in various situations. In particular the goals were:

- Evaluation of the practicality (expenditure of time, handling, ease of use) and application in a therapeutic setting.
- Evaluation of acceptance of the system by therapists and patients.
- Evaluation of the feedback produced by the system.

The study was performed in the rehabilitation clinic Valens, Switzerland. 15 physical therapists were given a short introduction to the Backtrainer system. After this introduction, the therapists were free to use the system at their discretion during their regular therapy sessions with patients that suffer from chronic back pain or that have undergone back surgery. Both, patients and therapists were asked to fill out questionnaires asking about their subjective satisfaction with the system. In addition, the software is equipped with extensive logging to provide objective data about the use of the system.

The study was performed for a period of five weeks. The following discussion of the results focuses on those aspects that relate to the software of the Backtrainer system. In depth analysis of the overall study is subject of further research.

During the period under investigation, the system has been in use between 1 and 2 hours per day. 26 patients have performed therapy sessions with the Backtrainer. They have performed a total of 248 exercises distributed as follows: mobilizing (17%), stabilizing (51%), game (32%). The average duration of an exercise was about 2 minutes with the 90% quantile at 5 minutes (8 minutes for the game exercise). Since the game exercise can be considered a mobilizing exercise, the distribution between the two types of exercises is just about half and half.

We have received filled-out questionnaires from 23 patients and 12 therapists. Tables 1, 2, and 3 show the answers to selected questions. In general, the feedback provided by the Backtrainer is considered helpful by both patients and therapists, and it matches the visual observations of the therapists. Patients generally indicate that when training individually, it is more fun to train with the Backtrainer. We will have to further investigate into the reason for the few less favorable answers though. Therapists indicate that they can use the Backtrainer to measure therapy progress. The evaluation of the software by the therapists is favorable.

**Table 1.** Patient questionnaires (n = 23). Encoding: 4 Completely agree, 3 Agree somewhat, 2 Disagree somewhat, 1 Completely disagree.

| Question | Encoded Answers [# answers] | | | | Statistics | |
|---|---|---|---|---|---|---|
| | 4 | 3 | 2 | 1 | $\bar{x}$ | $\sigma$ |
| Feedback from the Backtrainer was helpful | 15 | 8 | 0 | 0 | 3.7 | 0.5 |
| Feedback is easy to understand | 16 | 5 | 2 | 0 | 3.6 | 0.6 |
| Makes it easier to perform exercises on your own | 10 | 7 | 2 | 2 | 3.2 | 1.0 |
| Independent training is more fun with the Backtrainer | 9 | 9 | 3 | 1 | 3.2 | 0.8 |

**Table 2.** Physical therapist questionnaires (n = 12). Encoding: 4 Completely agree, 3 Agree somewhat, 2 Disagree somewhat, 1 Completely disagree.

| Question | Encoded Answers [# answers] | | | | Statistics | |
|---|---|---|---|---|---|---|
| | 4 | 3 | 2 | 1 | $\bar{x}$ | $\sigma$ |
| Feedback from the Backtrainer was helpful for patients | 8 | 4 | 0 | 0 | 3.7 | 0.5 |
| The feedback matched my observations | 6 | 3 | 2 | 0 | 3.4 | 0.8 |
| It is possible to measure therapy progress with the Backtrainer | 3 | 7 | 2 | 0 | 3.1 | 0.6 |
| Patients are motivated to use it | 6 | 6 | 0 | 0 | 3.5 | 0.5 |

**Table 3.** Software evaluation by physical therapists (n = 12). Encoding: 4 Completely agree, 3 Agree somewhat, 2 Disagree somewhat, 1 Completely disagree.

| Question | Encoded Answers [# answers] | | | | Statistics | |
|---|---|---|---|---|---|---|
| | 4 | 3 | 2 | 1 | $\bar{x}$ | $\sigma$ |
| Overall impression is very good | 0 | 12 | 0 | 0 | 3.0 | 0.0 |
| The software fulfills its task | 4 | 8 | 0 | 0 | 3.3 | 0.5 |
| The software matches my expectations and habits | 4 | 7 | 1 | 0 | 3.3 | 0.6 |

## 6    Conclusions and Future Work

We have developed a system to support physiotherapy of low back pain. We found that this is possible by measuring the orientation of the lumbar spine at two points, with three degrees of freedom. The dynamic behavior and accuracy of commercially available inertial sensors are good enough for this application.

The layered software architecture that we developed has proven effective in integrating different hardware systems, and providing the flexibility needed for prototyping. The separation of code for the exercise concept into core, logic, and feedback is useful for providing various feedback modalities in a modular way.

The task analysis and the division of the interface into three areas, resulted in a system that was easy to use and matched the workflow of a typical therapy session. The distribution of the exercise types performed during the clinical evaluation shows that the distinction between mobilizing and stabilizing movements is fundamental and well reflected in practice.

The abstract visual feedback that we designed was considered helpful. Ambient feedback in the form of the Lightbulb proved to be a very useful addition to the computer screen in a real-life therapy setting. With regard to feedback and motivation, the study only provided some first hints though. Participants liked the game and the feedback, but there is further systematic investigation needed to answer our research questions in this area.

The usage patterns and the answers from the questionnaires from the clinical evaluation provide a stable foundation for the further development of the Backtrainer system. Since therapists did not have to follow a fixed protocol, but were free to use the Backtrainer when they saw a need, we interpret the numbers that we found as encouraging indication for the adoption of such a therapy support system.

The above results suggest future work for the elaboration of the system in several areas:

– Evaluate other feedback modalities (e.g. auditory, tactile), and other ambient devices (e.g. light emitting floor panels). Also the use of wearable 3D-Displays (Eyegoggles) will be evaluated.
– Explore a telemedical scenario in which the exercises performed by the patient at home can be evaluated by geographically distant physiotherapists to provide guidance for the patients.
– Evaluate "virtual reality" game-like feedback modalities to further raise motivational factors.

**Acknowledgements.** This work was supported by funding from the Swiss Innovation Promotion Agency CTI. The authors would like to thank Hocoma AG for their support.

# References

1. van Tulder, M., Koes, B., Bombardier, C.: Low back pain. Best Pract. Res. Clin. Rheumatol. 16, 761–775 (2002)
2. van Tulder, M., Koes, B., Bouter, L.: A cost-of-illness study of back pain in The Netherlands. Pain 62, 233–240 (1995)
3. Deyo, R., Weinstein, J.: Low back pain. N Engl. J. Med. 344, 363–370 (2001)
4. Abenhaim, L., et al.: The role of activity in the therapeutic management of back pain: Report of the Paris International Task Force on Back Pain. Spine 25, 1S–33S (2000)
5. Hayden, J., van Tulder, M., Tomlinson, G.: Systematic review: strategies for using exercise therapy to improve outcomes in chronic low back pain. Ann. Intern. Med. 142, 776–785 (2005)
6. Van Herp, G., Rowe, P., Salter, P., Paul, J.P.: Three-dimensional lumbar spinal kinematics: a study of range of movement in 100 healthy subjects aged 20 to 60+ years. Rheumatology 39, 1337–1340 (2000)
7. Jordan, K., Dziedzic, K., Mullis, R., Dawes, P.T., Jones, P.W.: The development of three-dimensional range of motion measurement systems for clinical practice. Rheumatology 40, 1081–1084 (2001)
8. Dunne, L.E., Walsh, P., Smyth, B., Caulfield, B.: Design and evaluation of a wearable optical sensor for monitoring seated spinal posture. In: ISWC, pp. 65–68 (2006)
9. Lee, R.Y., Laprade, J., Fung, E.H.: A real-time gyroscopic system for three-dimensional measurement of lumbar spine motion. Medical Engineering & Physics 16, 817–824 (2003)
10. Zhengyi, Y., Griffith, J., Leung, P., Pope, M., Sun, L., Lee, R.: The accuracy of surface measurement for motion analysis of osteoporotic thoracolumbar spine. In: 27th Annual International Conference of the Engineering in Medicine and Biology Society, IEEE-EMBS 2005, pp. 6871–6874 (2005)
11. Mörl, F., Blickhan, R.: Three-dimensional relation of skin markers to lumbar vertebrae of healthy subjects in different postures measured by open MRI. European Spine Journal 15, 742–751 (2006)
12. Engels, L., Leloup, T., Warzée, N.: Imaging technologies for avoiding back pain at work. In: Proceedings First Symposium of the IEEE/EMBS Benelux Chapter, pp. 235–238 (2006)
13. Sucar, L.E., Leder, R.S., Reinkensmeyer, D.J., Hernndez, J., Azcrate, G., Casteeda, N., Saucedo, P.: Gesture therapy - a low-cost vision-based system for rehabilitation after stroke. In: Azevedo, L., Londral, A.R. (eds.) HEALTHINF (2), INSTICC - Institute for Systems and Technologies of Information, Control and Communication, pp. 107–111 (2008)
14. Huang, H., Ingalls, T., Olson, L., Ganley, K., Rikakis, T., He, J.: Interactive multimodal biofeedback for task-oriented neural rehabilitation. In: 27th Annual International Conference of the Engineering in Medicine and Biology Society, IEEE-EMBS 2005, pp. 2547–2550 (2005)
15. Mankoff, J., Dey, A.K., Hsieh, G., Kientz, J., Lederer, S., Ames, M.: Heuristic evaluation of ambient displays. In: CHI 2003: Proceedings of the SIGCHI conference on Human factors in computing systems, pp. 169–176. ACM, New York (2003)

# Multi-analytical Approaches Informing the Risk of Sepsis

Femida Gwadry-Sridhar[1], Benoit Lewden[1],
Selam Mequanint[1], and Michael Bauer[2]

[1] Lawson Health Research Institute, I-THINK Research Lab*, London, Ontario, Canada
femida.gwadry-sridhar@lhsc.on.ca
[2] Department of Computer Science, University of Western Ontario,
London, Ontario, Canada
bauer@uwo.ca

**Abstract.** Sepsis is a significant cause of mortality and morbidity and is often associated with increased hospital resource utilization, prolonged intensive care unit (ICU) and hospital stay. The economic burden associated with sepsis is huge. With advances in medicine, there are now aggressive goal oriented treatments that can be used to help these patients. If we were able to predict which patients may be at risk for sepsis we could start treatment early and potentially reduce the risk of mortality and morbidity. Analytic methods currently used in clinical research to determine the risk of a patient developing sepsis may be further enhanced by using multi-modal analytic methods that together could be used to provide greater precision. Researchers commonly use univariate and multivariate regressions to develop predictive models. We hypothesized that such models could be enhanced by using multiple analytic methods that together could be used to provide greater insight. In this paper, we analyze data about patients with and without sepsis using a decision tree approach and a cluster analysis approach. A comparison with a regression approach shows strong similarity among variables identified, though not an exact match. We compare the variables identified by the different approaches and draw conclusions about the respective predictive capabilities,while considering their clinical significance.

## 1  Introduction

Sepsis is defined as infection plus systematic manifestations of infection [6]. Severe sepsis is considered present when sepsis co-exists with sepsis-induced organ dysfunction or tissue hypo-perfusion [6]. Sepsis can result in mortality and morbidity, especially when associated with shock and/or organ dysfunction [3]. Sepsis is usually associated with increased hospital resource utilization, prolonged intensive care unit (ICU) and hospital stay, decreased long-term health related quality of life and an economic burden estimated at US $17 billion each year in the United States alone [3,5,13,15]. In Canada, there are limited data on the burden of severe sepsis; however, costs in Quebec may be as high as $73M per year [9], which contribute to estimates of total Canadian cost of approximately $325M per year. Patients with severe sepsis generally receive their

---

* Thanks to Corey Hilliard for her assistance in the preparation of this document.

A. Fred, J. Filipe, and H. Gamboa (Eds.): BIOSTEC 2009, CCIS 52, pp. 394–406, 2010.
© Springer-Verlag Berlin Heidelberg 2010

care in the ICU. A multicentre study of sepsis in teaching hospitals found that severe sepsis or septic shock is present or develops in 15% of ICU patients [2]. However, diagnosing sepsis is difficult because there is no "typical" presentation despite published definitions for sepsis [1,10]. In the Canadian Sepsis Treatment And Response (STAR) registry (mix of teaching and community hospitals across Canada), the total rate for severe sepsis was 19.0%. Of these, 63% occurred after hospitalization.

With advances in medicine there are now aggressive goal oriented treatments that can be used to help these patients [4,12,14]. If researchers were able to predict which patients may be at risk for sepsis they could start treatment before all diagnostic tests - such as blood cultures - are available and potentially result in a reduced risk of mortality and morbidity. Therefore, methods that can be developed to help with the early diagnosis of patients who either present with sepsis or develop sepsis in hospital are needed.

A variety of analysic techniques can be used to identify relationships among a set of measured variables or quantities. We hypothesized that analytic methods currently used in clinical research to determine the risk of a patient developing sepsis may be further enhanced by using multi-modal analytic methods that together could be used to provide greater precision. Researchers commonly use univariate and multivariate regressions to gather information about variables that are associated with the dependent variable, which in this case is whether the patient contracted sepsis or not. However, sometimes these models are constrained as we either use univariate analysis to guide our decision on which variable to include or rely on the literature to guide the variable selection. Earlier work had looked at the use of regression techniques to develop a linear predictive model or mortality and length of stay, but not sepsis [11].

In this paper, we consider the use of decision tree analysis and cluster analysis. Decision trees are interesting since they provide a prescriptive approach for arriving at a decision with an associated probability. In contrast, cluster analysis takes a holistic approach to partition the data into disjointed sets. We were interested in using these approaches to identify the key variables or variable sets that could potentially be used to predict the likelihood of sepsis or not having sepsis in patients admitted to an ICU.

## 2  Data in Study

We obtained data that was collected from 12 Canadian intensive care units that were geographically distributed and included a mix of medical and surgical patients [11]. Data were collected on all patients admitted to the ICU who had an ICU stay greater than 24 hours or who had severe sepsis at the time of ICU admission. Patients who were not anticipated to obtain to receive active treatment were excluded.

Hospitals collected a minimum data set on all eligible patients admitted to the ICU. This included demographic information and data about their admission, source of admission, diagnosis, illness severity, outcome and length of ICU and hospital stay. Illness severity scores were calculated using data obtained during the first 24 hours in the ICU [7,8]. All patients were subsequently assessed on a daily basis for the presence of infection and severe sepsis.

The management of severe sepsis requires prompt treatment within the first six hours of resuscitation [6]. Experts in critical care agree that the literature supports early goal-directed resuscitation which has been shown to improve survival in patients presenting to emergency rooms with septic shock [6].

## 2.1 Ethical Review, Funding and Data Ownership

The study was approved by the University of Western Ontario Research Ethics Board. Participating institutions submitted the study to their review process if local approval was required. All activities were compliant with the privacy and confidentiality practices of the participating institutions and the Federal and Provincial governments of Canada. Eli Lilly Canada provided a research grant to London Health Sciences Centre to support trial coordination, data collection, data management and data analysis. The investigators and sites retained control and responsibility for data collection, analysis and interpretation. Data is owned by and resides with London Health Sciences Centre.

## 2.2 Structure of Paper

The following two sections (3 and 4) describe our approaches to computing the decision tree and performing the cluster analysis. Section 5 describes our use of these approaches on the analysis of our data set. Section 6 then considers variables that have emerged as important from our prior regression analysis and these two approaches. We conclude with some directions for future work.

## 3    Decision Tree Approach

In data mining and machine learning, a decision tree is a predictive model, that is, a mapping from observations about an item to conclusions about its target value. In these tree structures, leaves represent classifications and branches represent conjunctions of features that lead to those classifications. The machine learning technique for inducing a decision tree from data is called decision tree learning, or (colloquially) decision trees.

A decision tree is made from a succession of nodes, each splitting the dataset into branches. Generally, the algorithm begins by treating the entire dataset as a single large set and then proceeds to recursively split the set. Three popular rules are typically applied in the automatic creation of classification trees. The Gini rule splits off a single group of as large a size as possible, whereas the entropy and twoing rules find multiple groups comprising as close to half the samples as possible. The algorithms construct the tree from the "top" down until some stopping criteria is met. In our current approach, we have used the gain in entropy in order to determine how to best create each node of the tree.

## 3.1 Entropy

In order to define information gain precisely, we used a measure commonly used in information theory, called entropy, that characterizes the "purity" (or, conversely, "impurity") of an arbitrary collection of examples. Generally, given a set S, containing only positive and negative examples of some target concept (a so-called two-class problem), the entropy of set S relative to this simple, binary classification is defined as:

$$Entropy(S) = -p_p \log_2 p_p - p_n \log_2 p_n \; . \tag{1}$$

where $p_p$ is the proportion of positive examples in $S$ and $p_n$ is the proportion of negative examples in $S$. In all calculations involving entropy we define $0 \log 0$ to be 0.

One interpretation of entropy from information theory is that it specifies the minimum number of bits of information needed to encode the classification of an arbitrary member of S (i.e., a member of S drawn at random with uniform probability).

If the target attribute takes on c different values, then the entropy of S relative to this c-wise classification is defined as

$$Entropy(S) = \sum_{i=1}^{c} -p_i \log_2 p_i \; . \tag{2}$$

where $p_i$ is the proportion of S belonging to class $i$. Note that if the target attribute can take on c possible values, the maximum possible entropy is $\log_2 c$.

## 3.2 Information Gain

Given entropy as a measure of the impurity in a collection of training examples, we can now define a measure of the effectiveness of an attribute in classifying the data. The measure we will use, called information gain, is simply the expected reduction in entropy caused by partitioning the examples according to this attribute. More precisely, the information gain, $Gain(S, A)$ of an attribute $A$, relative to a collection of examples $S$, is defined as

$$Gain(S, A) = Entropy(S) - \sum_{v \in Value(A)} \frac{|S_v|}{|S|} Entropy S_v \; . \tag{3}$$

where $Values(A)$ is the set of all possible values for attribute A, and $S_v$ is the subset of $S$ for which attribute $A$ has value $v$ (i.e., $S_v = s \in S | A(s) = v$). Note the first term in the equation for $Gain$ is just the entropy of the original collection $S$ and the second term is the expected value of the entropy after $S$ is partitioned using attribute A. The expected entropy described by this second term is simply the sum of the entropies of each subset $S_v$, weighted by the fraction of examples $\frac{|S_v|}{|S|}$ that belong to $S_v$. $Gain(S, A)$ is therefore the expected reduction in entropy caused by knowing the value of attribute $A$. Put another way, $Gain(S, A)$ is the information provided about the target attribute value, given the value of some other attribute. The value of $Gain(S, A)$ is the number of bits saved when encoding the target value of an arbitrary member of $S$, by knowing the value of attribute $A$.

The process of selecting a new attribute and partitioning the training examples is now repeated for each non-terminal descendant node in the tree, this time using only the training examples associated with that node. Attributes that have been incorporated higher in the tree are excluded, so that any given attribute can appear at most once along any path through the tree. This process continues for each new leaf node until either of two conditions is met:

1. Every attribute has already been included along this path through the tree, or
2. The training examples associated with this leaf node all have the same target attribute value (i.e., their entropy is zero).

Some of the variables in the data set are continuous variables, such as temperature. These require a somewhat special approach. This is accomplished by dynamically defining new discrete-valued attributes that partition the continuous attribute value into a discrete set of intervals. In particular, for an attribute $A$ that is continuous-valued, the algorithm can dynamically create a new Boolean attribute $A_c$ that is true if $A < c$ and false otherwise. The only question is how to select the best value for the threshold $c$. This is done by selecting values for the threshold based on the existing values of the attribute $A$ and computing the gain. The threshold $c$ that produces the greatest information gain is then chosen.

## 4   Cluster Analysis

Clustering is the classification of objects into different groups, or more precisely, the partitioning of a data set into subsets (clusters), so that the data in each subset (ideally) share some common trait - often proximity according to some defined distance measure. Data clustering is a common technique for statistical data analysis, which is used in many fields, including machine learning, data mining, pattern recognition, image analysis and bioinformatics. The computational task of classifying the data set into k clusters is often referred to as k-clustering.

Data clustering algorithms can be hierarchical. Hierarchical algorithms find successive clusters using previously established clusters. Hierarchical algorithms can be agglomerative ("bottom-up") or divisive ("top-down"). Agglomerative algorithms begin with each element as a separate cluster and merge them into successively larger clusters. Divisive algorithms begin with the whole set and proceed to divide it into successively smaller clusters. Hierarchical clustering builds (agglomerative), or breaks up (divisive), a hierarchy of clusters.

### 4.1   K-Means Clustering

The K-means algorithm assigns each point to the cluster whose center (also called centroid) is nearest. The center is the average of all the points in the cluster - that is, its coordinates are the arithmetic mean for each dimension separately over all the points in the cluster. The main advantages of this algorithm are its simplicity and speed which allows it to run on large datasets. Its disadvantage is that it does not yield the same result with each run, since the resulting clusters depend on the initial random assignments. It minimizes intra-cluster variance, but does not ensure that the result has a global minimum of variance.

### 4.2 Modified K-Means

For our cluster analysis we developed a slightly modified version of the K-means clustering algorithm. The regular algorithm was not useful as we needed to introduce a distance measure relevant to our problem. Our data are grouped into categories, which mean that the values that a variable could have are in "orthogonal planes", that is, each variable was essentially a discrete point in a space which has its own dimension and base (the dimension of each variable is simply the number of categories available). For example, if a variable can take 4 distinct values then our base for this space becomes $(1, 0, 0, 0)(0, 1, 0, 0)(0, 0, 1, 0)(0, 0, 0, 1)$. These 4 vectors are orthogonal to each other and therefore are a base for this particular space.

In a normal space each point can be expressed using a combination of its base's vector. In our case each point will only be able to take the value of one of its basis vectors. We need a measure so that the distance between a point from category 1 and category 2 is the same as from category 1 and 4 (our categories are equally spaced from each other). We use a distance measure roughly equivalent to the norm 2 of the cross product of 2 vectors. Basically, this produces a distance measure in which the distance between 2 points will be 0 if they are in the same category and 1 otherwise. This way, we can quantify the distance between points without introducing a hierarchy between categories.

Our basic algorithm then behaves pretty much as the standard K-means algorithm:

1. Initialization: We create N centroids for clusters by choosing the first data point as the first centroid. Then we calculate the distance between the other points and this centroid value and choose as second point the one furthest away (based on the distance calculated above). Then, we choose the furthest point from the first and second centroids as the third one. This is continued until we have N centroids.
2. Assign each remaining data point to the closest of the N clusters.
3. We recalculate the centres of each cluster. The centre needs to minimize the distance from itself and each point. Therefore, we minimize the distance between the centre and each point by assigning to it the most common category among the cluster to which it was assigned.
4. Then, return to Step 2, until no further changes result.

## 5 Comparative Analysis

In this section we compare the results of the decision tree and cluster analysis methods to the results obtained using regression techniques. In particular, we were interested in which variables were identified as key in determining sepsis in the approaches.

### 5.1 Regression Analysis

Previous work had focused on the analysis of the data using regression techniques. From a multivariate logistic regression a number of variables emerged as significant in the model (computed using SAS 8.2). The model was very accurate in being able to classify patients not likely to get sepsis (99%) and reasonably accurate at predicting

**Table 1.** Logistic Regression Model

| Variables | P value | Exp(B) |
|---|---|---|
| Anaerobea culture | .122 | .317 |
| Abdominal diagnosis | .000 | 15.027 |
| Blood diagnosis | .000 | 3.574 |
| Lung diagnosis | .000 | 10.360 |
| Other diagnosis | .000 | 8.492 |
| Urine diagnosis | .000 | 7.280 |
| Chest X-ray and purulent sputum | .000 | 2.756 |
| Gram negative culture | .047 | .679 |
| Gram positive culture | .001 | .533 |
| Heart rate > 90bpm | .000 | 16.933 |
| No culture growth | .000 | .103 |
| PaO2/FiO2 < 250 | .000 | 12.305 |
| pH < 7.30 or lactate > 1.5 upper normal with base deficit > 5 | .141 | 1.242 |
| Platelets < 80 or 50% decrease in past 3 days | .000 | 5.665 |
| Respiratory rate > 19, PaCO2 < 32 or Mechanical ventilation | .000 | 8.866 |
| SBP < 90 or MAP < 70 or Pressure for one hr | .000 | 9.963 |
| Abdominal culture | .259 | 1.872 |
| Blood culture | .000 | 2.311 |
| Lung culture | .724 | .932 |
| Other site culture | .614 | .869 |
| Urine culture | .100 | 1.450 |
| Temperature < 36 or > 38 | .000 | 8.246 |
| Urinary output < 0.5 mL/kg/hr | .000 | 3.166 |
| WBC > 12 or < 4 or > 10% bands | .000 | 6.281 |
| Yeast culture | .011 | .492 |
| Constant | .000 | .000 |

patients that were likely to get sepsis (66%). The variables in the model are summarized in Table 1. As one can see, the variables are not weighted by clinical significance using this method.

As indicated, we were interested in exploring whether decision tree analysis techniques and cluster analysis techniques could provide additional or at least complementary insight beyond the regression model and help establish the relative importance of the variables.

## 5.2  Decision Tree Analysis

The decision tree analysis yielded 9 distinct paths that led to a determination of sepsis with high probability. Table 2 displays the variables that appeared in at least one of these 9 paths and whether the evaluation of that variable along the path was positive ($Y$) or negative ($N$). Absence of either a $Y$ or a $N$ means that the variable did appear in the path.

**Table 2.** Decision Tree Analysis

| Variables | Path | | | | | | | | |
|---|---|---|---|---|---|---|---|---|---|
| | 1 | 2 | 3 | 4 | 5 | 6 | 7 | 8 | 9 |
| Lung diagnosis | | | N | Y | | N | N | N | Y |
| Chest X-ray and purulent sputum | | | Y | | | | | | |
| Temperature < 36 or > 38 > 10% bands | Y | | Y | N | Y | N | Y | Y | |
| WBC > 12 or < 4 or | Y | | | Y | | Y | | | Y |
| No culture growth | | | Y | | N | Y | Y | Y | Y |
| Heart rate > 90bpm | N | Y | Y | Y | Y | Y | Y | Y | Y |
| SBP < 90 or MAP < 70 or pressors for one hour | N | N | N | N | Y | Y | Y | Y | Y |
| PaO2/FiO2 < 250 | N | N | Y | Y | | | | | |
| Urinary output < 0.5 mL/kg/hr | N | N | | | | | | | |
| pH < 7.30 or lactate > 1.5 upper normal with base deficit > 5 | N | N | | | | | | | |
| Respiratory rate > 19, PaCO2 < 32 or Mechanical ventilation | | | Y | | | Y | | | Y |
| Other diagnosis | | | | | | | Y | Y | |
| Abdominal diagnosis | | | | | | | N | N | Y |
| Platelets < 80 or 50% decrease in past 3 days | Y | Y | | | | | | | |

## 5.3  Cluster Analysis

For this initial study, we chose to have the number of clusters created for Sepsis and non-Sepsis patients the same. We settled on eight clusters for each based on the approach described in Section 4. Eight clusters seemed to provide a good "clustering" of the patient records without creating clusters that were too small. Determining the optimal number of clusters is also an area for further study.

For each variable in a cluster, we computed the percentage of patients in that cluster having that variable "true". Essentially, the higher the percentage the more likely that variable being true occurred in each of the patients assigned to that cluster. For this paper, we considered clusters formed from a sample of 10,000 patients. Two Sepsis and two non-Sepsis clusters are presented in Table 3 along with a list of the variables. For each of the clusters and each of the variables, the percentage of patients in that cluster with the variable evaluating to "true" is reported. For example, for patients assigned to Sepsis Cluster 1, 93.1% of those had the variable *Temperature < 36 or > 38* as "true".

In looking at these example clusters, we can note several things:

- For both Sepsis clusters, the variables *Temperature < 36 or > 38, Heart rate > 90bpm, Respiratory rate > 19, PaCO2 < 32 or Mechanical ventilation, SBP < 90 or MAP < 70 or pressors for one hour*, occurred frequently (greater than 50%).
- Sepsis Cluster 2, also had variables *WBC > 12 or < 4 or > 10% bands, Urinary output < 0.5 mL/kg/hr, PaO2/FiO2 < 250, pH < 7.30 or lactate > 1.5 upper normal with base deficit > 5"* occurring frequently.
- For non-Sepsis Cluster 1, only the variable *No culture growth* occurred frequently, i.e., since we were considering only when variables held, this would seem to be a good indicator of a patient not developing sepsis. Note that in Sepsis Cluster 1,

**Table 3.** Example of Variables from Cluster Analysis

| Variable | Sepsis | | Non-Sepsis | |
|---|---|---|---|---|
| | Cluster 1 | Cluster 2 | Cluster 1 | Cluster 2 |
| | % patients | % patients | % patients | % patients |
| Chest X-ray and purulent sputum | 0.164 | 0.203 | 0.018 | 0.117 |
| Temperature < 36 or > 38 | 0.931 | 0.878 | 0.047 | 0.117 |
| Heart rate > 90bpm | 0.948 | 0.824 | 0.024 | 0.522 |
| Respiratory rate > 19, PaCO2 < 32 or Mechanical ventilation | 0.991 | 0.973 | 0.127 | 0.745 |
| WBC > 12 or < 4 or > 10% bands | 0.224 | 0.824 | 0.059 | 0.147 |
| SBP < 90 or MAP < 70 or pressors for one hour | 0.724 | 0.730 | 0.017 | 0.060 |
| Urinary output < 0.5 mL/kg/hr | 0.181 | 0.635 | 0.011 | 0.031 |
| PaO2/FiO2 < 250 | 0.164 | 0.716 | 0.006 | 0.028 |
| pH < 7.30 or lactate > 1.5 upper normal with base deficit > 5 | 0.147 | 0.595 | 0.005 | 0.012 |
| Platelets < 80 or 50% decrease in past 3 days | 0.060 | 0.000 | 0.000 | 0.003 |
| No culture growth | 0.276 | 0.000 | 0.966 | 0.000 |
| Gram negative culture | 0.259 | 0.095 | 0.012 | 0.745 |
| Gram positive culture | 0.267 | 1.000 | 0.009 | 0.119 |
| Yeast culture | 0.026 | 0.014 | 0.006 | 0.086 |
| Anaerobea cultrue | 0.009 | 0.014 | 0.001 | 0.007 |
| Other site culture | 0.095 | 0.203 | 0.008 | 0.114 |
| Urine culture | 0.155 | 0.095 | 0.021 | 0.235 |
| Blood culture | 0.276 | 0.351 | 0.021 | 0.154 |
| Lung culture | 0.138 | 0.378 | 0.022 | 0.468 |
| Abdominal culture | 0.026 | 0.068 | 0.003 | 0.023 |
| Lung diagnosis | 0.155 | 0.162 | 0.028 | 0.083 |
| Urine diagnosis | 0.052 | 0.041 | 0.003 | 0.026 |
| Other diagnosis | 0.103 | 0.108 | 0.010 | 0.031 |
| Blood diagnosis | 0.069 | 0.027 | 0.000 | 0.013 |
| Abdominal diagnosis | 0.069 | 0.135 | 0.009 | 0.013 |

some patients assigned to that cluster also showed *No culture growth*. For Sepsis Cluster 2, none did.

- Non-Sepsis Cluster 2 held patients with *Heart rate > 90bpm, Respiratory rate > 19, PaCO2 < 32 or Mechanical ventilation*, and *Gram negative culture holding*. Of these the *Heart rate* and the *Respiratory rate* were also important in the Sepsis clusters. This suggests that by themselves these two variables may not be valid determinants of whether a patient will or will not develop Sepsis.

## 6   Discussion

Our research focused on exploring whether decision tree and cluster analysis techniques could provide additional or at least complementary insight into a regression model approach. We initially looked at the variables that appeared to be the most *"important"* in each of the techniques. Since the way in which a variable was determined *"important"* varied with the technique, we simply ranked the variables by importance within each of the approaches. At this stage of the research we are interested in a qualitative comparison of the variables identified by the different approaches.

For the regression model, We considered any variable with a beta coefficient (Exp(B)) (see Table 1) greater than 1 as *"important"*.

For the decision tree, we were interested in some intuitive determination of variables which where *"important"*. One notion of *"important"* is based on a variable's position in a tree; intuitively the root variable is most important since it the first variable evaluated and determines which half of the tree to consider. Using this idea, then for each variable $v$, in a tree $T$, we compute its *"importance"*, $I_T(v)$, of as follows:

$$I_T(v) = \frac{\sum_{n \in T, n=v} level(n)}{c} \, . \tag{4}$$

where $n$ is any node in the tree that represents $v$, and $level(n)$ is the level in the tree, i.e., how far from the root (where a variable at the root has a value of 0), and $c$ is the number of times that $v$ occurs in the tree. In this case, the variables with the lowest overall values are considered the most *"important"*. Based on this computed *"importance"* we ranked the variables; variables with the same values received the same ranking.

For the cluster analysis approach, we considered variables that occurred with a frequency of 50% within a particular cluster to be *"important"* to that cluster. We then looked at the number of times a variables was considered *"important"* in any cluster (Sepsis or non-Sepsis). That resulting value determined which variables were deemed *"important"*.

We present a summary of the variables deemed *"important"* in each of the approaches in Table 4 based on the order of importance from the regression analysis. A $\sqrt{}$ in a column indicates that the particular variable is has been deemed *"important"* for the particular method. Nine variables appear *"important"* in all three approaches: *Heart rate > 90bps, PaO2/FiO2 < 250, Lung diagnosis, SBP < 90 or MAP < 70 or pressors for one hour, Respiratory rate > 19, PaCO2 < 32 or Mechanical ventilation, Temperature < 36 or > 38, WBC > 12 or < 4 or > 10% bands, Urinary output < 0.5 mL/kg/hr, pH < 7.30 or lactate > 1.5 upper normal with base deficit > 5*. These variables would appear to have clinical importance in predicting who might develop sepsis or not.

Other variables may also play important roles. Seven variables appeared as *"important"* in the regression and decision tree approaches, but not in the cluster analysis (at least based on the way we determined *"important"*). Two variables were deemed *"important"* in the decision tree and cluster analysis approaches but not in the regression. These may be important as well, though may also be model dependent. Finally, some variables appeared as *"important"* in only one of the approaches, while two variables (*Yeast culture, Anaerobea culture* - X's in Table 4) were not deemed "important" in any of the approaches. Interestingly, *Heart rate > 90bpm* was deemed important in all analyses.

This has implications for practice since clinicians want to apply models of risk at the bedside. Often it is not feasible to collect data on 20 variables, such as those we found

**Table 4.** Important Variables in Respective Models

| Variable | Regression | Decision Tree | Cluster |
|---|---|---|---|
| Heart rate > 90bpm | √ | √ | √ |
| Abdominal diagnosis | √ | √ | |
| PaO2/FiO2 < 250 | √ | √ | √ |
| Lung diagnosis | √ | √ | √ |
| SBP < 90 or MAP < 70 or pressors for one hour | √ | √ | √ |
| Respiratory rate > 19, PaCO2 < 32 or Mechanical ventilation | √ | √ | √ |
| Other diagnosis | √ | √ | |
| Temperature < 36 or > 38 | √ | √ | √ |
| Urine diagnosis | √ | √ | |
| WBC > 12 or < 4 or > 10% bands | √ | √ | √ |
| Platelets < 80 or 50% decrease in past 3 days | √ | √ | |
| Blood diagnosis | √ | √ | |
| Urinary output < 0.5 mL/kg/hr | √ | √ | √ |
| Chest X-ray and purulent sputum | √ | √ | |
| Blood culture | √ | | |
| Abdominal culture | √ | | |
| Urine culture | √ | √ | |
| pH < 7.30 or lactate > 1.5 upper normal with base deficit > 5 | √ | √ | √ |
| No culture growth | | √ | √ |
| Other site culture | | √ | |
| Lung culture | | | √ |
| Gram positive culture | | | √ |
| Gram negative culture | | | √ |
| Yeast culture | X | X | X |
| Anaerobea culture | X | X | X |

in the regression or cluster and models that are easy to use to either rule out patients who are not at risk of sepsis or those who are at risk would be more useful. To test any model we have to ensure that it is reliable and valid. Here we have shown with the 30 patient accuracy test that our tree is reliable and it approaches 100% validity. Our analysis also illustrates the value of multiple methods: 1) in our analysis, regressions can be used to provide a broad estimate of risk, and 2) a more precise estimate in this case can be made using a decision tree. In a separate paper, we will compare a cluster analysis approach to a decision tree. This is outside the scope of this paper. A valid approach for future research is the comparison of cluster analysis, decision trees and regression analysis. Finally, there is a lot of interest in clinically important variables and clinically important differences among variables. This concept, which is well articulated in comparing outcomes measures such as quality of life, requires future inquiry beyond traditional markers of "importance".

# 7  Conclusions

Multiple methods of analyzing clinical data provide different perspectives on assessing the risk of disease. To develop robust models researchers may want to consider regression to get a broad perspective on the risk and utilize decision trees to provide more parsimonious models.

This study has several strengths. This was a prospective observational cohort and the determination of sepsis used standard criteria. The large sample size provided a large number of variables that we could use for our analyses. Future research will now entail determining the precision of the decision tree and cluster analysis methods as well as exploring finer-grained elements of both approaches, e.g. constraining variables to be used, number of clusters. We did not have an opportunity to test other methods such as Bayesian methods or neural networks, which we hope to do in the future.

# References

1. American College of Chest Physicians/Society of Critical Care Medicine Consensus Conference Definitions for sepsis and organ failure and guidelines for the use of innovative therapies in sepsis. Critical Care Medicine 20, 864–874 (1992)
2. Alberti, C., Brun-Buisson, C., Burchardi, H., Martin, C., Goodman, S., Artigas, A., Sicignano, A., Palazzo, M., Moreno, R., Boulme, R., Lepage, E., Le Gall, R.: Epidemiology of sepsis and infection in ICU patients from an international multicentre cohort study. Intensive Care Med. 28, 108–121 (2002)
3. Angus, D.C., Linde-Zwirble, W.T., Lidicker, J., Clermont, G., Carcillo, J., Pinsky, M.R.: Epidemiology of severe sepsis in the United States: Analysis of incidence, outcome, and associated costs of care. Critical Care Medicine 29, 1303–1310 (2002)
4. Bernard, G.R., Vincent, J.L., Laterre, P.F., LaRosa, S.P., Dhainaut, J.F., Lopez-Rodriguez, A., Steingrub, J.S., Garber, G.E., Helterbrand, J.D., Ely, E.W., Fisher Jr., C.J.: Efficacy and safety of recombinant human activated protein C for severe sepsis. New England Journal of Medicine 344, 699–709 (2001)
5. Brun-Buisson, C., Doyon, F., Carlet, J., Dellamonica, P., Gouin, F., Lepoutre, A., Mercier, J.C., Offenstadt, G., Regnier, B.: Incidence, risk factors, and outcome of severe sepsis and septic shock in adults: A multicenter prospective study in intensive care units. French ICU Group for Severe Sepsis. JAMA 274, 968–974 (1995)
6. Dellinger, R.P., Levy, M.M., Carlet, J.M., Bion, J., Parker, M.M., Jaeschke, R., Reinhart, K., Angus, D.C., Brun-Buisson, C., Beale, R., Calandra, T., Dhainaut, J.F., Gerlach, H., Harvey, M., Marini, J.J., Marshall, J., Ranieri, M., Ramsay, G., Sevransky, J., Thompson, B.T., Townsend, S., Vender, J.S., Zimmerman, J.L., Vincent, J.L.: Surviving Sepsis Campaign: International guidelines for management of severe sepsis and septic shock. Critical Care Medicine 36, 296–327 (2008)
7. Knaus, W.A., Draper, E.A., Wagner, D.P., Zimmerman, J.E.: APACHE II: a severity of disease classification system. Critical Care Medicine 13, 818–829 (1985)
8. Knaus, W.A., Wagner, D.P., Draper, E.A., Zimmerman, J.E., Bergner, M., Bastos, P.G., Sirio, C.A., Murphy, D.J., Lotring, T., Damiano, A.: The APACHE III prognostic system: Risk prediction of hospital mortality for critically ill hospitalized adults. Chest 100, 1619–1636 (1991)
9. Letarte, J., Longo, C.J., Pelletier, J., Nabonne, B., Fisher, H.N.: Patient characteristics and costs of severe sepsis and septic shock in Quebec. Journal of Critical Care 17, 39–49 (2002)

10. Levy, M.M., Fink, M.P., Marshall, J.C., Abraham, E., Angus, D., Cook, D., Cohen, J., Opal, S.M., Vincent, J.L., Ramsay, G.: SCCM/ESICM/ACCP/ATS/SIS International Sepsis Definitions Conference. Critical Care Medicine 31, 1250–1256 (2003)

11. Martin, C., Priestap, F., Fisher, H., Fowler, R.A., Heyland, D.K., Keenan, S.P., Longo, C.J., Morrison, T., Bentley, D., Antman, N.: A prospective, observational registry of patients with severe sepsis: The Canadian Sepsis Treatment And Response (STAR) Registry. Critical Care Medicine (in press)

12. Minneci, P.C., Deans, K.J., Banks, S.M., Eichacker, P.Q., Natanson, C.: Meta-analysis: the effect of steroids on survival and shock during sepsis depends on the dose. Ann. Intern. Med. 141, 47–56 (2004)

13. Pittet, D., Rangel-Frausto, S., Li, N., Tarara, D., Costigan, M., Rempe, L., Jebson, P., Wenzel, R.P.: Systemic inflammatory response syndrome, sepsis, severe sepsis and septic shock: Incidence, morbidities and outcomes in surgical ICU patients. Intensive Care Medicine 21, 302–309 (1995)

14. Rivers, E., Nguyen, B., Havstad, S., Ressler, J., Muzzin, A., Knoblich, B., Peterson, E., Tomlanovich, M.: Early goal-directed therapy in the treatment of severe sepsis and septic shock. New England Journal of Med. 345, 1368–1377 (2001)

15. Salvo, I., de, C.W., Musicco, M., Langer, M., Piadena, R., Wolfler, A., Montani, C., Magni, E.: The Italian SEPSIS study: Preliminary results on the incidence and evolution of SIRS, sepsis, severe sepsis and septic shock. Intensive Care Medicine 21(suppl. 2), S244–S249 (1995)

# Author Index